Building and Facilities Maintenance Management

Building and Facilities Maintenance Management: An Innovative Operational Approach is a new introductory textbook aimed at students who are studying building maintenance and asset management for the first time as part of their undergraduate or foundation course. The book plugs a gap in the current curriculum and literature which must now seek to educate emerging professionals about the significance of the operational phase of a built asset's lifecycle, if the challenges of the day are to be met head on.

Each chapter is supported by seminar questions and formative assessment tasks to allow the learner to assess their own progress through the text and covers:

- The nature and importance of building maintenance for the successful strategic operation of an estate
- contracting and procurement of maintenance works
- risk assessment in maintenance management
- surveying for common maintenance issues and defects
- the relationship between maintenance management and environmental sustainability
- and maintenance and conservation of historic buildings

The book will help readers enhance their awareness, understanding and breadth of knowledge of the issues around improving organisational maintenance management. A handy guide for emerging professionals, this practical book is also ideal reading for anyone studying facilities management, building surveying, construction management, real estate and property management.

Dr Jason Challender MSc PhD Eng. D.Hon FRICS FAPM FAHE has acquired 35 years 'client side' experience in the UK construction industry and procured numerous successful major construction programmes during this time. He is Director of Estates and Facilities at the University of Salford, a member of its Senior Leadership Team and responsible for overseeing a large department of approximately 350 estates and construction-related staff. He is also a construction researcher who has published seven books and twelve academic journal and conference papers in recent years, all of which have been dedicated to his studies around construction and engineering management. He has attended many national and international construction and institutional conferences as a guest speaker over the years and is a visiting professor.

Robert Challender is employed at CS2 Charted Surveyors in Manchester, where he is serving an apprenticeship in building surveying. The building surveys featured in the book are those which he has undertaken himself. He is currently studying for a BSc degree in building surveying at the University of Salford.

Building and Facilities Maintenance Management

An Innovative Operational Approach

Jason Challender and Robert Challender

Routledge
Taylor & Francis Group
LONDON AND NEW YORK

Designed cover image: Getty Images

First published 2025
by Routledge
4 Park Square, Milton Park, Abingdon, Oxon OX14 4RN

and by Routledge
605 Third Avenue, New York, NY 10158

Routledge is an imprint of the Taylor & Francis Group, an informa business

British Library Cataloguing-in-Publication Data
A catalogue record for this book is available from the British Library

ISBN: 978-1-032-41587-1 (hbk)
ISBN: 978-1-032-41554-3 (pbk)
ISBN: 978-1-003-35880-0 (ebk)

DOI: 10.1201/9781003358800

Typeset in Times New Roman
by Apex CoVantage, LLC

Contents

Foreword

This text by Jason and Bobby is a welcome and valued addition to the literature on asset and maintenance management. Jason has been able to write it from the unique perspective of being in a very senior client-side position responsible for awarding work to construction supply chains, and from his doctorate award and subsequent visiting professorship, and Chair of the Industry Advisory Board at the University of Bolton. It is important that we keep this subject area extremely prominent in our minds and in our actions. Jason will not mind that I describe him primarily as a practitioner, and it is books from practitioners that we all want to read, since they are often of most interest and easy to apply in our lives. The text admirably addresses the complexity of the subject area of maintenance management in a thoughtful way, using case studies and the insights of colleagues to give real-life examples and scenarios. I know Jason and Bobby have written this book with the intention to incentivise improved ways of working. In this way hopefully it will inspire some of his readers to question their current practices, with a view they would otherwise not have done if they had not read of his experiences. It is a long journey ahead, so I believe that this book which promotes better collaboration and technical advancement in the built environment will prove a useful intervention for best practice. Accordingly, I hope this book inspires readers to take a different perspective on the procurement of maintenance management strategies and in conducting building surveys and will encourage transformational real change in practice.

Thank you, Jason and Bobby, for taking us some distance on that journey.

Dr Peter Farrell MSc CEng, FRICS, FCIOB, FCABE, SFHEA

Acknowledgement

Thanks also goes out to Richard Jackson, a long-time friend of Professor Challender, for his support in the past for previous publications.

Figures

Chapter 1

Introduction

> Proficient maintenance management strategies if developed, harnessed, and adopted correctly can make a significant contribution to the corporate goals of organisations in terms of efficiency, reputation, and the challenge of attracting and motivating talent.
>
> —Professor Jason Challender (2024)

1.1 Introduction to the book

With reference to the previous quotation there has in recent years been a stronger focus on maintenance management responsibilities and strategies aligned to businesses. Accordingly, the subject matter of what this book refers to as *Building and Facilities Maintenance Management: An Innovative Operational Approach* has become a hot topic and one that businesses need to pay special attention to if they want to succeed. For this reason, the book will explore many different aspects and themes around maintenance management, specifically within the context of the built environment. It will highlight some of the problems and challenges that are currently being experienced in the sector on a national and international level. It is hoped that this will stress the importance, relevance, and maintenance management in the workplace. The book will then explore and articulate many different innovative initiatives and measures to addressing some of these challenges. It is hoped that these can be introduced within estates management departments, particularly where they are overseeing large property portfolios, and therein represent a model of good practice for the future of the built environment. The following sections will hopefully give some context and background to the discussions around maintenance management, alongside the justification and aims and objectives for the book.

1.2 Context to the significance, relevance and importance of maintenance management

The increasing use of private finance for the delivery of major public infrastructure together with the publications of various policy pronouncements as part of the UK government's Industry Strategy for Construction 2025 has caused a paradigm shift in the construction industry's view of the maintenance and upgrading of existing buildings, with asset management no longer seen as the 'Cinderella' of the construction industry. As a result, higher education providers have started to place an increased focus on the longer-term view of construction projects advocated in Wolstenholme's critical review of the construction industry on behalf of Constructing Excellence. In this report he attested construction projects cannot be seen in isolation from the operational phases of completed buildings (Wolstenholme *et al.* 2009).

DOI: 10.1201/9781003358800-1

The management of building maintenance operations continues to be an important area of practice and indeed plays an important role in the education of undergraduate and postgraduate learners. Despite this, most previously published textbooks in this field have failed to support this important area of academic endeavour and have not adequately supported then needs of a significant proportion of undergraduate students. This problem has been exacerbated by the evident trend in recent academic publications to focus solely on the technological aspects of maintenance at the determent of the wider curriculum. As a result, a clear gap exists within the existing literature base, which the book is designed to fill.

1.3 Overarching rationale, synopsis and justification; the aim and objectives for writing the book

Building and Facilities Maintenance Management: An Innovative Operational Approach is concerned with how we make improvements and positively contribute to existing estates management practices in terms of providing safe and sustainable environments to work in. Furthermore, maintenance management should be focused on improving estates strategies, to create thriving and aspiring, motivational and learning workplaces. For this reason, the book covers the significance and nature of building maintenance alongside asset planning strategies, programmes, policies, and systems. The focus and relevance in these important areas is related to all organisations involved with the procurement or life cycle of buildings from planning through design, construction, operations and facilities management.

The aim of the book is to provide the foundations for advance thinking and shape future practices in maintenance management and articulate how there can be a vast array of methods, motives, and rationales for achieving more successful outcomes on both an organisational and industry-wide basis. In this pursuit there are clearly many unexplored challenges and dilemmas to grapple with and it is recognised that we are at the start of a largely unexplored journey. Such challenges include leadership, statutory responsibilities and regulations, health and safety, motivational management, and collaborative working.

The improvements in practice in the aforementioned key areas will be drawn from case studies that the lead author, Professor Challender, has been personally involved with across many other knowledge fields. These include other industries such as manufacturing and retail and involve the cross pollination of ideas from a broad range of collaborating stakeholders. In this regard, previous examples of what works best in practice will be given to pave the way forward for the advancement of applied knowledge as a new and clear agenda for change in the built environment. The cultural changes in the built environment, especially in the UK over the last thirty years, including significant improvement in health and safety, can lead us to be optimistic that further reforms can be made possible around asset maintenance management. This may be assisted by changes through legislative, leadership and training and education initiatives to achieve the same shifts for maintenance management in those areas previously referred including sustainability, human resources, motivational management and collaborative working.

1.3.1 The need to rethink the concept of maintenance management

Maintenance management is not a concept that is new. Moreover, its roots have gone back thousands of years to the times of our ancient forefathers in maintaining religious temples, places of burial and landmarks e.g. Egyptian pyramids. In recent years, however, the world has been witness to many different challenges in maintaining buildings and facilities which are deemed safe, compliant, and conducive to user needs. The failures and scandal around the use of aluminium composite cladding (ACM) that has been deemed unsafe and potentially hazardous cladding for buildings, following the Grenfell Tower Disaster in London of 2017, presents one notable example of poor practice.

Although over many years business organisations have focused mainly on strategies linked to making increased profits and creating value for shareholders, in recent years companies are realising that they need to focus on nonfinancial strategies alongside their economic goals. This involves behaving ethically in providing good, well-maintained and legally compliant working environments for employees which in turn can raise their reputations and in turn increase their success in the marketplace. This is supported by Orlitzky *et al.* (2003) who found from their research that a positive correlation exists between corporate financial performance and environmental/social performance. Furthermore, Chye Koh and Boo (2004) found that organisational commitment to providing good standards of accommodation can have a positive relationship with job satisfaction and career success of employees. Companies that embraced asset and maintenance strategies around social and environmental benefits found that this raised the motivation levels of employees. In these instances, they found that staff loyalty and willingness to suggest organisational improvements were prevalent where motivation levels are high. In addition, such policies were found to reduce absenteeism and staff turnover which are very costly for organisations. The evidence would therefore suggest that asset and maintenance strategies underpinned by moral considerations for staff and the communities they work in could have far reaching positive effects on organisational success. It is for this reason that the book will articulate the various estates measures that organisations in the built environment should be progressing to achieve more successful maintenance management outcomes. Such measures will revolve around maintenance management policies, programmes and plans alongside strategies linked to leadership, human resources management, environmental sustainability, procurement, and motivation of employees.

1.3.2 The fragmentation and bespoke nature of the built environment sector in the UK

In the UK nearly half of all building works associated with new construction, maintenance, refurbishment, adaptions, and alterations are undertaken by approximately 200,000 small contractors. Some of the workforce in this regard, especially those who undertake smaller jobs, are often individuals with limited and potentially insufficient experience. Because of such inexperience there may be cases where their knowledge and awareness of issues relating to regulations, compliance, best practice, health and safety and repair/renewal considerations are lacking. Furthermore, in the context of the built environment, which is predominantly associated with fragmented, complex, and potentially confrontational practices, this can create a dilemma for the sector. It is, therefore, perhaps not surprising that there have been many reported cases in the press of examples of malpractice and contractors' breaches of regulatory standards and codes of practice. Some of these cases have resulted in litigation proceedings brought by their clients, which has resulted in reputational damage for the industry. In addition, these practices have frequently led to less than acceptable project outcomes in terms of value for money, delays, and poor-quality build and maintenance standards.

The book will respond to the increasing demand for practical, and industry-aligned, best practice in the built environment, covering asset and maintenance management. More specifically the book will address how existing maintenance standards can be pragmatically applied to professional practice from an organisational perspective and provide case studies and example scenarios aligned to modern-day requirements. Furthermore, the book will provide coverage of those real-world situations where the minimum regulatory requirements continue to require practitioners to demonstrate professional judgement and decision-making in keeping estates compliant.

1.3.3 The book's unique and outstanding features

In previous studies into maintenance management, very little attention has been focused on tailoring decision-making and everyday practices around corporate strategies and processes of business planning. Their attention has instead been concentrated on only the property benefits rather than the business benefits for employers, employees, and customers and in areas such as gaining operational efficiencies. In some cases, silo mentalities have emerged around estates issues and maintenance of assets and a clear disconnect from operational planning and core business activities. This has created a dilemma for organisations where departments and teams do not collaborate and share knowledge, leading to problems and challenges in achieving long-term goals and aspirations. As the name of the book suggests, it will seek to adopt an innovative and alternative approach to maintenance management and therein infill the current gaps in knowledge. In doing so it will examine how buildings and estates can be regarded as business enablers rather than expensive assets which burden companies with repairing liabilities. In this regard the book will make recommendations to overcome poor and outdated traditional working practices and behaviours in the pursuit of more successful business outcomes.

There have been relatively few books which have been written on the specific subject of encouraging and incentivising appropriate maintenance strategies around environmental sustainability. The book has responded to this in Chapter 10, specifically looking through the lens of how organisations can improve their environmental credentials in the process of carrying out their maintenance programmes. Other gaps that the book has targeted include the inspection and building survey processes associated with maintenance management. To respond to this, building pathology will be covered in Chapter 4 and Chapters 6 and 7 devoted to defect diagnosis when carrying out surveys. In addition, a practical guide which Professor Challender developed for students will be presented in Chapter 9, which is designed to streamline and assist in building surveying processes and methodologies. The book will also draw on case studies from the lead author's experiences with projects that he has overseen over the last 35 years in industry. This will include a range of different project types including restoration works, and Chapter 5 is devoted to maintenance and conservation of historic buildings in this regard.

The book will take a slightly different course and approach from previously published works by asking some very fundamental questions:

- How can improvements in maintenance management positively promote good standards, behaviours and practices and influence the success of corporate business strategies?
- What is the importance and role of long-term asset planning in allowing organisations to realise their future visions and aspirations for their businesses?
- How can maintenance management strategies be fully integrated into operational planning and help organisations perform more successfully, efficiently and reduce disruption to business activities?
- How can asset management benefit businesses for their owners, customers, and employees?
- Why are environmental considerations so relevant for businesses today and how can these by integrated within maintenance management strategies?

In consideration of the aforementioned questions, the book's objectives are:

- To use case studies to look at how successes, failures and key risks can be influenced by maintenance management practices.
- To identify how adopting policies and delivering strategies linked to maintenance management can assist organisations in the built environment gain more efficiencies and achieve other successful outcomes for business owners, customers, and employees.
- To identify a set of clear guidelines, national or international, to support the adherence and regulation around maintenance management.
- To form the basis of a practical toolkit for guidance and teaching of maintenance management in the built environment.
- To be the standard reference for businesspeople in understanding how maintenance management affects organisational risk.

The book is intended to assist students, academics, construction-related practitioners and clients in their awareness, breadth of knowledge and comprehension of the issues around maintenance management considerations, with the overarching aim of delivering more successful outcomes. This is felt to be particularly important as, in previous studies into maintenance management in the built environment, very little attention has been focused on giving practical advice. The book has sought to infill the literature gaps through examination of traditional roles of maintenance and facilities teams and through providing guidance on potential improvement measures.

Case studies and practical examples have been included to assist readers on how theoretical perspectives can be applied to real-life construction projects and scenarios which involve overcoming problems, challenges, and dilemmas. This has responded to academic calls for greater insight into how leadership around maintenance management can be created, mobilised, and developed and more understanding of the resultant positive effects and impact that can be generated therein.

There will be frequent reference to construction practitioners' views and opinions throughout the book and these have been sought through qualitative research carried out in 2023 from a small sample of semi-structured interviews. Participants from these interviews included clients, design consultants, main contractors and subcontractors. Notwithstanding this premise, the majority of the book is largely based on the 'lived' experiences of the authors. This is intended to pass on to readers the benefit of their experiences and knowledge in dealing with a multitude of different maintenance projects in the past.

Understanding the risks posed for breaches of health and safety and statutory standards in the built environment is a growing area as the tolerance for project failure reduces. Contemporary books have in the past sought to externalise maintenance management strategies to traditional practices and codes of conduct which regulate them. This book takes a different perspective and is unique in considering maintenance management as an integral and valuable component of normal business activities. In doing so it creates a practical baseline for awareness of all asset-related considerations and therein smoothing the transition between traditional teaching and embedding standards, values and principles within maintenance management. Accordingly, it is a starting point for standard maintenance practices to be developed and integrated within organisations to support and de-risk their business activities.

The value proposition of this book is that it will hopefully be read, understood, and accepted by businesspeople as their main guidance and reference tool for reducing risks associated with adverse disruption from maintenance operations or noncompliance with statutory regulations.

Accordingly, this is a book written for businesspeople by businesspeople based on sound theory (how to do it) and sound practice (lessons derived from case studies). It is considered unique in that it represents a comprehensive and wide-ranging analysis of best maintenance practice in the built environment as well as other industries and sectors and what can be learned from them. The book will take established and widely accepted business management practices and models and align them with maintenance strategies. From this perspective, it does not seek to introduce new or adapt established maintenance management systems and processes but simply to use best practice models to co-exist alongside them. The book also investigates the area of environmental sustainability in raising 'green' accreditation and status for businesses. These have become more prominent and important over recent years with the advent of reputational benefits for companies pursuing environmental agendas.

Although the research was undertaken in the UK, and all findings are likely to therefore have best fit with the UK construction industry, the overall knowledge and understanding to be provided by this book will have international relevance. Other countries seeking to develop similar approaches to the UK will be able to utilise the book, with consideration of how the findings fit with their own understanding in practice.

It is worth acknowledging that the lead author, Professor Challender, has gained over 35 years' experience of construction and maintenance management from both a practitioner and academic perspective. From this, the book has drawn on both academia and practice, and it seeks from both these perspectives to prove an important insight into an area which has long been problematic for the built environment.

1.4 Previous publications on maintenance management

Over the last 20 to 30 years a diverse range of books have been published focused specifically on the topic of maintenance and asset management. Unfortunately, however, a number of these texts have not undergone periodic refreshment, leaving today's learners with a limited range of core and recommended reading for the increasing array of modules which seek to address the whole life cycle of built assets. As a result, the review of existing textbooks considers books published from the mid-1980s up to current time.

The market is led by three arguably seminal texts: Chanter and Swallow's *Building Maintenance Management* published in 2008, Seeley's *Building Maintenance* published in 1987 and finally Wordworth and Lee's *Building Maintenance Management* which was updated for a 4th education by Wordsworth in 2001. Whilst these texts adopt very different positions in the nature of maintenance, Chanter and Swallow's arguably sector-leading text addresses all aspects of maintenance including dimensions, planning, organisation, procurement and post-contract management and finally the role of knowledge management within maintenance organisations. Despite the comprehensive coverage of these fundamental issues, the text has really failed to deal with the increasingly important areas of carbon resilience and information management. It is now essential that clients need to consider the viability of retrofitting micro generation and renewable technologies to existing structures during major maintenance activities. The sophistication of the text, however, suggests its primary market would be higher-level learners, such as those nearing the completion of their degree and professionals preparing for professional assessment by bodies such as the RICS and CIOB. As a result of this complexity the text is inaccessible for those learners who are being introduced to the field of building maintenance as part of intermediate awards such as higher national's and foundation degrees or indeed those learners studying maintenance as part of the earlier stages of their surveying degree programmes.

Despite adopting a similar approach to the topic of maintenance management to that documented previously, Seeley's text on building maintenance provides both a wider view of the topic, whilst also considering the wider implications, with extensive coverage relating to the role of the Quantity Surveyor in the execution of building maintenance. Indeed, the book provides an extensive coverage of the quantification, costing, procurement and contractual management aspects of building maintenance. Alongside this, the text provides the students (particularly building surveying students) with an extensive commentary on the pathology issues relating to existing structures, including identification and potential remedial works. The latter chapters of the text are devoted to the planning and management of maintenance activities, including coverage of maintenance plans, the structure and operation of maintenance organisations and the day-to-day management of maintenance teams. Whilst this text provides better alignment with the needs of the learners alluded to previously, the material is over 20 years old; as a result the books fails to acknowledge legislative, procedural and technological changes not influencing the maintenance of built assets in the 21st century.

A number of these omissions are corrected in the later fourth edition of *Lee's Building Maintenance Management* written by Wordsworth in 2001 and published by Blackwell Science. As with Seeley's work, this book is well suited to the student market, alluded to earlier. Unlike Seeley's text, however, Wordsworth provides a less comprehensive analysis of the pathology and financial management aspects of building maintenance, instead favouring to adopt a similar model to that used by Chanter and Swallow in that the book provides a more succinct coverage of the various facets of maintenance management. However, the text provides a far more accessible coverage of the various topic areas than those provided by Chanter and Swallow, making it clearly suitable for the undergraduate student market targeted by this proposal. Yet despite its eminent suitability to the market sector covered by this proposal, the text is lacking in currency. It fails to review the role of framework agreements, key performance indicators and other advancements in the procurement of maintenance activities. Nor does the text evaluate the increasingly important area of sustainability, sustainable maintenance, carbon resilience and the growing retrofit market.

In addition to these main texts devoted to the maintenance management of buildings, a range of periphery texts have been published in recent years including Riley and Cotgrave's (2011) *The Technology of Refurbishment and Maintenance*, Gahlot and Sharma's (2006) *Building Repair and Maintenance Management*, Highfield and Gorse's (2009) *Refurbishment and Upgrading of Buildings*, and finally Nick's (2009) *The Handbook of Sustainable Refurbishment.* In most cases, these texts have adopted a strong focus on the technological aspects of building maintenance and refurbishment (Riley and Cotgrave 2011; Nick 2009; Gahlot and Sharma 2006; Highfield and Gorse 2009) with the texts collectively addressing the building pathology and technologies aspects of the refurbishment and maintenance processes including façade retention, demolition techniques and the underpinning of buildings. Despite the extensive information provided by these texts, they fail to address the comprehensive nature of building maintenance. Instead, they appear to be targeting more specialist modules such as building pathology and the higher-level advanced technology modules demanded as part of the RICS and CIOB accreditation processes.

The books reviewed previously indicate that whilst the management of building maintenance operations continues to be an important area of practice and indeed plays an important role in the education of undergraduate and postgraduate learners, the three seminal texts published in this field are currently failing to support this important area of academic endeavour. With the texts either focusing on the higher levels of academic attainment or being so dated (Seeley 1987) they are failing to adequately support the needs of a significant proportion of undergraduate students.

A problem exacerbated by the evident trend in recent academic publications to focus solely on the technological aspects of maintenance at the determent of the wider curriculum (Nick 2009; Wolstenholme *et al.* 2009; Riley and Cotgrave 2011). As a result, a clear gap exists within the existing literature base, which the book proposed would fill.

1.5 Value proposition of the book

Some previous publications have successfully considered aspects of maintenance including its dimensions, planning, organisation, procurement, and post-contract management and finally the role of knowledge management within maintenance organisations. Despite the comprehensive coverage of these fundamental issues, most of these previous works, as previously referred to, have not considered the increasingly important areas of carbon resilience and information management which the book will address. This is an important and topical aspect as it is now essential that clients need to consider the viability of retrofitting micro generation and renewable technologies to existing structures during major maintenance activities. For this reason, Chapter 10 has been devoted to environmental sustainability in asset and maintenance management. In addition, the book will provide a much needed area which has previously been overlooked, in how building assets can help rather than hinder organisational performance.

The majority of previous books around building maintenance have been targeted predominantly on higher-level learners, including those nearing the completion of their degree and professionals preparing for professional assessment by bodies such as the RICS. As a result of this complexity these textbooks have not always been accessible for those learners who are being introduced to the field of building and facilities maintenance management. For this reason, the book has been designed to accommodate those students undertaking intermediate awards such as higher national's and foundation degrees or indeed those learners studying maintenance as part of the earlier stages of their surveying degree programmes.

If this rebalancing of the curriculum is to be achieved successfully it is essential any new 'asset management focused modules' are underpinned by current and topical textbooks. As a visiting professor delivering, examining, or externally moderating such modules, the book's lead author has become increasingly aware of the severe limitations associated with existing textbooks, which have failed to meet the needs of current and future learners. In consideration of this deficiency and with an ever increasing need to bridge this knowledge gap, it is hoped that the book can provide the impetus and material that can be utilised by its readers. In this regard it is intended that this textbook will sit alongside, and support Chanter and Swallow's *Building Maintenance Management* published in 2007, which is arguably focused on more advanced learners, potentially who are studying building maintenance and asset management for the first time as part of their course. In addition, the book could be conducive to those working towards chartership with one of the professional bodies or experienced practitioners working in the field of built asset management. For this reason, some chapters of the book will be supported by hypothetical examples and real-life case studies from the authors' experiences which are designed as best practice guides.

The book provides students undertaking building surveying courses or modules with an extensive commentary on the pathology issues (Chapter 4) relating to existing structures, including identification and potential defects and associated remedial works (Chapters 6 and 7). The book will also cover the planning and management of maintenance activities, including coverage of maintenance plans, the structure and operation of maintenance organisations and the day-to-day management of maintenance teams.

1.6 Research sources

The research for the book has been mostly derived from a combination of sources which include the authors' own experiences, and interviews with a wide range of construction professionals and literature. In addition, the authors have conducted a literature review of many of the aspects and subject matters covered in each chapter, and utilised various academic sources which are included in the references.

1.7 Target audience of the book

This textbook has been designed to sit alongside and support Chanter and Swallow's seminal text, which is arguably focused on more advanced learners, those working towards chartership with one of the professional bodies or experienced practitioners working in the field of built asset management. Notwithstanding this premise, it has also been aimed at Higher National, Foundation degree and Honours degree students, who are studying building maintenance and asset management for the first time as part of their course.

In addition to students, the book will be designed to assist estates and facilities practitioners in their awareness, understanding and breadth of knowledge of the issues around improving organisational maintenance management.

Most chapters will be supported by case studies from the authors' own experiences. As such it represents a practical guide for students, academics and practitioners that can develop into a common framework on how to initiate, procure and manage more collaborative maintenance management strategies and working philosophies.

The book will not be designed as a holistic course textbook albeit it could be worthy inclusion in a recommended reading list for courses related to estates management and construction procurement. As such, it is not intended solely as a practitioner guide. Rather, the book aims to cross this divide and provide useful insight to both academics and practitioners in developing their understanding of the topic area.

The research findings will be presented to prioritise use by professional practitioners and therein provide a practical guide for adherence to best practices as a toolkit, for improvements in asset management in the built environment.

1.8 Structure of the book

The book has been structured into 14 chapters and covers the subject of maintenance management from many different angles and perspectives but specifically in the context of the built environment. A brief outline of the themes of each chapter are detailed next:

Chapter 1: Introduction

Introduction, justification, aims and objectives, value proposition, previous publications, readership and brief description of the areas covered throughout the book.

Chapter 2: Significance and nature of building maintenance

Definitions of maintenance
Concept of maintenance and refurbishment

Implications and impact of building deterioration and failures
Age, type and use of buildings as influential factors on scope and extent of maintenance
Delivering acceptable standards of maintenance
Ensuring statutory compliance
Business continuity and disaster recovery planning
Building life cycle as applied to maintenance
Concepts around planned preventative and reactive maintenance
Longevity and obsolescence
Changing the nature of asset and facilities management

Chapter 3: Maintenance and asset planning strategies, programmes, policies and systems

Maintenance strategies, policies, systems
Managing building maintenance programmes
Identifying and delivering on maintenance needs. Functional suitability and use of space
Maintenance and asset management plans
Relationship between different maintenance management functions
Statutory compliance
Expectations and standards
Reactive maintenance management and estates help desks
Prioritising maintenance and backlog maintenance
Managing the maintenance process
Financial management, budgets and cost control

Chapter 4: Common maintenance issues and managing defects

Exploring the pathology of building maintenance issues
The importance of understanding the nature and effect of agents that can lead to building defects
Mechanisms of decay, failure and deterioration
Dilemmas associated with repair or renewal decisions
Managing the remedial works process to address maintenance issues and defects
Measures to mitigate and prevent defects
Subcategories of settlement and subsidence
Subcategories of movement and cracking
Subcategories of corrosion and decay
Dampness and water penetration in buildings

Chapter 5: Maintenance and conservation of historic buildings

Complexities, challenges and dilemmas associated with historic buildings
Developing robust maintenance management systems and policies for conservation
Statutory protection afforded to various types of heritage buildings
Planning (Listed buildings and Conservation Areas) Act 1990 and the Town and Country Planning Act 1990
Conservation areas
Role of Heritage England and local authority conservation officers

Enforcement action for breaches and failures to obtain statutory approvals
Potential conflict between conservation and meeting current standards
Case studies for two different projects in which the author was personally involved: former Stockport Union Workhouse and Gorton Monastery

Chapter 6: Common internal maintenance issues, defects and remedial works

Problems and challenges pertaining to internal defects
Influence of quality, use, condition, and age of buildings in defect analysis
Categories of different internal defects including:
Rising dampness
Condensation
Water penetration through buildings
Timber decay
Presence of asbestos and other hazardous materials
Wall, floor, and roof defects
Defective electrical, gas and water installations
Underlying causes, descriptions and diagnosis of internal defects
Alternative repair and remedial works to address different defects
Misdiagnosis of building defects

Chapter 7: Common external maintenance issues, defects and remedial works

Problems and challenges pertaining to internal defects
Importance of linking external and internal defects
Categories of different external defects including:

- Brickwork defects, pointing, cracking, dampness
- Differential settlement and subsidence
- Window and door defects
- Flat and pitched roof defects
- Defective rainwater goods
- Presence of hazardous and potentially flammable materials
- Defective external areas and hardstandings

Correct investigation and diagnosis of external defects
Misdiagnosing the underlying causes of building problems
Alternative repair works for different building types and scenarios

Chapter 8: Surveying existing buildings

Background to conducting building surveys
The methodical process of undertaking building surveys
Challenges and obstacles when undertaking building surveys
The importance of building investigation works (intrusive and nonintrusive) for conducting surveys

Different types of building surveys and specialist inspections
Managing the remedial work process

Chapter 9: An innovative project to assist building surveying students: designing a handbook as a practical teaching and learning guide for surveying buildings

Practical tool to guide students through the different stages of surveying buildings
Introduction and justification for the project
Planning and devising the project
Description of the project
Implementing the project
Use of the handbook for carrying out surveys
Feedback and evaluation

Chapter 10: The relationship between maintenance management and environmental sustainability

Philosophy of environmental sustainability as applied to maintenance management
Carbon resilience during retrofit/refurbishment works
Corporate social responsibility
Sustainability rating techniques including BREEAM and Passivhaus
Managing maintenance to reduce waste and use of nondegradable materials
Energy efficiency measures and renewable technologies in refurbishment and retrofit decision making

Chapter 11: Case studies related to embedding environmental sustainability into estates practices

Two case studies: A college and university, both in the North West of England, where the lead author was instrumental in developing new environmental policies

Chapter 12: The role of information technology and digital twins in asset and maintenance management

Importance of asset and maintenance information
Growing challenges that are faced by the built environment
Role of digitalisation
Background and context to the use of digital twins
Benefits of digital twin technologies

Chapter 13: Implications in practice for maintenance management in the built environment

Challenges from traditional poor and outdated maintenance management practices
The need to rethink the concept of maintenance management
Fully integrating asset management strategies into operational planning

Raising environmental sustainability credentials for organisations as part of asset and maintenance management
Improving health and safety in maintenance management
Raising awareness of how successes, failures and key risks within organisations can be influenced by maintenance management practices
Promoting collaborative working and partnering principles in maintenance management
Education, training and continued professional development (CPD)

Chapter 14: Summary, reflections and conclusions

Summary of the key issues raised throughout the book
Implications for future maintenance and estates strategies
Conclusions to the themes and subject areas
Recommendations for the future
Final reflections, overview and closing remarks

1.9 Summary

The UK government's Industry Strategy for Construction 2025 has caused a paradigm shift in the construction industry's view of the maintenance and upgrading of existing buildings. Accordingly, providing improved asset and maintenance management strategies has become a hot topic and one that businesses need to pay special attention to if they want to succeed. This book has responded to such needs brought about by the prominence of this area of practice, which has been neglected in previous years. *Building and Facilities Maintenance Management: An Innovative Operational Approach* is concerned with how we make improvements and positively contribute to existing estates management practices in terms of providing safe and sustainable work environments. The aim of the book is to provide the foundations for advance thinking and shape future practices in maintenance management and articulate how there can be a vast array of methods, motives, and rationales for achieving more successful outcomes on both an organisational and industry-wide basis.

There is a need to rethink the concept of maintenance management. In recent years companies are realising that they need to focus on nonfinancial strategies alongside their economic goals. This involves behaving ethically in providing good, well-maintained and legally compliant working environments for employees which in turn can raise their reputations and in turn increase their success in the marketplace.

The book will examine how buildings and estates can be regarded as business enablers rather than expensive assets which burden companies with repairing liabilities. In this regard it will make a recommendation to overcome poor and outdated traditional working practices and behaviours in the pursuit of more successful business outcomes. The book is intended to assist students, academics, construction-related practitioners and clients in their awareness, breadth of knowledge and comprehension of the issues around maintenance management considerations, with the overarching aim of delivering more successful outcomes.

The research for the book has been mostly derived from a combination of sources which include the authors' own experiences, and interviews with a wide range of construction professionals and literature. In addition, the authors have conducted a literature review of many of the aspects and subject matters covered in each chapter, and utilised various academic sources which are included in the references. Most chapters will be supported by case studies from the authors' own experiences. As such it represents a practical guide for students, academics and practitioners that can develop into a common framework on how to initiate, procure and manage more collaborative maintenance management strategies and working philosophies. The research findings will be presented to prioritise use by professional practitioners and therein provide a practical guide for adherence to best practices as a toolkit, for improvements in asset management in the built environment.

References

Chanter, B. and Swallow, P. (2008). *Building maintenance management.* John Wiley & Sons.

Chye Koh, H. and Boo, E.F.H. (2004). Organisational ethics and employee satisfaction and commitment. *Management Decision*, 42(5), pp. 677–693.

Gahlot, P.S. and Sharma, S. (2006). *Building repair and maintenance management.* CBS Publishers and Distributors.

Highfield, D. and Gorse, C. (2009). *Refurbishment and upgrading of buildings.* Spon Press.

Nick, B. (2009). *The handbook of sustainable refurbishment: Non-domestic buildings.* Routledge.

Orlitzky, M., Schmidt, F.L. and Rynes, S.L. (2003). Corporate social and financial performance: A meta-analysis. *Organization Studies*, 24(3), pp. 403–441.

Riley, M. and Cotgrave, A. (2011). *Construction technology 3: The technology of refurbishment and maintenance.* Macmillan International Higher Education.

Seeley, I.H. (1987). *Building maintenance.* Bloomsbury Publishing.

Wordsworth, P. and Lee, R. (2001). *Lee's building maintenance management.* London: Blackwell Science.

Chapter 2

Significance and nature of building maintenance

2.1 Introduction to chapter

The chapter will start by considering the different definitions of maintenance and related terms to fully understand their meaning and what they entail. It will then consider the importance of maintenance and the implications of building deterioration of buildings when repairs and remedial works are not undertaken to acceptable standards. In this regard it will discuss the impact of building failures and the detrimental effects that these can bring in terms of disruption to occupants and business operations.

The extent and degree of maintenance requirements will then be covered using practical examples, and how age, type and use of buildings can become influential factors. The chapter will then discuss the significance of setting standards as benchmarks for maintaining buildings and facilities and their importance for creating comfortable environments for occupants. This will encapsulate the different approaches to establishing building standards and the reasoning and justifications that underpin these. Considerations around fitness for purpose for different types and uses of buildings will then be analysed.

Managing risks around statutory compliance with regulations will be explained and why this is a crucial requirement for maintenance strategies. Implications for organisations around statutory noncompliance will then be articulated and the possible impacts of breaches and infringements. Health and safety will then be discussed from the perspectives of workplace regulations and in undertaking maintenance and construction and maintenance projects. This will cover the Construction Design and Management Regulations (CDM) 2015 for eligible new works. Furthermore, accreditation schemes such as the Construction Skills Certification Scheme will be referred to, and the ways that these have assisted companies in managing their health and safety duties and responsibilities.

The chapter will then move on to managing risks in workplaces through business continuity and disaster recovery planning. Examples of different types of risk will be given and how contingency planning for unexpected events, issues and failures can benefit organisations in crisis situations. A real-life case study will be presented from the experience of the lead author and an account of how business continuity planning assisted his organisation during the COVID-19 pandemic.

Planned preventative and reactive maintenance will then be covered, which will include an explanation of each and how these can be integrated into maintenance management programmes. The importance of life cycle analysis in considering future maintenance needs will be referred to, alongside decision-making around repairs or renewals. This will then lead into the notion of obsolescence and, using a real-life case study, how unforeseen economic and market external forces can influence the longevity of buildings.

DOI: 10.1201/9781003358800-2

2.2 The definition of maintenance

One definition for maintenance is presented as 'the act of keeping something in good condition by checking or repairing it regularly'. A similar definition comes from it being used as an uncountable noun: 'the maintenance of a building, vehicle, road or machine is the process of keeping it in good condition by regularly checking it and repairing it when necessary' (Collins Dictionary, 2024). An alternative definition is presented as 'the combination of all technical and associated administrative actions intended to retain an item in or restore it to a state in which it can perform its required functions' (Thomas 2013). There is, therefore, no universal definition for maintenance, but clearly there are common strands which exist to articulate its meaning in practice.

Within the umbrella of maintenance there are many different terms used commonly from an estates and built environment perspective and some of these, provided by Thomas (2013), are included in Table 2.1, with their respective definitions.

Table 2.1 Different terms used commonly from an estates and built environment perspective (Source: Thomas 2013)

Term or terminology	*Definition*
Building element	A building element is a part of a structure or building with a distinct function, such as a wall or roof. The sub-elements of a roof would include the structure, covering, insulation and drainage
Building pathology	Building pathology is the systematic treatment of building defects, its causes, its consequences, and its remedies
Condition-based maintenance	Preventative maintenance initiated from knowledge of the condition of an item derived from routine or continuous monitoring
Corrective maintenance	Maintenance carried out after a fault has occurred and intended to put an item into a state where it can perform a required function in a safe and efficient manner
Economic life	The estimated number of years until an item no longer represents the least expensive method of performing its function
Maintenance management	The organisation of maintenance within an agreed policy
Maintenance policy	A written statement issued by, or on the authority of, the client and which is acceptable to the building owner of the building
Planned maintenance	Maintenance organised and carried out with forethought, control, and the use of records to a predetermined plan. Preventative maintenance is always part of planned maintenance, whereas corrective maintenance may not be
Preventative maintenance	Maintenance carried out at predetermined intervals or corresponding to prescribed criteria and intended to reduce the probability of failure or the performance degradation of an item
Reactive (responsive maintenance)	Repairs carried out on an ad hoc basis as the need arises and which cannot be deferred for inclusion in planned maintenance programmes
Scheduled maintenance	Preventative maintenance carried out in accordance with an established schedule including time interval, number of operations or mileage
Sinking fund	A sinking fund is money invested to provide future funds to cover the cost of capital expenses which may be forthcoming in the future or long term

2.3 The importance of maintenance

On completion of buildings, building elements will start to physically deteriorate naturally owing to forces of nature and wear, tear and use by building occupants. As well as deterioration during the life of buildings, they can sometimes, especially when not maintained to the correct standard, fail unexpectedly from physical or performance perspectives. This can lead to malfunctions of building services or physical building defects. The types of failure and defects will be covered more in Chapters, 4, 6 and 7 which deal with building pathology and defects analysis. When buildings fail unexpectedly it can cause significant disruption and expense to building owners, and their occupants. In extreme cases buildings may have to be vacated whilst major repairs are carried out. Examples could include electrical, heating or fire alarm malfunctions or flooding from burst pipes or roof leaks. To avoid such failures, it is important that maintenance regimes are put in place from the date of practical completion of new buildings, when they are legally handed over to clients for their occupation. This will also ensure that buildings as valuable assets do not depreciate over time, and therein lose value. They should be considered as expensive assets and resources and maintaining the value of buildings is clearly important for their owners especially when they come to sell their properties. In a similar way, maintaining building elements in good condition is crucial for ensuring that the performance of buildings and those operations carried out within them are not compromised at any time. This may be particularly important for commercial or industrial buildings delivering services or manufacturing products, where any building issues could affect production processes and business outputs.

2.4 The extent and degree of maintenance required for buildings

The use and type of buildings can affect the maintenance regime deployed and, in this sense, there is 'no one size that fits all'. For example, in terms of use, a five-star hotel may require to be always maintained in a first-class condition, as this is an expectation of their customers. To achieve and maintain this standard, regular inspections may take place numerous times during the day by hotel staff with facilities personnel on hand to deal with any reported issues or complaints immediately. Conversely a building predominantly used for distribution or storage may not require the same urgency of attention to issues unless these are deemed to have health and safety implications or to avoid breaches in statutory regulations. Notwithstanding this premise, it is not just the uses of buildings that affect the extent and degree of maintenance requirements but building types. Accordingly, the type and nature of different buildings could equally determine the scope of maintenance requirements and liabilities. For example, older buildings and those which hold special architectural or historical interest, e.g. listed buildings, may demand more specialist maintenance than more conventional newer buildings. Maintenance of historic buildings will be covered in Chapter 5 of the book.

2.5 Establishing standards for buildings

It is essential to consider the standards by which building owners wish to maintain their estates. This is especially important for preparing asset management plans and maintenance programmes which will be discussed more in the next chapter. These standards will vary according to the type of business, building and use. The standards which are established can have a direct effect on the comfort and satisfaction of building occupants which in turn could affect the performance of businesses. For instance, if building environments were considered poor, with little or no natural lighting, heating or cooling problems, poor central facilities, lack of privacy and overcrowding then this could have detrimental effects on staff and customer morale and motivation. Furthermore, it could consequently affect productivity and service outputs as staff become demoralised and disillusioned with their places of work. In some cases, staff could leave organisations altogether in pursuit of more improved working environments elsewhere.

Even when standards are established, the processes of deterioration will constantly seek to change the condition of buildings for the worst, with a resultant decline in actual standards if not maintained correctly. At the same time expectations from building occupants and increased regulation will drive those acceptable standards to a higher level over time. This can become a dilemma for building owners as a gap between expected and actual standards opens up. If gaps grow sufficiently wide, buildings over time may become unusable and obsolescent. An example could be an old office building, where problems and associated cost of renewing all faulty and noncompliant building services may render any refurbishment or upgrading works financially unviable. This may result in buildings being earmarked for demolition, to make way for new buildings.

There can be different approaches to setting building standards. Building owners in some cases may wish to set standards very high to present the right image of their organisation. They may consider that having attractive buildings and facilities in pristine condition projects their businesses in a positive light to possibly attract new customers. In this regard buildings, as visual entities, can bring reputational benefits for companies, especially during processes associated with rebranding their businesses. However, not all building owners will adopt this policy around standards. Conversely, they may regard built assets as merely accommodation for conducting their core business functions within. In this sense they may seek to limit repairs, maintenance, and any improvements to a minimum, taking account of statutory regulations and health and safety legislation. In this scenario they may choose to only carry out urgent and unavoidable works. Standards may also vary where leasehold interests are in place where tenants are required only to maintain buildings over their life cycles to the standards prescribed by lease terms and initial schedules of condition. Any improvement works and upgrading may be regarded as an unnecessary expense especially when lease terms are short, and therein any benefits to tenants from building investments considered limited.

In addition to the standards associated with state of repair and physical condition, there are also expectations for building accommodation to be fit for purpose in terms of space and functional suitability to meet the expectations of building occupants. Satisfaction levels in such cases will not just be dependent on internal comfort levels but on the layout and configuration of space in terms of shape, size, and location. For instance, in an educational context the design of classrooms has been proven to assist with teaching and learning and improve morale and behaviours of staff and students. Accordingly good design could be related to providing adequate, appropriate, and functional space for educational organisations. For this reason, there are government design guidelines, as benchmarks, which set down recommended minimum levels of classroom space for teaching facilities, as m^2 area per student, for architects to adhere to. In addition, configuring desks in classrooms to give adequate visibility of teaching walls and projector screens could present another factor affecting the quality of learning. Other considerations may include the quality and scope of equipment, fixtures, and fittings in designated spaces e.g. computers and information technologies. This highlights the need to carefully consider the design of accommodation, for improved functional suitability, when remodelling space as part of maintenance, retrofit and refurbishment programmes.

2.6 Statutory compliance

Every organisation should be aware of the differentiation between essential and desirable maintenance. The former would include compliance with regulations and statutory requirements, to achieve minimum standards normally underpinned by established codes of practice, and where any breaches or infringements could be unlawful. This classification of work would normally be carried out at regular intervals and certification would be produced to confirm that it has been undertaken competently by qualified personnel. Managing risks around statutory compliance with regulations is therefore a crucial requirement for maintenance management. Any breaches or infringements could result in criminal prosecutions or civil actions being brought against organisations. This can not only be expensive in terms of litigation and enforcement action but potentially cause reputational damage to those companies affected. Conversely desirable works could include redecoration, window replacement, new finishes, and nonessential refurbishment programmes. Although this classification of works may not be required for statutory compliance, any works will still need to be carried out in a safe and competent manner and in adherence with health and safety legislation.

On new buildings, major refurbishment or where material alterations to properties are proposed in the UK, the building regulations would need to be complied with. This would involve submission and approval of building regulation applications, which would cover areas such as fire safety, access for disabled persons and structural stability. The requirements set down by building regulations are updated regularly and standards have been raised for areas such as sustainability and energy efficiency in recent years. Building regulation in the UK are not retrospective for existing buildings unless approval for material alterations and changes of use are sought. In addition, for new builds, changes of use and material alterations to the properties planning permission would need to be sought, unless exempt through permitted development.

2.7 Health and safety

In addition to the prescriptive regulations and statutory requirements referred to previously, there is a requirement by law for employers to maintain the health and safety of their employees at all times. The Health and Safety at Work etc. Act 1974 represents the main legislation in the UK. This act imposes duties to safeguard the health and safety of employees by their employers and to comply with all measures put in place. This is particularly important when carrying out maintenance programmes when arguably the level of risks to occupants and members of the public is greater. Accordingly, works needs to be carefully managed with use of method statements and risk assessment tools to ensure their health and safety at all times. In the UK the government agency that oversees all aspects relating to health and safety is the Health and Safety Executive (HSE).

On existing buildings and new builds, once occupied, there are a wide range of statutory regulations relating to health and safety of workplaces and these are included in Figure 2.1.

In consideration of new works related to carrying out construction and maintenance projects there is additional specific health and safety legislation and regulations which need to be complied with. Such regulation has responded to reported accidents and fatalities in previous years. According to the Health and Safety Executive (HSE), statistics have year on year shown that the construction industry represents one of the most dangerous industries in the United Kingdom and in 2019 alone, the UK suffered 30 construction worker deaths (Challender 2023). Concerted efforts to improve health and safety management when carrying out construction

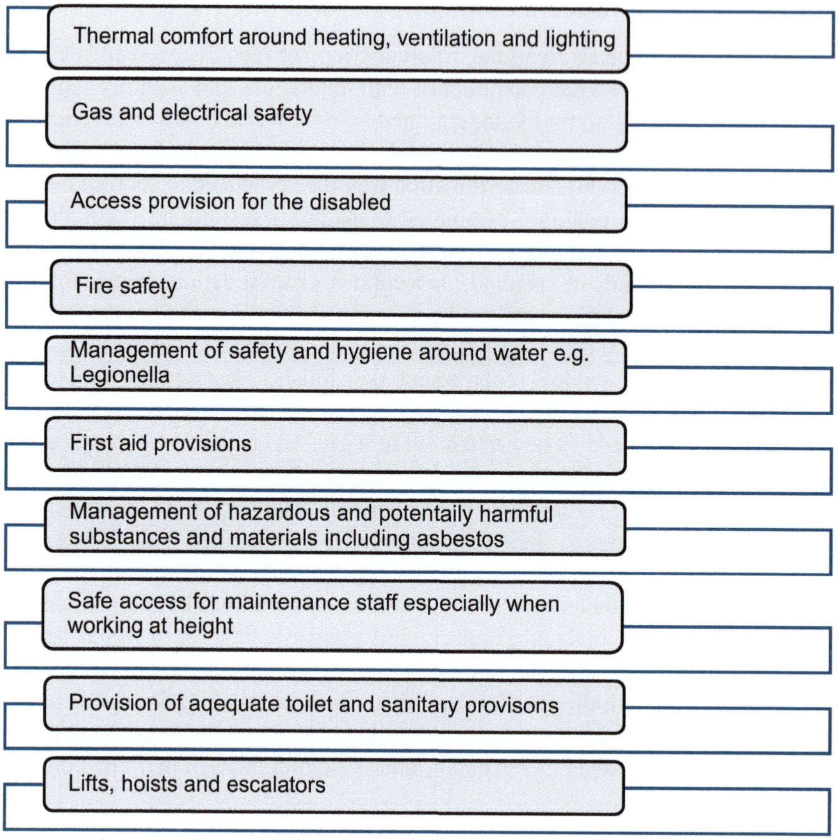

Figure 2.1 Statutory regulations relating to health and safety of workplaces

and maintenance works have been undertaken in the past. Measures have included having 'near miss' reporting and more rigorous procedures in place for ensuring that risk assessments and method statements are strictly adhered to for all works. Other improvement measures have included more stringent induction requirements and training for all workers carrying out building and maintenance works. In addition, some main contractors have introduced mandatory special safety accreditation requirements for all their supply chains. To obtain such accreditation has normally involved training of those staff that are required to work on construction and maintenance sites. Furthermore, in some cases, accreditation has become a prerequisite for the award of subcontractor work orders. One of the most common accreditation schemes in the UK of this kind is the CSCS (Construction Skills Certification Scheme) which involves individuals completing an online test to qualify as accredited members (Challender 2023). Furthermore, most construction and maintenance teams take adequate precautions in closely managing methods of work with safety in mind and ensuring that all personnel have the right personal protective equipment (PPE). In this way risks associated with potential injuries or fatalities can be reduced to an absolute minimum.

In addition to health and safety in the construction phases of projects, the UK government has introduced the Construction Design and Management Regulations (CDM) 2015 which

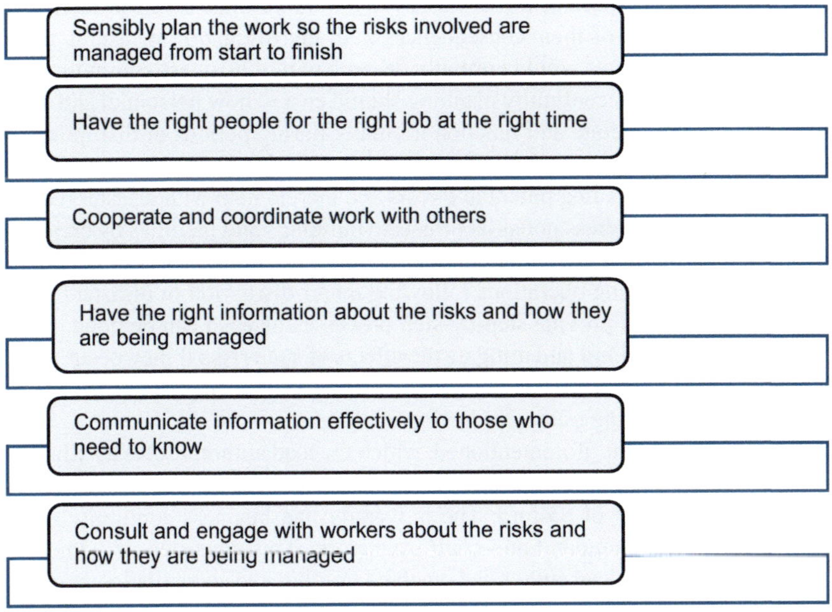

Figure 2.2 Requirements of the CDM Regulations

superseded the CDM Regulations 2007. The regulations describe the law that applies to the whole construction process on all projects from conception through to completion. They also dictate what each duty holder must or should do to comply with the law to ensure projects are carried out in a way that secures health and safety. According to Challender (2023), whatever the roles of individuals in construction and maintenance projects, the CDM Regulations aim to improve health and safety through the requirements in Figure 2.2.

2.8 Managing risks in workplaces, business continuity and disaster recovery planning

Addressing risks in the workplace, from an estate and facilities perspective, is an integral part of any good risk and maintenance management strategies. Estates can be sophisticated and complicated and can present endless risks from unexpected events which if not managed correctly can be extremely disruptive or disastrous. These issues could be related to maintenance projects or programmes or simply arise during normal business operations. The impacts from these risks and issues if not managed correctly could result in unacceptable costs being incurred, time delays, performance shortfalls, or closure of facilities in some cases. For this reason, organisations should be able to identify which buildings are critical to business operations and how responses can be planned in the event of unexpected issues arising.

It is crucial for estates teams to understand not only the specific risks to building users during construction and maintenance works, as previously articulated, but to have an awareness of other eventualities which could arise. These could include fires, floods, electrical failures, loss of heating and cooling, and information technology malfunctions e.g. cyberattacks. All these events could have severe implications for organisations in terms of disruption to their business activities. It is therefore important for risk assessments to be undertaken to identify potential estates issues, the severity of their consequences and any mitigation measures to limit their impact. These risk assessments would normally become part of business continuity and disaster recovery planning. Business continuity planning should ensure how personnel and assets can be protected and be able to operate and function normally during periods of disruption and crisis. They should be able to assist organisations in creating strategies, policies, systems, and procedures with the aim of overcoming potential threats and therein help with disaster recovery. In an estates context, these can address potential threats to buildings and facilities by creating systems of prevention and recovery if risks materialise. Having these processes in place can help businesses survive by maintaining operations following major disruption or disasters. They should be systematic in nature and provide step-by-step processes of what can be done to prevent the risks from materialising or limit and mitigate the effects of such risks if they occur. Having good business continuity and disaster recovery plans in place in this way can make the difference between companies remaining trading or having to close.

One estates example of the aforementioned, which the lead author, Professor Challender, was involved with, occurred during the COVID-19 pandemic, whilst he was employed as Director of Estates at the University of Salford. The university had business continuity and recovery plans in place through a command-and-control network. This consisted of 'Gold, Silver and Bronze Commands' and the lead author had received training and accreditation as a 'Gold Commander'. When the UK government in March 2020 issued a mandate for lockdown to control the spread of the COVID-19 virus, the business continuity plan was enacted to deal with the crisis from an operational perspective. He became at this point a leading senior officer of various teams which were created to oversee certain functions, and this included Chair of the Estates and Infrastructure Group. This group was commissioned to look at the implications of the lockdown on estates and facilities. It had to consider how the university's campuses could be safely secured and maintained during a period when most staff and students were working from home and teaching and learning was carried being out online. Considerations included making each campus safe and secure for some staff and international students who had to remain on site, and how basic services could still be maintained. The national and global crisis caused by the COVID-19 outbreak cross the world caused many organisations to close temporarily and, in some cases, permanently. Fortunately for the university their planning allowed them to conduct their business operations remotely and in a safe and secure manner for many months, which was generally regarded as a highly successful outcome. Other aspects of the university's business continuity plans included contingency arrangements for building closures from unforeseen risks such as fire, flooding, cyber and terrorist attacks. Mitigation through business continuity planning enabled the university to enter into a reciprocal contractual agreement with another local university to use their accommodation for teaching if future closures arise. This contingency planning was therefore designed to maintain teaching operations, from alternative accommodation if needs arise.

2.9 Planned preventative and reactive maintenance

According to Higham *et al.* (2021) maintenance can be described as unscheduled or planned maintenance and will become required when items become damaged or are not functioning at their full or required capacities. At these stages, repairs or renewals will be required. Wherever possible, 'planned preventative maintenance programmes' (PPMPs) can reduce the degree of 'reactive unplanned' maintenance. Planned preventative maintenance represents any form of maintenance on buildings or equipment that is scheduled and conducted on a regular basis. Its concept is predicated on the premise that maintenance teams should not wait for building elements to fail. Instead, plans are created which incorporate regular checks, servicing, and repairs to ensure that they remain in full working order and condition for as long as possible. Reactive maintenance refers to repairs or renewals performed after assets have broken down which can affect business operations and productivity. For this reason, reactive maintenance can be highly disruptive in nature. Notwithstanding this premise, there will always be a degree of reactive maintenance within any management planning as not all eventualities or failures of elements and components for buildings can be anticipated. For instance, buildings may become impacted by force majeure factors such as extreme weather conditions, where roofs and walls become damaged by heavy winds, or flooding may result from heavy rainfall. In other cases, electrical or mechanical installations and plant could fail unexpectedly, even with the highest standards of servicing and maintenance. Buildings can also suffer from vandalism or inappropriate use by building occupants which could necessitate remedial works to overcome. Accidental damage could also exist from building users or members of the public and vehicular damage would fall under this category. For all the aforementioned reasons organisations should have monies allocated within their maintenance budgets to deal remedial works from these eventualities. Notwithstanding the need to 'plan' for unplanned works, which appears to be a contradiction in terms, the degree of reactive maintenance should be reduced to a minimum. Instead, organisation should look to incorporate planned preventative maintenance (PPM) wherever possible.

Planned preventative maintenance methodologies utilise programmes for maintaining building elements to predetermined standards. Such programmes rely on investigating the physical condition of existing buildings and facilities through condition surveys. PPM should predict potential defects or faults which could arise from condition survey information. Thereafter planning, scheduling, and programming of remedial works to respond and address any existing or potential issues can be undertaken. Programmes to carry out any repairs, maintenance, and renewals as part of such remedial works need to be regularly updated and become part of 'rolling' annual implementation programmes. For this reason, planned preventative maintenance represents the scheduling and implementation of building maintenance activities which seek to prevent or reduce unexpected failures in the future. Accordingly, as the name suggests, planned preventative maintenance programmes are predicated on prediction of potential failures, defects, and issues to avoid them occurring in the first place. This is preferable rather than simply responding to the aftermath of building failures. Accordingly, they are considered to be 'prevention rather than cure' orientated.

As part of planned preventative maintenance regimes, building services should be regularly serviced and repaired by suitably qualified individuals. Other examples of preventative measures could include carrying out regular building surveys and undertaking roof repairs before roof coverings fail and inflict resultant rainwater damage to internal environments. Consequently, planned preventative maintenance should serve to anticipate future implications associated with the condition of building elements. This predictive approach is designed to avoid potential defects emerging and to minimise resultant damage and disruption associated with prolongation of any issues. Examples could include identifying early signs of timber decay before a major infestation of wet or dry rot or beetle attack is caused. Other examples could include addressing early signs of corrosion of reinforcement in concrete panels and structural frames affected by concrete cancer. Capturing problems earlier rather than later could assist the maintenance team in preventing the condition escalating into a major health and safety issue associated with spalling concrete. Early diagnosis and action should at the same time significantly reduce the extent of repairs required and avoid the requirement for major remedial works. In this way any potential faults can be anticipated before failures exist, which will serve to limit remedial works and avoid unacceptable levels of disruption to business operations and building users. This concept of prevention will be discussed further in Chapters 4, 6 and 7 of the book which cover building pathology and potential detrimental effects on buildings and potential defects.

Planned preventative maintenance can be based on 'life cycle analysis' principles which will be discussed more in Chapter 11. These principles will predict the likelihood of building failures or undesirable events occurring and thereby through risk assessment techniques plan and estimate what works are required to avoid them. This can sometimes present a daunting task for maintenance management in accurately programming associated timescales for implementation. Other challenges are associated with correctly estimating and costing the required works for future budgetary planning. Complications and difficulties arise as there are many different factors which will influence when planned preventative maintenance works need to be undertaken. Such factors could include the quality of design and construction, age of buildings, the quality and degree of repairs to maintain condition and the type and intensity of building use. For instance, older buildings, which are used intensively by members of the public every day, may require more works over shorter time spans than newer buildings used intermittently by their staff.

Sometimes risk assessment processes as part of PPM need to consider the potential consequences, impact, and severity of not carrying out maintenance works or other mitigation measures. They are sometimes presented in the form of a scoring matrix based on the severity of failure and their consequences.

2.10 Benefits of planned preventative maintenance

The proportion of unplanned to planned maintenance within any organisation should be carefully considered as part of maintenance management strategies. Auditors commissioned by the UK government recommended a ration of 25:75 unplanned to planned maintenance expenditure (Thomas 2013). Notwithstanding this government target, maintenance teams will be responsible for presenting and demonstrating evidence of efficiencies and other benefits from delivering a regime of planned maintenance investment. One of the main advantages of planned maintenance is that work can be programmed in advance which gives time for reflection and procurement of required contract works. In this way time is made available to consider methodologies for best value procurement and carrying out works more efficiently. Sometimes this can involve grouping works into programmes e.g. roof replacement, which can bring financial cost savings from economies of scale. It also gives maintenance teams the time to decide on the best means to procure contract works, via tendering or through established existing term or framework contracts. It is usually more cost-effective to undertake planned maintenance work in this way rather than reactive works which normally relies on quick responses to isolated failures. One could also argue that the planned maintenance can be more carefully organised and managed to reduce disruption to end users, a minimum alongside avoiding any known risks.

2.11 Decision-making around repairs or renewals

Planned maintenance should not only be regarded as responding to future failures but can look at obsolescence where future parts for old equipment, plants and machinery may become difficult to source. Decisions may be taken then to replace building elements earlier rather than later where future parts availability present problems for repairs to be undertaken. In addition, decision-making to renew elements of buildings may be predicated on improving the environmental sustainability credentials of buildings. Examples of planned maintenance in this regard could include a programme to renew light fittings within the building to replace old tungsten filament lighting with new LED fittings. Maintenance managers can present business cases to offset the capital costs associated with new LED lighting against energy efficiency savings generated in the future.

Other potential environmental sustainability improvements could include replacing wall cladding or roofing systems with newer alternatives that increase building performance around energy efficiency. In such scenarios, cost benefit analysis as part of business cases would assess and evaluate the relative costs and pros and cons of repair or renewal alternatives. The benefits of repair could be to reduce initial capital costs, but the longer-term implications could be higher running costs associated with poor heat loss through building envelopes. Conversely the increased capital investment costs of renewal of cladding and roofing systems would have to be measured against future benefits of lower running costs and improved environmental credentials for buildings. This exercise can sometimes present a dilemma for organisations. Finance directors have tendencies to think short term rather than longer term and therein are more orientated to reduce initial capital costs at the expense of increased ongoing operational revenue costs. To 'turn the tide' on this dilemma, maintenance managers should therefore present justifications for recommendations around longer term considerations and use life cycle analysis to unpin their assertions in business cases. Life cycle analysis will be discussed later in the book.

2.12 Obsolescence of buildings

The definition of obsolescence varies between sources but according to Oxford Languages (2024) it is the process of becoming obsolete or outdated and no longer used. It can also be described as a loss in the utility or value of property that results over time from intrinsic limitations or external circumstances. It is generally accepted that there are four different types of obsolescence, and these include physical deterioration and obsolescence brought about by function, location, and economic forces. The following real-life example of the latter two types of obsolescence can hopefully articulate how location and economic considerations can influence the life of buildings.

Case study of British Midland Airways Central Training Facility, West Drayton Middlesex

The lead author was employed by the airline to procure a new central teaching facility close to Heathrow Airport for provision of all training services including cabin crew, pilots, and customer services representatives. With available land for development being in short supply within the boundaries of Heathrow Airport, a site was identified by Professor Challender on a small industrial estate within a short drive from the airport. Closeness to the main airport was essential for the airline in transporting staff, mostly based at Heathrow Airport, as its operational hub, over to the new facility. The new building, shown in Figure 2.3, comprised four flight simulator pods, a large atrium containing a large aircraft fuselage for cabin crew evacuation training and many different classrooms for teaching and learning.

The design and construction of the building was successfully completed to an excellent standard and the facility was internally regarded as a centre of excellence. It was visited by many international airlines interested in training their staff and went on to win many prestigious awards. The overall project costs in 1997 when the building was completed amounted to approximately £30m. The level of quality in construction, technology and flexibility incorporated were all designed to future proof the building for many decades. Unfortunately, although still in pristine condition, the building's fate was affected by external factors after only 13 years in operation. The global airline industry had suffered from the aftermath of the terrorist attack in New York on 11th September 2001, coupled with a trend for budget fares in lieu of business travel. This dramatically affected the airline which could no longer afford to run the facility and the building was sold shortly after its closure in 2010. At the same time the site had become a strategic site for distribution with its close reach of Heathrow Airport. From an economic perspective the building was no longer viable for its current use with the demand for airline training highly diminished and land values for distribution-related uses rapidly increasing. Accordingly, the building was demolished by the new owners, and the site developed as a large distribution centre for freight to and from Heathrow Airport.

Figure 2.3 Photographs of British Midland Airways Training Centre

The previous example demonstrates that even with the best forward planning, buildings can be affected by unforeseen external forces which can render them obsolete. Accordingly, obsolescence can sometimes be extremely difficult to control or forecast as processes linked with external market forces can be random in nature. This represents a common dilemma for organisations in developing new proposals and maintaining their existing estates in trying to predict the future of their operations and asset holdings.

Determination of time horizons for the life of buildings, sometimes referred to as life cycles, is something that can be daunting for maintenance teams. Notwithstanding this premise, it is important to consider the expected life expectancy of buildings for continuity, planning of new facilities or relocation of businesses to alternative sites. For instance, take the hypothetical example of a business operating in the South East of England. Their premises may have a relatively short term to run before demolition becomes necessary which could be because of its age, condition, functionality, and appearance. The building owners may be incurring high operational and maintenance costs in retaining the building in its present form, and it may no longer portray the image they want to project for their customers and the public. They may therefore decide that there is a need for a new building to create a modern image and branding for their business. In addition, they may consider that it could be more cost-effective to relocate to the North West of England where land, capital and operating cost are generally lower. Notwithstanding these longer-term aspirations, their plans may be affected by the estimated longevity of their current building, as they may wish to stay in their existing premises for as long as economically possible. For this reason, they need to be able to plan, sometimes for several years in advance, for such a strategic move to a new location to take place. If they have established that the economic life of retaining their existing accommodation is relatively short, e.g. five years, then they are very unlikely to want to invest in any significant maintenance expenditure, with imminent building demolition in mind. However, they would still be required by law to undertake maintenance works deemed urgent, and possibly linked to potential breaches of statutory requirements and health and safety issues. In this way they can plan to run down their facilities in a controlled and planned way to avoid any unnecessary maintenance expenditure, prior to demolition. Obsolescence of buildings therefore represents an important consideration for any business in understanding the parameters and issues for planning new facilities and accommodation.

2.13 Summary of chapter

There is no universal definition for maintenance, but clearly there are common strands which exist to articulate its meaning in practice. As well as natural deterioration processes, buildings, when not maintained to the correct standard, can fail unexpectedly from physical or performance perspectives. This can lead to malfunctions of building services or the emergence of physical building defects which can cause significant disruption to users and expensive repairs for building owners. To avoid such failures and ensure that buildings as valuable assets do not depreciate over time, it is important that maintenance regimes are put in place. Such regimes can be affected by use and type of buildings and examples were presented in this chapter to explain the different scenarios e.g. listed buildings.

It is essential to consider the standards by which building owners wish to maintain their estates. These standards can have a direct effect on the comfort and satisfaction of building occupants which in turn could affect the performance of businesses. There can be different approaches to setting building standards and some building owners may wish to set standards very high to present the right image of their organisations, and thereby attract new customers.

Conversely others may regard built assets as merely accommodation for conducting their core business functions within and therefore seek to limit repairs, maintenance, and any improvements to a minimum. In addition to the standards associated with state of repair and physical condition, there are also expectations for building accommodations to be fit for purpose in terms of space and functional suitability to meet the expectations of building occupants.

Essential maintenance includes compliance with regulations and statutory requirements and any breaches or infringements could be unlawful and possibly result in litigation and enforcement action. Moreover, desirable maintenance refers to works of a non-statutory nature including redecoration, window replacement, new finishes, and nonessential refurbishment programmes.

The Health and Safety at Work etc. Act 1974 requires employers to maintain the health and safety of their employees at all times. This is particularly important when carrying out maintenance programmes when arguably the level of risks to occupants and members of the public is greater. Accordingly, work needs to be carefully managed with use of method statements and risk assessment tools to ensure health and safety is maintained at all times. There have been concerted efforts to improve health and safety management when carrying out construction and maintenance programmes. Improvement measures have included 'near miss' reporting, accreditation such as the Construction Skills Certification Scheme and the introduction of new regulations including the Construction Design and Management Regulations (CDM).

Estates can be sophisticated and complicated and can present endless risks from unexpected events which if not managed correctly can be extremely disruptive or disastrous. The impacts from these risks and issues if not managed correctly could result in unacceptable costs being incurred, time delays, performance shortfalls, or closure of facilities in some cases. Examples of risks have been given in this chapter which have included fires, floods, electrical failures, loss of heating and cooling, and information technology malfunctions e.g. cyberattacks. It is therefore important for risk assessments to be undertaken to identify potential estate issues, their severity, and any mitigation measures to limit their impact. These risk assessments would normally become part of business continuity and disaster recovery planning.

Maintenance can be described as unscheduled or planned maintenance. Planned preventative maintenance programmes (PPMPs) can reduce the degree of 'reactive unplanned' maintenance. They incorporate regular checks, servicing, and repairs to ensure that buildings remain in full working order and condition for as long as possible. Furthermore, they should predict potential defects or faults which could arise from condition survey information to plan, schedule, and programme remedial works more effectively. This predictive approach is designed to avoid potential defects emerging and to minimise resultant damage and disruption associated with the prolongation of any issues. Examples to assist readers have been provided in this chapter which have included defects associated with timber decay and concrete cancer. Benefits of PMPs can include the ability to group works into programmes e.g. roof replacement, which can bring financial cost savings from economies of scale.

Renewal as opposed to repair decisions may be made when buildings or some of their component elements are approaching obsolescence. There are four different types of obsolescence, and these include physical deterioration and obsolescence brought about by function, location, and economic forces. A case study was provided in this chapter where a relatively new building procured by the lead author was adversely affected by unexpected economic forces leading to its demolition. Accordingly, predicting time horizons for the life of buildings, referred to as life cycle analysis, is something that can be daunting for maintenance teams.

The next chapter will follow on from the concepts raised and introduce delivery strategies associated with maintenance policies, programmes, and systems.

References

Challender, J. (2023). *Organisational ethics in the built environment.* Blackwell-Wiley.

Higham, A., Challender, J. and Watts, G. (2021). *Introduction to built asset management.* Blackwell-Wiley.

Oxford Languages (2024). Available from https://languages.oup.com/oxford-languages-branding-resources/ (accessed 23rd June 2024)

Thomas, M. (2013). *Managing building pathology and maintenance construction managers' library.* Leonardo da Vinci.

Chapter 3

Maintenance and asset planning strategies, programmes, policies and systems

3.1 Introduction to chapter

The chapter will begin by introducing maintenance strategies and policies and how organisations seek to maintain their buildings and estates, keeping them operationally fit for purpose and retaining their assets' values. Within this area, it will elaborate on the mechanisms by which these strategies and policies are derived and implemented alongside establishing the levels, and timescales of maintenance needs as part of the maintenance programmes. This will then lead into managing building maintenance programmes together with the quality assurance measures which can be undertaken to measure and audit their performance. From there the chapter will then move on to the area of asset management and use hypothetical examples of threats and opportunities that organisations are faced with when competing in competitive market environments. Accommodation and space management considerations arising from these market forces will be examined and how organisations are faced with alternative choices in their decision-making. Asset management plans and building condition surveys will then be covered alongside statutory compliance and health and safety. This part of the chapter will also assess utilisation of space, satisfaction levels of building occupants, operational and functional performance, and suitability of buildings and facilities.

The chapter will then move on to balancing maintenance budgets, and the need to consider prioritisation of certain works over others. This will include the different considerations and implications of deferring certain works and any effects that this could have on business operations and continuity, occupier satisfaction and deterioration of buildings. Again, hypothetical examples will be presented to assist readers in contextualising the possible impacts of delaying certain essential repair and maintenance works.

Concepts and practices in establishing maintenance requirements will then be introduced. This will cover the various categories of urgency that work can be encapsulated within depending on noncompliance, condition, health and safety and other considerations. This will then lead into discussions around prioritisation which is an important aspect of maintenance planning. In this regard the influence of risk assessments, especially around health and safety, will be examined. Practical difficulties using examples will then be presented to explain why organisations may decide to reprioritise certain works over others. Following on from this, procurement of maintenance works will be introduced and factors which could influence whether repair works are undertaken by in-house teams or externally outsourced. Pros and cons of each approach to procurement will be articulated with the aid of hypothetical examples for practical contextualisation.

The chapter will then move on to discussions around maintenance programmes and problems associated with backlog maintenance. Finally, the various issues around defect liability periods

DOI: 10.1201/9781003358800-3

for newly completed construction and maintenance contracts will be examined, from perspectives of responsibilities and liabilities for clients and main contractors.

3.2 Maintenance strategies and policies

As part of organisational planning, all medium to large companies should have what are commonly referred to as '*strategic corporate business plans*'. These normally include statements relating to vision, mission, short- and long-term aims, goals and objectives alongside action plans and approaches to delivering these. In these plans there should be aspects that relate to their building and estate assets. This could relate to their organisation aspirations to acquire or dispose of properties or leasehold interests, expand, or contract their estates, or undertake any major redevelopment and master-planning programmes. They should also, from the perspective of real estate planning, include a section on property maintenance strategies and any programmes or policies that support them.

Building maintenance strategies and policies should include statements of how organisations seek to maintain their buildings and estates and keep them operationally fit for purpose whilst retaining their assets' values. A clear set of objectives should be articulated together with the methodologies and approaches for how the maintenance team plans to deliver them. In this way, maintenance strategies and policies should become frameworks for implementation of maintenance operations. Building maintenance policies should list each building asset, their market and book values and give a clear indication of their longevity and life spans. Sometimes what is referred to as an impairment loss can occur where the book value of an asset is greater than its market value. In such a case finance departments within organisations need to record the new value and write off the difference as part of their financial and governance accounting and regulations.

Building maintenance strategies and policies should follow a well-structured format which would normally include the following areas:

- Budgets and budgetary controls
- Health and safety
- Use of maintenance contractors
- Directly employed maintenance staff within the organisation

In addition to the aforementioned, they should also articulate the scope, level, and timescale of maintenance needs as part of the maintenance programme and the quality control measures to competently execute maintenance and servicing operations.

3.3 Managing building maintenance programmes

Managing maintenance programmes relies on processes related to compiling maintenance plans with associated allocated timescales for implementation, alongside quality assurance measures which are planned to be undertaken to measure and audit their performance. In preparing this level of information the following stages in Figure 3.1 would need to be carried out.

In carrying out the stages in Figure 3.1, maintenance teams will be required to possess skills, knowledge and expertise relating to information gathering, possibly from historical data. It will also require them to have experience in planning resources, budget setting and procurement of maintenance and works contracts. They will also be required to have supervisory skills in

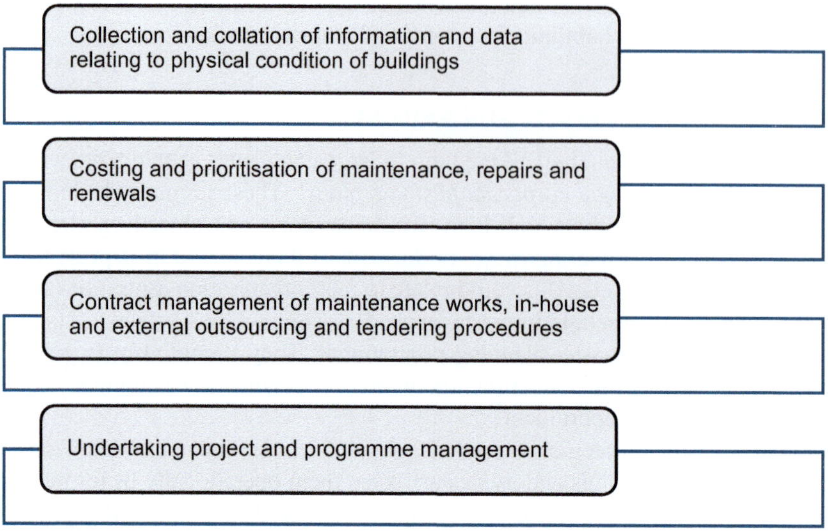

Figure 3.1 Stages required to manage building maintenance programmes

overseeing works, both by in-house maintenance staff and external contractors to ensure they are carried out to the required standard. Furthermore, there will be expectations for them to assess which maintenance works would be more cost-effective to carry out in-house, with directly employed maintenance tradespersons, or to outsource through external contractors. In the latter case consideration would be required whether to tender individual projects separately or utilise 'term' or 'framework' contractors that may have been appointed to undertake a programme of works over a designated period of time.

3.4 Asset management

Asset management in property terms should ensure that buildings and land provide a higher level of benefits to organisations in meeting their business aims and objectives. Accordingly, property asset management is a business process which should be very much linked to the finite resources available to organisations. Its purpose is to achieve the best possible utilisation of real estate assets and in doing so minimise the opportunity costs in tying up resources in buildings and land. Moreover, those valuable resources should be deployed in the business activities of organisations in helping them perform successfully, grow and attain good profit levels and returns for shareholders. An example could be a manufacturing company which would reap more operational benefits from investing in new machinery rather than carry out property repairs. The new machinery could increase their production and turnover, whereas the same expense incurred on replacing their roof would not improve their operational position. Clearly one could argue that if the roof was defective and leaking this could hamper operations and therefore repairs were essential for productivity reasons and business continuity.

Organisations are constantly faced with threats and opportunities from competitive market conditions. To respond to these challenges, they are frequently faced with making changes and devising new business propositions. In most cases these propositions will require business cases to be prepared which detail the associated costs and the anticipated benefits to be generated. It is usual for such plans to include some investment linked to estates and property assets. For example, if a company has a desire to expand then this proposition may require more operational space. This additional space could be achieved in different ways, through leasing more accommodation, extending their existing buildings, or moving to alternative larger premises. All these options carry different costs and benefits, and asset management can assist them in analysis and evaluation as part of decision-making processes. Accordingly, asset management strategies are frequently amongst the key considerations in organisations when preparing their business plans. They should contain all the requirements of the estate for conducting 'business as normal' or any alternative business proposition put forward. The key benefits for preparing and delivering asset management plans in this way, as identified by Thomas (2013), are included in Figure 3.2.

In addition to the benefits in Figure 3.2, asset management plans will identify real estate shortfalls in terms of delivering corporate aims and objectives, normally focused on core business activities. These can include shortage of space, unsuitable or inflexible accommodation, surplus space or building defects which could be restricting business operations. Asset management plans should address these factors and devise different alternative propositions and opportunities that could deliver benefits to assist business operations. Sometimes this would involve redevelopment, reconfiguration of buildings and facilities or complete rationalisation of

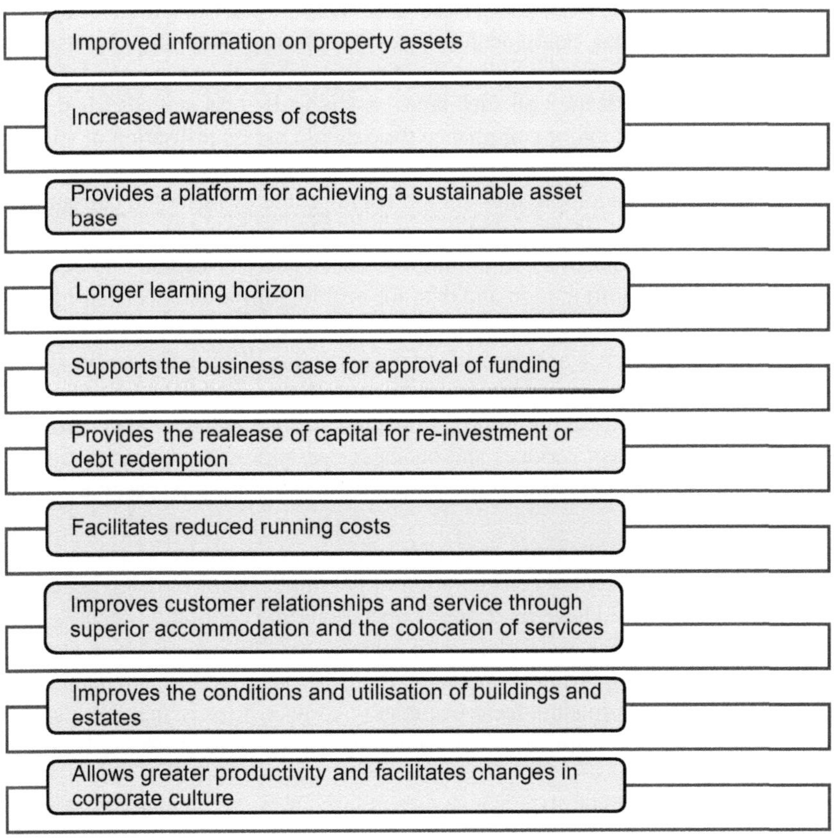

Figure 3.2 Key benefits for preparing and delivering asset management plans

the whole estate portfolio. In the latter scenario, it would entail preparing an estates master plan across different phases to deliver benefits in a staged delivery programme to reduce disruption whilst maintaining business continuity.

Property asset management relies on having facilities management regimes in place which are both effective and responsive. These should underpin the core activities of businesses rather than be regarded as separate activities in their own right. They should include areas such as maintenance management, environmental sustainability, energy efficiency, procurement of services, project management, information technology, security, waste management and cleaning. Property asset management should be centred around aims and objectives related to preserving the condition and interests of property assets over time, complying with statutory requirements, maintaining the health and safety of users and satisfying requirements and expectations by building owners and occupants. Property asset management plans should not be regarded as stagnant documents but conversely should provide a framework for constant review as business environments change and allow for review and upgrading as required to support organisations.

3.5 Surveys as part of asset management

In preparing asset management plans, it is necessary for organisations to undertake condition surveys of all their buildings and estates portfolios. These surveys should establish the state and condition of all elements of building structures, fabric, electrical and mechanical services, internal spaces, external areas, equipment, fixtures and fittings. They should consider statutory compliance of various elements including building services against current legislation. This will involve checking certification on such areas as legionella, gas and electrical certification. As well as condition and statutory compliance they should assess utilisation of space, satisfaction levels of building occupants, operational and functional performance and suitability of buildings and facilities. In addition, they should appraise and evaluate the current cost of running and maintaining estate assets including existing and predicted maintenance costs, rents, potential acquisitions and disposals, and building values and any capital depreciation. In this way, they can provide the information and data for enabling informed decision-making on any estates planning. This can include redevelopment, relocation, acquisition and disposal strategies alongside the long-term sustainable use of estate portfolios. This information should enable decision-making around changes or investment in estates to maintain them as fit for purpose and 'future proofed'. In some cases, the information can facilitate business plans around delivery of more effective and efficient services and business operations. Furthermore, it should assist in maintaining the health, safety and well-being of building users and preserve the asset value of estates.

In balancing budgets, maintenance managers will need to consider prioritisation of certain works over others. This will entail consideration of the implications of deferring certain works and any effects that this could have on business operations and continuity, occupier satisfaction and building deterioration. For example, a roof renewal may be regarded as a main priority if water penetration is affecting business operations, causing ongoing resultant damage to internal elements, and receiving complaints from occupiers. A lower priority may be allocated to the replacement of a major piece of plant, which albeit inefficient due to old technology, is currently operating satisfactorily.

Normally asset management plans are presented in the form of databases and spreadsheets. Each element may be listed separately and dashboards or scores allocated against its condition, performance and its level of priority if remedial works are required. Costs and timescales associated with remedial, renewal or mitigation works would normally be apportioned to each element. This information would become part of a maintenance programme and used for budget setting year on year, which was previously referred to in Chapter 2. Aspects of environmental sustainability may be considered for elements that may benefit from retrofit improvement works or replacement to make them become more energy efficient. An example could include insulation measures to walls and roofs to reduce heat loss and therein improve energy conservation. Environmental sustainability and relationships with maintenance management will be discussed further in Chapter 10.

3.6 Establishing maintenance requirements

The process by which maintenance requirements are established is based on expected levels of needs. These maintenance needs are predicated on methodologies, data and other information generated from the following sources in Figure 3.3.

Building condition surveys will normally be carried out at intervals normally ranging from five years to ten years. They should address the physical condition of all elements of buildings including structure, fabric, finishes and building services. The state of condition for all elements is normally categorised according to the findings of surveyors conducting the inspections. These can range between very poor to very good and all the categories in between. In addition to outlining the state of condition of each element, the surveys should make recommendations as to necessary remedial works to address any issues of disrepair. Any items deemed to be very poor, or poor will clearly require remedial works whereas those classed as good or very good, by definition, may not require any attention. Those classified as acceptable or satisfactory may require some remedial works to counter further deterioration. The survey reports should also address elements of noncompliance with statutory regulations and make provision for upgrading or renewal works where deemed necessary. Any health and safety compliance issues should be addressed in a similar way. The information and data from these surveys is compiled as schedules and on databases which will be the basis for maintenance plans and programmes.

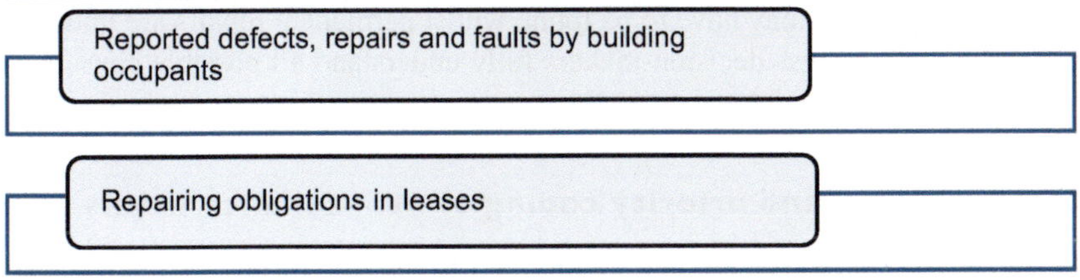

Figure 3.3 Information sources to establish maintenance needs and requirements

Normally the repair works recommended in condition surveys will be categorised into minor or major works of renewal, repair, redecoration, refurbishment or adaption. Irrespective of which category they are placed within, each element of work should be fully quantified, prioritised and costed. Reference is sometimes made to life expectancy of elements which can affect the order of priority they are designated.

3.7 Prioritising and classifying maintenance works

Having conducted the building condition surveys, it is necessary to make assessments relating to the urgency of carrying out any remedial works. Prioritisation is therefore an important aspect of maintenance planning and one which can sometimes require surveyors to use their knowledge of building pathology to predict rates of deterioration. If building elements are predicted to deteriorate faster than others owing to their condition, they are more likely to receive higher levels of prioritisation, to avoid further damage occurring to buildings. An example could be leaking internal water vessels which left unattended over a period of time could cause further issues associated with dampness, corrosion and timber decay. Prioritisation is not only linked to condition and consequential damage, as it can be influenced by compliance with regulation and statutory requirements. For instance, if an electrical installation cannot be certified as it does not comply with current regulations then upgrading and/or renewal must become a priority. Other highly prioritised works could include those where health and safety risks or dangerous conditions have become apparent, with consequential risk to life e.g. presence of hazardous materials such as asbestos. In these circumstances compliance with the Health and Safety etc. Act of 1974 would deem urgent remedial works or action to be required to avoid such risks.

Conversely if works are deemed to be required but organisations for financial reasons cannot undertake them then their prioritisation could be classified as lower in favour of more affordable maintenance works. This can occur with major items of expenditure where the costs are regarded as unaffordable and prohibitive. Examples could include renewal of major pieces of equipment or plant e.g. boiler replacement where companies feel they can defer the works and simply operate with a heating system which is both inefficient and unreliable. Another example would be roof renewal works where business owners may decide to delay in favour of carrying out further patch repairs. In these two scenarios, maintenance teams should make building owners or their senior leadership teams aware of the implications and risks associated with deferring remedial works. When considering delaying the boiler replacement, in the example given, this could result in higher energy costs, unreliable levels of heating and in a worst-case scenario the sudden unscheduled failure of the heating plant. Similarly in the example of the required roof renewal, patch repairs may fail to prevent water penetration through the roof covering which could lead to internal damage. In both these cases possible disruption to building occupants and business operations may be unavoidable. Furthermore, it is not inconceivable that temporary alternative accommodation may have to be found whilst permanent repairs are undertaken. It is therefore crucial that business decision-makers fully understand all possible consequences and risks of deference of such maintenance works.

3.8 Categorisation and priority coding of maintenance works

In prioritising maintenance works it is preferable for minimum timescales to be designated for each classification of priority. Where works are related to statutory compliance and adherence with health and safety they should be deemed urgent and unavoidable. Measures to address

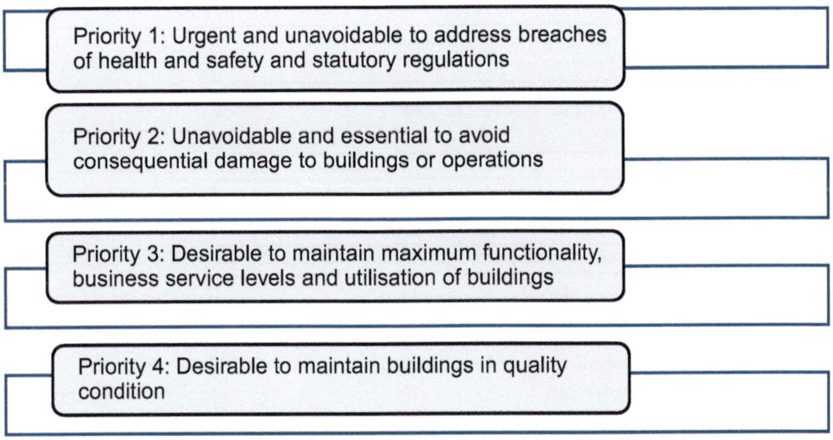

Figure 3.4 Different levels of priority for repair and maintenance works

them could mean that remedial works are required to be completed within as little as a week. Failure to attend to them in a timely manner can lead to unnecessary risks to end users and prosecution in some cases. Conversely if works are classified as essential but not urgent then repairs or renewals within a year may be acceptable. To record levels of priority as part of building condition surveys it is best practice to code all required works. Priority levels could be presented to depict works as shown in Figure 3.4.

This prioritisation can apply to both reactive and planned maintenance works. For reactive works, where faults are reported from building users to maintenance teams via help desks, it would normally be the systems administration teams that would classify the level of prioritisation, based on the information they have received. To enable administrators to make these valued judgements, sufficient information on faults from building users would need to be provided. To assist them and furnish them with all the relevant information, help desks such as CAFM should prompt users to explain as much as possible about the urgency of issues being reported.

3.9 Procurement of maintenance works; in-house or externally outsourced

As referred to earlier in the chapter, maintenance works can be delivered by in-house maintenance teams made up of maintenance managers and directly employed tradespersons. Alternatively, maintenance can be completely outsourced to specialist companies that specialise in providing 'whole service', 'turnkey solutions' for their clients. Each of these solutions can have advantages and disadvantages for businesses. The latter solution may suit those smaller companies who do not have the financial resources to invest in their own teams or feel that building maintenance goes far and beyond their core businesses. Alternatively, the former solution might be more conducive and suitable for those companies that already operate building services as part of their core business operations. This is not to say that companies must opt for this 'all or nothing' position where they must choose from each scenario. In most medium to large businesses there is usually a balance of both in-house and contracted maintenance works. From the lead author's position in running a large university estates department, this may entail employing

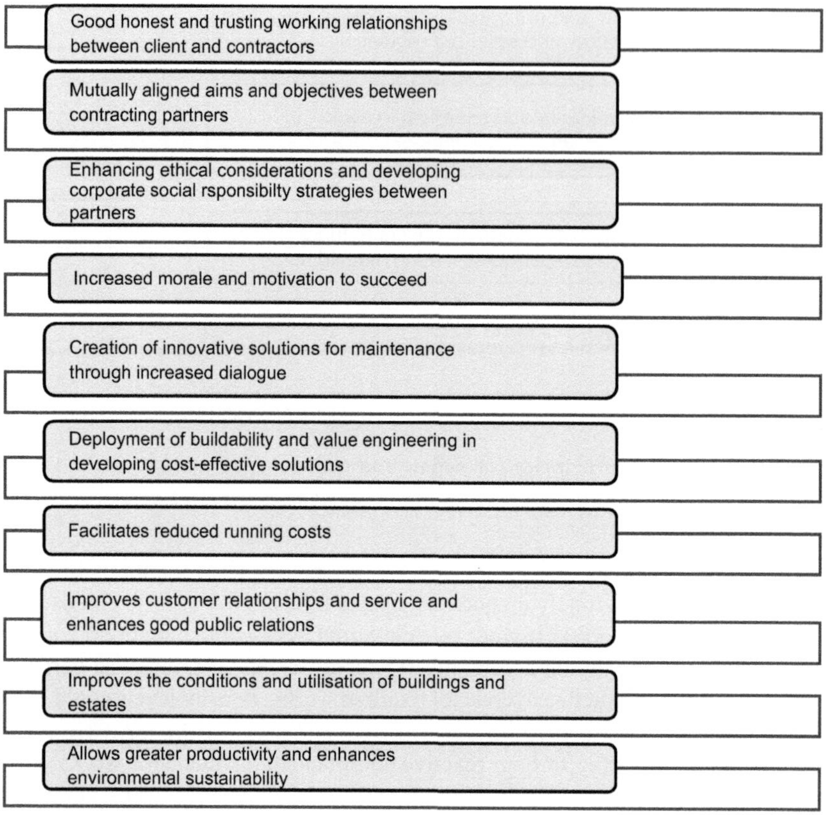

Figure 3.5 Benefits and successful outcomes from partnering (Challender *et al.* 2019)

a maintenance management team, to oversee planned maintenance and having in-house tradespersons to carry out more minor day to day repairs. However, where larger specialised works are required such as roof renewal programmes then external contractors would logically be appointed for reasons of scale, resources and competency. These external contractors could be procured through existing established framework contracts or alternatively through separate tendering procedures. In the former case framework or 'term' contracts would involve some form of partnership between clients and their appointed contractors. This could entail colocation of workplaces in some cases, established open book accounting procedures and 'gain share, pain share' financial mechanisms. According to Challender *et al.* (2019), these strategic partnering relationships can bring about many benefits and successful outcomes for clients and contractors in the past, some of which are detailed in Figure 3.5.

Other advantages for partnering between clients and framework contractors could be associated with formulating standardised solutions to common maintenance problems, better deployment of supply chain knowledge and expertise and improved quality control. All these factors can assist organisations in achieving best value for money and obtaining better maintenance outcomes.

3.10 Maintenance programmes

Estates and facilities teams within organisations should carefully programme for various planned works to be carried out. Such works may have been determined by building condition surveys undertaken which itemised the required remedial actions and set down timescales for their completion. As referred to earlier in the chapter, it is usual for such maintenance programmes to be on five-year cycles to spread the costs and ideally budgets and sinking funds set for each year to finance these works. An example of such a maintenance programme is contained in Appendix 1. Beyond five years it can be difficult for estates teams to predict the correct level of maintenance required owing to many different variables which could emerge in the intervening period. In addition, finance departments within organisations normally operate their budgetary planning on five-year cycles, so it makes sense for estates planning to utilise the same time parameters.

Maintenance programmes represent a way that organisations can better oversee their asset management. They can be regarded as enabling tools to maintenance teams in forward planning works into the future. They are also designed to avoid cases where buildings become deteriorated to the point where their condition adversely brings about failures of their structures, fabric or services. In such cases there could be consequential risks posed to the health and safety of building occupants. Maintenance management should be about addressing known building issues from reactive maintenance as identified in the previous chapter. This could relate to a programme of window replacement where windows from building survey records are deemed to be in poor physical and operational condition. In addition, the best estates planning is related to planned preventative maintenance predicated on avoidance of issues emerging in the first instance.

3.11 Backlog maintenance

Maintenance backlogs occur when repairs or renewal works fail to keep pace with the condition of buildings as they deteriorate, or are alternatively linked with statutory regulations e.g. legionella testing and certification. Accordingly, backlog maintenance represents a programme of maintenance works or tasks which are deemed essential and have been scheduled or are planned to be carried out. The essential nature of these works might be related to defects such as water penetration through roof coverings which if left unattended could grow progressively worse causing damage to internal elements and finishes.

Backlog maintenance can normally be put in the context of value in how much expenditure needs to be invested to bring buildings or facilities back up to an acceptable state of repair and condition. This would apply where property estates assets have fallen below a defined acceptable standard relating to their condition. Sometimes this could entail renewal decisions predicated on replacing elements of buildings such as roofs which are deemed to be beyond economical repair. In addition, there could be some aspect of improvement works where organisations seek to enhance buildings in areas such as energy efficiency and environmental sustainability.

The reasoning for not carrying out essential repairs may vary between organisations as there may be practical difficulties associated with unacceptable disruption brought about by the work and maintaining business continuity. However, in most cases, backlog maintenance is caused by funding difficulties where maintenance budgets have had to be reduced to take account of prevalent financial difficulties within organisations. Where financial resources are severely impacted it is not unusual for organisations to only carry out those works that are deemed urgent by law. This scenario may include only those works for statutory compliance e.g. gas and electrical certification or those linked to health and safety, e.g. risk of tripping or persons falling from building faults.

Completing backlog maintenance works in a proactive manner is essential for any maintenance regime. Failure to address them in a timely manner can sometimes lead to them being constantly being pushed back year after year to a point where the scope of backlog grows to an unacceptable level. This can sometimes occur when organisations are performing badly and normal operation costs including paying staff salaries takes priority over maintenance costs. The only possible exception for allowing backlog maintenance to grow year on year is when buildings are scheduled for demolition. In these cases, organisations may make decisions to run down facilities prior to end of life rather than waste resources and undertake unnecessary investment in keeping them in pristine condition.

3.12 Defects liability period

Normally in the first year following completion of a new building or refurbishment there is a *defects liability period* whereby the main contractor is responsible for faults or latent defects which may become apparent in the occupation period. Typical examples may include electrical faults, cracking plasterwork, uplifting carpet tiles, mechanical plant misfunctions and any water penetration through the building envelope. In such cases it is the responsibility of building owners, normally through their maintenance teams, to log faults and issues and report these through the main contractor's representations for corrective action. The main contractor would then be given a reasonable amount of time to undertake any repairs or renewal works accordingly. This liability, however, does not cover accidental damage or wear and tear by building occupants. Furthermore, it excludes any building failures that are deemed to have emanated from poor building maintenance. A scenario could be where there was a roof leak and water penetration through the roof covering had caused resultant internal damage to finishes and building services. If it was established that the cause of the water penetration resulted from blocked valley gutters and building maintenance teams' inability to clean them, then it would be more than likely the responsibility of building owners. A similar scenario could be where there is a failure of new mechanical plants or installations, resulting from the absence of any specialist maintenance contracts or servicing plans being in place. In this case it could be argued that the failure has resulted from the negligence of the building owners' facilities teams in not having an appropriate maintenance regime in place.

When looking at sophisticated building services in any building, it is not always black and white who is responsible during defects liability periods for faults and defects, especially when very expensive pieces of mechanical equipment e.g. a chiller plant becomes damaged. It is sometimes a source of contention where main contractors will argue that maintenance, albeit put in place, may not have been sufficiently scoped or competently undertaken. In extreme cases blame could be apportioned for the damage being a direct result of the wrong type of maintenance works being carried out by facilities teams or their appointed contractors. For this reason, certainly during defect liability periods for buildings, it is best practice to employ the original installation subcontractors to undertake all maintenance in connection with their original installations. This establishes a single-source responsibility approach to repairing liabilities and thus avoids dilemmas of who is responsible in the event of failures.

3.13 Summary of chapter

Strategic corporate business plans articulate the vision, mission, short- and long-term aims, goals and objectives of organisations but also should relate to building and estate assets. Accordingly building maintenance strategies and policies should include statements of how organisations seek to maintain their buildings and estates and keep them operationally fit for purpose whilst retaining their assets' values. A clear set of objectives should be articulated together with the methodologies and approaches for how the maintenance team plans to deliver them. In this way, maintenance strategies and policies should become frameworks for implementation of maintenance operations.

Maintenance programmes rely on compiling maintenance plans with associated allocated timescales for implementation, and quality assurance measures. This involves planning resources, budget setting and procurement of maintenance and works contracts. Furthermore, they will need to consider which maintenance works would be more cost-effective to carry out in-house, with directly employed maintenance tradespersons, or to outsource through external contractors.

Asset management in property terms should ensure that buildings and land provide the highest level of benefits to organisations in meeting their business aims and objectives. It should also achieve the best possible utilisation of real estate assets and in doing so minimise the opportunity costs in tying up resources in buildings and land. Asset management should include areas such as maintenance management, environmental sustainability, energy efficiency, procurement of services, project management, information technology, security, waste management and cleaning.

In preparing asset management plans, it is necessary for organisations to undertake condition surveys of all their building and estates portfolios. These surveys should establish the state and condition of all elements of building structures, fabric, electrical and mechanical services, internal spaces, external areas and equipment, fixtures and fittings. They should consider statutory compliance and involve checking certification on such areas as legionella, gas and electrical certification. Furthermore, they should appraise and evaluate the current cost of running and maintaining estates assets including existing and predicted maintenance costs, rents, potential acquisitions and disposals and building values and any capital depreciation. In balancing budgets, maintenance managers will need to consider prioritisation of certain works over others. There should be consideration of the implications of deferring certain works and any effects that this could have on business operations and continuity, occupier satisfaction and building deterioration.

The process by which maintenance requirements are established is based on expected levels of needs. Information sources to capture these needs emanate from building condition surveys, known building requirements for statutory compliance, reported defects, repairs and faults by building occupants and repairing obligations in leases. Building condition surveys will normally be carried out at intervals normally ranging from five years to ten years. They should record the physical condition of all elements of buildings including structure, fabric, finishes and building services. Building elements should be categorised to reflect their state of condition and the level of urgency for any recommended remedial works. Normally the repair works recommended in condition surveys will be categorised into minor or major works of renewal, repair, redecoration, refurbishment or adaption. Irrespective of which category they are placed within, each element of work should be fully quantified, prioritised and costed. Prioritisation is therefore an important aspect of maintenance planning and one which can sometimes require surveyors to use their knowledge of building pathology to predict rates of deterioration. However, it is not only linked to condition and consequential damage, as influenced by compliance with regulation and statutory requirements. In prioritising maintenance works it is preferable for minimum timescales to be designated for each classification of priority. Where works are related to statutory compliance and adherence with health and safety they should be deemed urgent and unavoidable. Procurement of maintenance works, either in-house or externally outsourced, is a major consideration for maintenance teams. Decision-making around this could be predicted by the size and nature of businesses. For instance, smaller companies who do not have the financial resources to invest in their own teams or feel that building maintenance go far and beyond their core businesses.

Maintenance programmes can be regarded as enabling tools for facilitating maintenance teams in forward planning works into the future. They should be designed to avoid cases where buildings become deteriorated to the point where their condition adversely brings about failures of their structures, fabric or services. Backlog maintenance represents a programme of maintenance works or tasks which are deemed essential and have been scheduled or planned to be carried out. Backlog maintenance can normally be put in the context of value in how much expenditure needs to be invested to bring buildings or facilities back up to an acceptable state of repair and condition.

Normally in the first year following completion of a new building or refurbishment there is a *defects liability period* whereby the main contractor is responsible for faults or latent defects which may become apparent in the occupation period. This liability, however, does not cover accidental damage or wear and tear by building occupants or any building failures that are deemed to have emanated for poor building maintenance. Finally, single-source responsibility for defects and faults in this period, by using the same installation contractors and subcontractors for servicing and maintenance, can avoid issues and disputes over liabilities during this period.

References

Challender, J. with Farrell, P. and McDermott, P. (2019) *Building collaborative trust in construction procurement strategies*. Blackwell-Wiley.

Thomas, M. (2013). *Managing building pathology and maintenance construction managers' library*. Leonardo da Vinci.

Chapter 4

Common maintenance issues and managing defects

4.1 Introduction to chapter

This chapter of the book is largely based on the previous work of the lead author which featured in Higham, A., Challender, J. and Watts, G (2021) *Introduction to Built Asset Management*. It will focus on the pathology of buildings with particular emphasis on building degradation and will describe the ways and means of identifying and addressing potential defective elements and defects within buildings. In this pursuit, it will examine the different methodologies for diagnosis of defects and their root cause (rather than just the symptoms of defects) to assist in identifying the most appropriate type of repair.

An analysis of the underlying conditions and mechanisms which lead to defects and building failures result from poor design and quality of build during the construction stage or refurbishment will be covered from dampness to structural movement, corrosion and timber decay. Furthermore, the role of maintenance across the life cycles of buildings will then be discussed and the influence that this has on their preservation and deterioration. The debate around this important area will then extend into the examination of planned and reactive policies for maintenance management of building estates.

The challenges associated with identifying the underlying causes of building failures will be presented where one or more possible different diagnoses may be initially debated. In this context the requirements for further investigation will be examined in correctly determining the underlying root cause of problems more accurately and confidently. The different categories of building failures which lead to the emergence of defects will then be discussed and the mitigation measures that should be undertaken to reduce the risk of such failures. In addition, the individual forces at work leading to different types of defects and how these can vary dependent on climate and the geographical locations around the world will be articulated. Furthermore, the difficulties and challenges of correctly identifying faults and defects to building services such as heating, lighting, fire alarms, cooling and other forms of gas and electrical installation will be explained, and practical examples provided.

The importance of understanding the nature and effect of agents that can lead to building defects will be examined in detail including human misuse of buildings and deliberate vandalism, coupled with electromagnetic, mechanical, chemical, building use, thermal and biological related forces and effects at work. The wrong use of materials and components and design defects will be discussed and how these can develop into latent construction defects in the future and therein introduce maintenance liabilities for building owners. Thereafter the magnitude of managing remedial works and the associated disruption that this can bring for building users will then be investigated with reference to different examples in practice. Dilemmas associated with repair or renewal decisions will be introduced and debated in detail and the overall effects

DOI: 10.1201/9781003358800-4

that decision-making can have on future maintenance and life cycle costings analysis. This will entail a discussion around capital cost (Capex) versus operational cost (Opex) perspectives. Following on from this, the various influences of different factors on the repair or renewal decision-making will be covered including capital funding, building longevity and potential disruption from proposed remedial works.

Managing the remedial works processes to address maintenance issues and defects will be examined with a focus on best practice and achieving the correct level of supervision and quality control throughout the contract period. This will introduce quality control and best practices in the election of the most appropriate design teams and contractors to attain the necessary skills, experience and competency levels required to undertake projects successfully. Finally measures to mitigate and prevent defects will be covered and the importance of allowing access provisions in buildings to enable and carry out regular maintenance inspections as part of planned preventative maintenance programmes.

4.2 Exploring the pathology of building maintenance issues

The pathology of building maintenance issues revolves around the investigation, inspection and reporting of building defects. According to Thomas (2013) building pathology is defined as 'the systematic treatment of building defects, its causes, its consequences and its remedies'. The principle aims and objectives around building pathology are related to the need to comprehend the processes of building degradation. In this way tools and methodologies can be introduced to identify potential defective elements and defects within buildings. This should facilitate solutions to be devised to eliminate or decrease their detrimental effects and therein avoid unnecessary or unforeseen maintenance expenditure.

Building pathology is related to the school of applied sciences and for this reason it is universally regarded as a specialist area in its own right. It is founded on investigation work to identify the source or sources of building problems. Thereafter it is concerned with diagnosis of the underlying technical issues which could be causing the problems and then determines the extent and specification of the required remedial works. The findings from this investigation and diagnosis work should identify specific components, elements or areas of buildings where problems are arising and the underlying defects or malfunctions that are responsible for them. In this regard such building problems may relate to individual building systems or alternatively groups of systems. For instance, water penetration in a building could emanate from a flat roof covering and could be caused by a single tear in a roof covering (single building system) or a combination of different envelope defects including the roof and walls (group of systems).

The term pathology has only been applied to buildings relatively recently and was previously only used in conjunction with the health sector when describing medical problems in people and diagnosing types of condition and disease. In this context it allowed medical practitioners an understanding of particular health complications and recommend a course of medicine or procedure for address. Buildings and people in terms of pathology have many aspects in common. The condition of a building is not always correlated with its age, on the basis that some newer buildings can have more defects and signs of wear and tear than an older building, as a result of poor design, workmanship and quality of build. A building's condition could also result from neglect and a lack of routine maintenance. In the same way the health of a person is not always related to how old they are but heavily influenced by such aspects as lifestyle, exercise, diet and maintenance of their well-being.

Normally when considering the process of building pathology, it will require suitably qualified specialists to undertake inspections and building surveys. Building pathologists will try to identify, assess, eliminate or control (through a managed process) building defects and their causes. In the UK building pathologists are normally building surveyors and members of the Royal Institution of Chartered Surveyors (RICS). However, if a particular building defect is related to major complicated structural problems, then a consultant structural engineer and possible member of the Institute of Structural Engineers (IstructE) may be appointed. Normally these structural engineering professionals are more specialised in dealing with structural issues such as these and can more proficiently prescribe the most appropriate remedial works. In a similar way if a building defect is related to mechanical and electrical issues, possibly related to heating, ventilation or cooling problems, then a suitably qualified mechanical and electrical engineer may be appointed. These individuals are normally members of the Chartered Institute of Building Services Engineers (CIBSE) and are more specialised in dealing with issues such as these and can more proficiently prescribe remedial works to building services.

4.3 Context to the discussion on building defects

A building defect could be described as a fault, deficiency or imperfection in a building which causes it to not properly function in terms of appearance, performance or usage. When building surveyors are determining remedial works, following diagnosis of defects, it is important for them to focus on the root cause of the problem rather than simply address the symptoms of the defect. An example could be structural movement cracking of external walls, which require structural foundation remedial works to address differential movement and subsidence, rather than simply repointing of cracked mortar joints. In this way it should ensure that repairs are not temporary superficial 'fixes' but conversely present long-term solutions.

When investigating, diagnosing and specifying remedial works to address defects building surveyors should be mindful through their knowledge, experience and training of the underlying conditions and mechanisms which lead to defects and building failures. Certain building failures result from poor design and quality of build during the construction stage or refurbishment. Other defects may simply arise due to natural forces associated with weathering and normal use. In the latter case, examples could include weathering of timber cladding that has been open to the elements for many years causing shrinkage cracking and rot in certain instances. Clearly the degree of maintenance that buildings receive over their life cycles will strongly influence the degree to which they are preserved or deteriorate. Planned maintenance policies in this regard aim to preserve buildings to a perpetually good standard and involve routine frequent interventions in this regard. Where maintenance policies are more reactive in nature, this is where problems can be exacerbated. For instance, there may be water penetration through a roof covering that goes unchecked for a number of months or years owing to a lack of planned maintenance and reactive maintenance only occurs once the problem becomes apparent. In such cases hidden water damage may have resulted in dry rot of roof timbers which is a much bigger problem to resolve than the initial roof leak.

Diagnosis of building defects is not always straightforward as there may be more than one force at play in terms of multiple sources of different problems. In such cases this can sometimes complicate normal lines of investigation and enquiry and present difficulties in determining the cause and associated remedial works. Complex investigations can arise in the following circumstances:

1. Wrongly specified materials and construction detailing. Roof abutment detailing and basement waterproofing is commonplace in this defect category, and it may be many months before water penetration can be first identified.

2. Correct materials are specified but detailed in an unsuitable way for the nature of the product, component or material. By way of an example, the correct roof flashing detailing for lead may be very different to that used with a different roofing material such as bituminous felt.
3. Wrong installation of products, components or material, albeit specified and detailed correctly. For this reason, it is important to not always rely on building contractors to complete installations without a supervising officer, which could be the architect or architectural technologist inspecting to confirm adherence to construction drawings and other design documentation.
4. Wrong assessment of conditions associated with the design which could include exposure to the weather or structural loading. An example of this could be the specification and design of steelwork that does not meet accepted factors of tolerance and building regulations.

When considering the typology of defects in buildings one should distinguish between different types of problems. According to Addleston (1989, as cited in Thomas 2013) these can be categorised by type and cause into i) settlement and subsidence, ii) condensation, dampness and water penetration, iii) loss of bond and adhesion, iv) decay and corrosion and v) movement and cracking. Using these five categories, they can be broken down into subcategories and these are illustrated in Figures 4.1, 4.2, 4.3, 4.4 and 4.5. Clearly the extent of each of the defect categories will be largely dependent on the different forms of construction. In the UK, house building tends to be heavily geared to traditional forms of construction with brick and block cavity external walling and pitched tiled roofs. Conversely in Scandinavia the predominant domestic construction comprises timber-framed and -clad buildings of a lightweight, and sometimes modular nature. Alternatively in some parts of Europe such as Spain and France casting concrete structures for housing and commercial buildings is predominantly the norm.

When we consider building services such as heating, lighting, fire alarms, cooling and other forms of gas and electrical installation, it can be more problematic to categorise defects and faults owing to the specialist nature of technologies in some of these systems. Figure 4.6 does however attempt to illustrate some of the types of failures, defects and faults in mechanical and electrical systems.

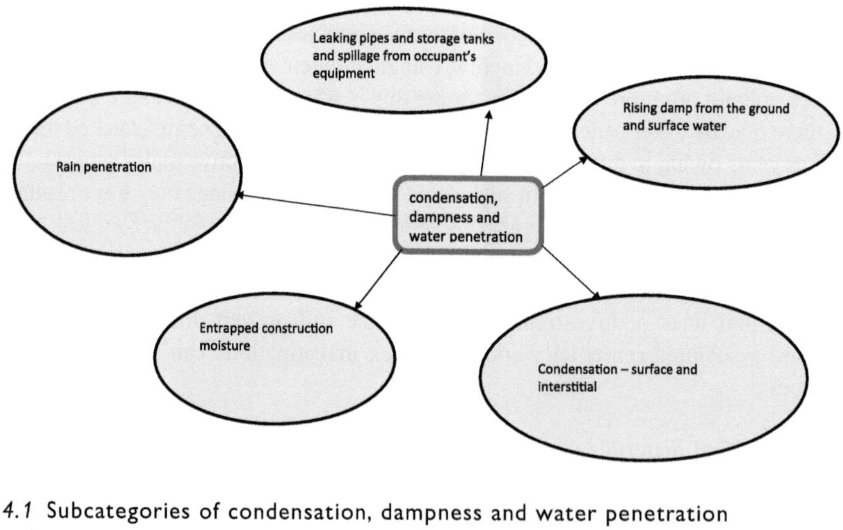

Figure 4.1 Subcategories of condensation, dampness and water penetration

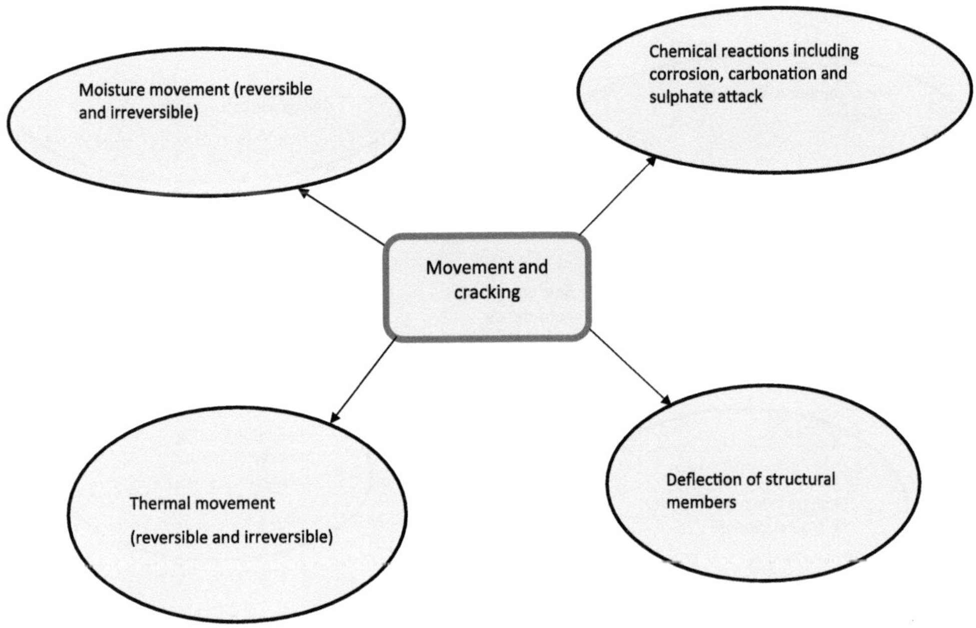

Figure 4.2 Subcategories of movement and cracking

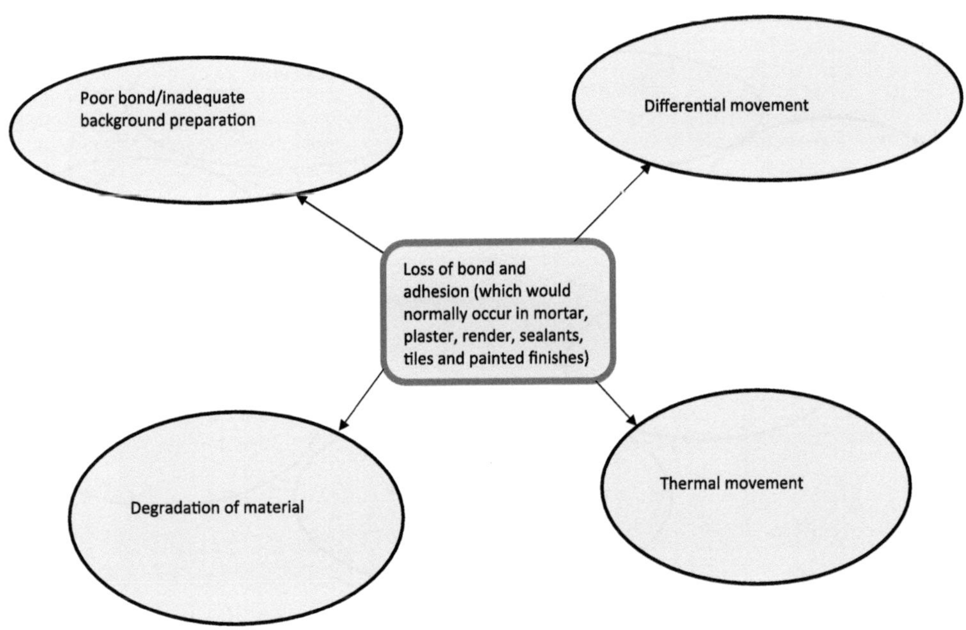

Figure 4.3 Subcategories of loss of bond and adhesion

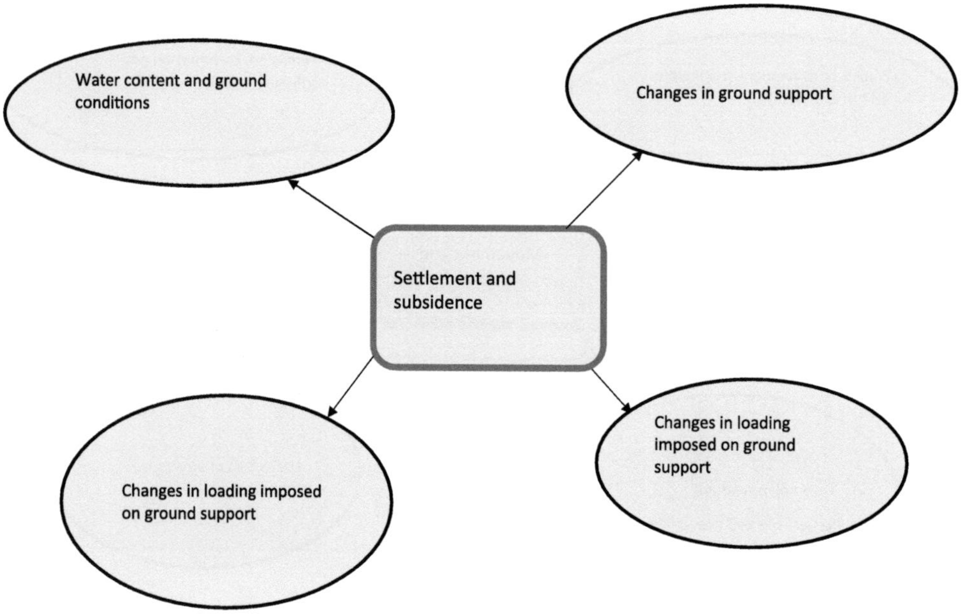

Figure 4.4 Subcategories of settlement and subsidence

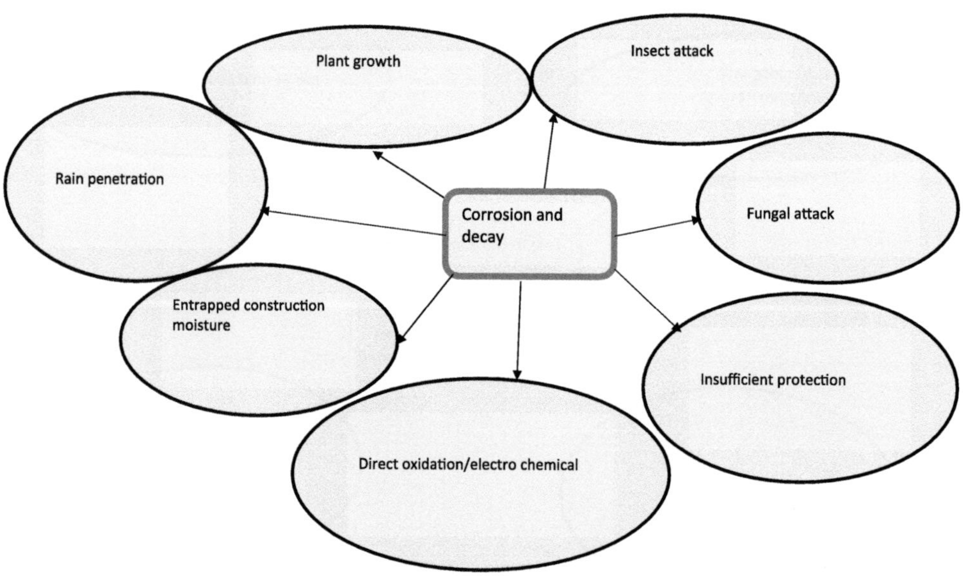

Figure 4.5 Subcategories of corrosion and decay

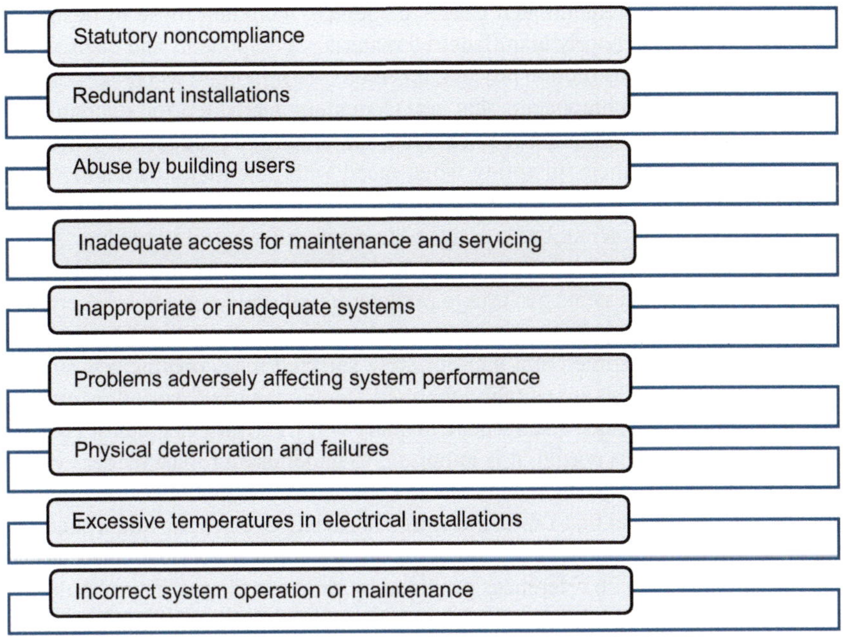

Figure 4.6 Possible defects and faults in mechanical and electrical systems

4.4 The importance of understanding the nature and effect of agents that can lead to building defects

The causes and agents at work that can introduce some of the defects mentioned can be categorised into different bands and groups and these include electromagnetic, mechanical, chemical, building user, thermal and biological forces and effects. Examples of building user causes can range between deliberate vandalism to accidental or wrongful misuse. Conversely, an example of chemical agents at work could include oxidation, acid attack and water damage to materials. One of the main causes of defects can be a result of poor design and simply not allowing for thermal or moisture movement in building elements and components. An obvious example could be where long lengths of brick walling are constructed without adequate movement joints. Thermal expansion of structures such as brickwork without provision for movement can cause stresses with these structures leading to cracking and potential structural failures. Different building materials and components have far-ranging coefficients of expansion when heated. It is for this reason that when components and materials come together to form building elements an understanding of the nature of the materials and how they will react with each other is critical. The example of the Liverpool's Catholic Cathedral, known now as the Metropolitan Cathedral of Christ the King, shortly after it was built in the 1980s presented a classic example of when a roof covering designed for the building was insufficiently designed and detailed without sufficient provision for movement joints. Thermal movement of the lead roofing system was not fully understood with inadequate measures installed to cater to such movement in the materials. This caused water penetration through the roof covering and ultimately led to its failure, resulting in an expensive and long-drawn legal proceedings by the diocese of Liverpool against the architects for the costs of remedial works.

Notwithstanding the aforementioned causes of defects including those of design, the most common defects arise from poorly manufactured materials, components and elements alongside poor workmanship in the construction process. It is therefore important wherever possible to only use materials and building components that are adequately accredited for standards or quality. Normally to attain such accreditation requires these materials and products to be robustly tested in lab conditions to verify their suitability when faced with potentially damaging climatic and environmental conditions. In the UK accreditations for materials can come from British Standards whereby an accreditation number is assigned to a particular building product. Accreditation and the use of quality, tested components are particularly important for mechanical and electrical installations to ensure that systems perform to predetermined standards and meet prescribed levels of longevity and durability. To combat poor workmanship in the build process it is essential to only appoint suitably qualified and appropriately selected main contractors, subcontractors and specialist contractors who are suitably qualified for the nature of work they are required to carry out. Every project is unique and bespoke in many ways, carrying degrees of complexity and wide-ranging values. For this reason, it is important to undertake a robust tender evaluation and selection process to ensure that the most suitable contractors are appointed for a particular project, and only use contractors that can demonstrate knowledge, expertise and a successful track record for similar types of work. Checks should be made wherever possible on previous projects that they have undertaken with references provided by previous clients. This should ensure that a reputable contractor is appointed with all the required levels and degrees of competence. In the selection process the nature of the proposed project should be considered against the skills and expertise of those contractors who are being considered. After all, a contractor that may be suitable to construct a £200k house refurbishment will seldom be appropriate for a £50m new build and vice versa. In addition to selecting the right contracting team it is also vitally important to have arrangements in place to monitor and inspect works completed to ensure the right level of quality is being achieved. Normally this task would fall to members of the design team and they should cross-reference the works against design drawings and specifications to confirm close adherence with the contract documentation. Any deviation from the design drawings and specification should be reported to the supervising officer on the project which in normal circumstances would be the client's project manager. The project manager would then inform the contractor of the noncompliance and potentially defective works and request that the matter be addressed.

Another common cause of premature failure and defects is a general misunderstanding and lack of maintenance of buildings. Although most organisations have planned preventative maintenance systems in place which seek to address maintenance issues before they become more serious defects there are still those who will only perform works on a reactive basis. This can lead to minor repairs becoming major repairs and is particularly common in vacant and derelict buildings. It is not unusual by way of an example for a simple roof leak in a building that has been left unattended and unchecked for a long period of time to have caused problems associated with dampness and dry rot in roofs and walls, potentially requiring major remedial works. Wherever possible architects and engineers should design buildings to be as maintenance free as possible. This involves not only specifying robust building products and materials but allowing for access for maintenance. An example could be to allow at the design stage access gantries to roofs to inspect, repair and clean roof coverings and roof lights. The same safe and easy access could be used to maintain roof sited mechanical and electrical plants including ventilation and cooling systems. It is where such simple and safe access is not designed into buildings where problems can arise with occupiers and owners not wanting to buy or hire expensive access plants or scaffolding to reach those areas that are hard to get to.

4.5 Dilemmas associated with repair or renewal decisions

Other dilemmas can be introduced when deciding on the right level of remedial works to address defects and maintenance issues. The repair or replace question frequently arises when considering building components and elements and how they interact and work with each other. It is a similar scenario to car maintenance in sometimes considering whether a patch repair will be sufficient over time or whether it's merely storing up a bigger problem for a later time. An example could be an old roof covering that is clearly showing obvious signs of wear and tear. The question arises: should one continue to arrange patch repairs on the roof covering when water penetration occurs or renew the roof covering entirely? Patch repairing old roofs can sometimes be cumbersome and disrupt other areas of the existing roof further from the actual roofing operations. This can lead to more costs and disruption being incurred over a longer period of time than simply replacing the roof and curing the problem once and for all. Furthermore, patch repairs on old roofs rarely can be guaranteed but normally new roof coverings carry a minimum level of ten years' guarantee, materials and labour, which can give building owners some added assurance for future maintenance costs in the future. Other dilemmas could relate to old mechanical and electrical installation, e.g., gas boilers that could carry substantial maintenance liabilities and future high energy costs associated with inefficient and outdated technologies. A decision might be made in these scenarios to replace such plants in these circumstances with more reliable, cost-effective and carbon-friendly technologies. One factor which could influence the decision to renew plants with more energy-saving technologies, possibly involving renewable technologies, could be the potential of government grant funding. In the UK there have been many schemes such as the Green Grant Fund which was set up to encourage householders to install energy primary and secondary environmentally sustainable measures. These included replacement of single-glazed windows and doors for double- and triple-gazed alternatives alongside other measures including the installation of cavity walls and loft insulation. In addition, the UK government announced a £144m 'Cleantech Funding' in 2020. This was designed as part of the government's green agenda to meet its zero carbon targets by the prescribed date of 2050, and with the aim of helping to promote clean growth projects, develop new technologies, create new jobs, drive productivity and tackle climate change. Other considerations for the repair or renewal question might be dependent on the historical status of a particular building. If a building is listed, it would be seen to be more acceptable where possible to conserve and repair individual elements of its composition rather than trip out historic fabric and replace it with new ones. Conservation officers would rarely allow whole elements to be replaced without a strong case for doing so and evidence to show that these are beyond repair. In the case of timber decay and particularly dry rot, in the past this has necessitated considerable strip-out and removal of historic fabric. However, in recent years preservation treatments have improved which has allowed some infected timbers to be retained and strengthened without risking spread of dry rot to other areas.

Sometimes remedial works do not solely look to repair or renew a particular element back to its former state but to improve and adopt the design to make it function more effectively. One example could be a roof that has limited overhang or insufficiently sized rainwater gutters and downpipes to prevent rainwater cascading down facades, causing dampness issues. In these cases, consideration should be given to undertaking improvement and upgrading works to avoid historical problems becoming future maintenance issues. Other similar cases which seek to address design issues in the original build could include removing substandard materials and components. The Grenfell Tower tragedy in 2010 where flammable cladding was found to have resulted in the building becoming ablaze and the death of 70 people has become a landmark case on poor design and quality control. Many owners and leaseholders are left with the ongoing problems associated with the cost of replacing inappropriate cladding at what can amount to a significant level of expenditure for retrofitting new nonflammable alternatives.

There are many factors which can influence the surveyors' recommendations on the renewal or replacement question, but whichever option they select should be arrived at through a carefully considered mechanism and one ideally which the building owner should be consulted on. After all it would be pointless and counterproductive for a surveyor to recommend a large programme of building replacement if the building owner has plans to relocate or simply cannot afford the expenditure that a large refurbishment/redevelopment would require. Through consultation with the building owners a programme of short-term temporary repairs may serve their purposes better than a large-scale/long-term renewal programme. Accordingly, there are both technical skills and intuition required on the part of surveyors in optimising the choice of the most economic and effective recommended repair/renewal solution. Notwithstanding this premise it is always the client, which is normally the building owner or their representative, that will make the final choice depending on their future plans, aspirations, resources and budget. They will however be influenced by the surveyor's technical report and so it is important that this clearly set out the different options, and the benefits, disadvantages and implications of each one. In a scenario that the repair or renewal costs are disproportionately high to the value of a building, the client may decide to demolish and redevelop, change the use of the building or simply sell the building and relocate elsewhere.

In certain instances, building owners may choose to do nothing when faced with a major defect and simply live with the effects and tolerate the defect and its effects if not too disruptive or damaging to their business operations. One example could be a basement cellar is letting in water owing to defective tanking. Any remedial works could be considered to be too expensive and disruptive to justify the benefits of having the basement area be watertight. Accordingly, the building owner may allow the groundwater to seep through and leave to drain naturally or alternatively pump the water away.

4.6 Managing the remedial works process to address maintenance issues and defects

When the remedial works are agreed upon between the client and their supervising officer/surveyor overseeing the work, a brief outlining the works should then be prepared with budget costings and a proposed plan of implementation. The implementation plan will normally have a programme of timelines for different stages of the procurement process from start to finish. For instance, there will be a design period whereby tender documents are prepared including drawings, specifications, specialist reports. For structural works this will normally entail the preparations of structural calculations for any repairs or reinforcement works and which will form part of a building regulations submission. Sometimes quantities will need to be prepared to assist tendering contractors in formulating prices for measured works. The next stage will be to tender the works, and this may involve selecting a handful of preferred contractors who are deemed suitable for the nature and value of the works. A prequalification process may take place to formulate a 'short list' of suitable contractors and this may consist of each potential contractor completing a questionnaire and providing supporting info to support their application for inclusion on the tender list. The works will normally then be tendered and evaluated on predetermined criteria by a panel of individuals agreed upon between the client and the supervising officer/surveyor responsible for overseeing the works. Once a contractor is selected contracts will be signed based on the agreed tender price and a programme of delivery. Contractors will be responsible for providing evidence of all insurances (employer's liability and professional indemnity normally) alongside all the health and safety documentation including risk assessments and method statements.

The works will be closely managed by the supervising officer, normally the surveyor who is overseeing the works, and this will involve frequent inspections of the works and certification at various stages of the project. It may be necessary to have reviews of the work if there are unforeseen areas of work that cannot be adequately specified owing to access difficulties and/or opening up works via destructive intrusive investigation. In some instances, building contracts for repairs and remedial works can be phased into two separate phases. Phase one could consist of works to access areas and provide the means for inspection of elements that would otherwise be inaccessible and not visible. Phase two would then consist of the remedial works themselves having had the benefit of phase one to allow further investigation and properly specify the proposed works. If the works cannot be fully quantified and are therein unable to provide fixed prices by tendering contractors, then another option is to award the contracts on the basis of agreed upon rates for carrying out various repairs. This could include a schedule of rates for different elements of work that can be quantified as the work proceeds and a valuation for the amount of work undertaken agreed upon at the end of the contract. Whichever building contract and procurement route adopted, the contract particulars should reflect the nature of the work and include all the aspects included in Figure 4.7.

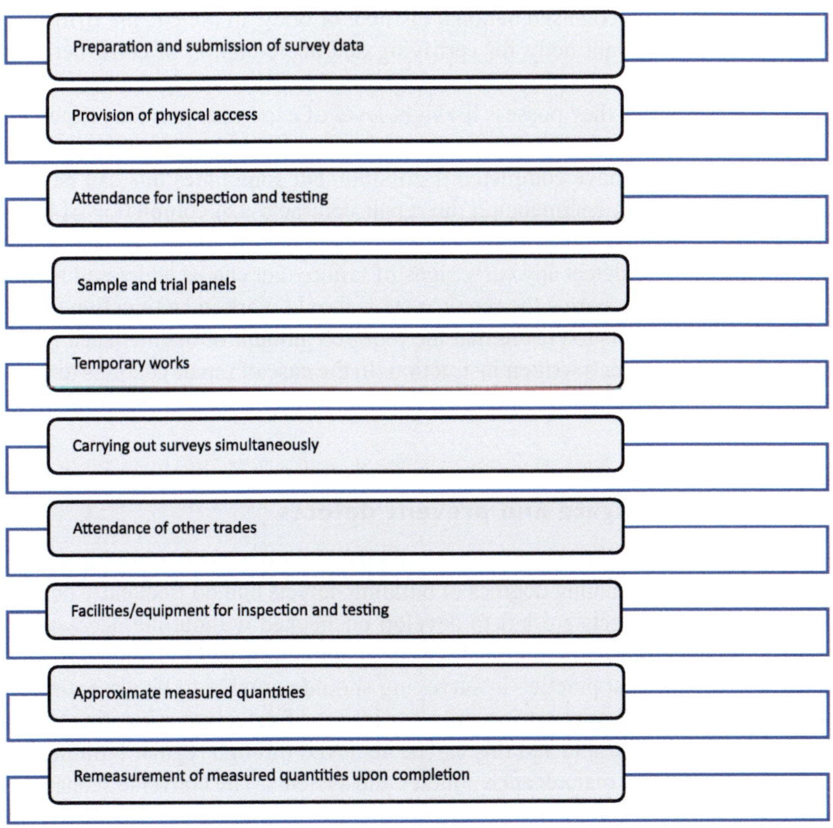

Figure 4.7 Contract particulars that should be included in contract for remedial works

Sometimes trials and tests of areas of repair will be undertaken to evidence that a particular repair method is working correctly before being applied over a much larger area. An example of a façade repair contract could involve destructive removal of a small area of repair material to ascertain its adhesion to the underlying wall fabric. If the repair compound is relatively easy to remove by chipping off, then this might deem the repair system inadequate owing to a loss of adhesion and a new method may need to be devised. Other tests may be related to waterproofing where a small area of roof or walling is water pressure treated to ensure that the new material is fully waterproof before being deployed across a whole roof or wall area. In this way it is far easier to test small areas and make adjustments where required to the design or specification of the repair system rather than assume a system will work and discover later there are widespread problems. Such trial areas are required sometimes to be undertaken at height or in areas that are difficult to access in other ways. In such cases temporary scaffolding, cherry pickers or scissor lifts may be required to provide access.

During the remedial works it is essential to have levels of supervision to reflect the nature and complexity of the works. On specialist works a competent and specialist supervisor may have to be commissioned to undertake the inspections and to test the areas of repair where appropriate or deemed necessary. Faulty remedial works can sometimes result from poor workmanship or materials and good supervision should avoid the need for further remedial works. In addition, materials and repair systems should only be used where they are tried, tested and preferably accredited to a recognised national institute or body. In the UK the British Standards Institute would be the relevant body for certifying and accreditation of construction products and materials. In addition, contractors who are appointed to undertake such repair works should be selected on the basis that they possess the right level of expertise, previous experience and a good track record of undertaking similar works. References should be obtained and inspections of previous work that they have completed if possible, but sometimes this can be challenging owing to time constraints in commencing the repair contracts. On completion of the remedial works, period inspections should be carried out to ensure that the repair system is performing adequately and possibly to detect any early signs of failure that can be addressed before further failure emerges. In many scenarios the repair system should work in conjunction with building owners' maintenance systems to ensure that the required amount of maintenance is carried out as per the repair manufacturer's written instruction. In the case of repair coatings to wall facades or roof coverings this could involve specialist cleaning and/or periodic recoating at prescribed intervals.

4.7 Measures to mitigate and prevent defects

Notwithstanding the array of information that is available for diagnosis and addressing building defects there are still wide-ranging degrees of building defects that do frequently occur. In some more severe cases such defects are left to develop unchecked without the necessary remedial works which subsequently causes more damage, e.g., dampness and timber decay. In a similar way to the health sector, best practice in surveying should continue to be disseminated around prevention of defects as well as cure. Certainly, avoidance of defects to mitigating measures and strategies are normally preferable and this can be achieved through regular building surveys as part of planned preventative maintenance policies and systems. The converse scenario would be reactionary maintenance whereby works are only planned when resultant damage from defects has been identified. An example could be water penetration through roof coverings, causing damage to ceilings and other finishes which can incur more costs and increased disruption for building occupiers.

In addition to having adequate and effective maintenance management policies in place, defects can be reduced through good designs and specifications, which are geared to achieving longevity for buildings. Furthermore, ensuring quality control during the construction period for new buildings and refurbishment work can avoid latent defects being stored up for emergence at later stages. Guidance on maintenance planning and achieving performance of buildings through design and construction is widely available through professional bodies such as the Royal Institution of Chartered Surveyors (RICS) alongside trade bodies, manufacturers and academic institutions. Such guidance is underpinned by a good knowledge of building pathology coupled with a comprehensive understanding of the science of the components, elements and materials which make up the structure and fabric of buildings. The science in this regard will be related to factors such as maintaining dimensional stability and a comprehension of the different ways that building components and materials move when subjected to varying degrees of temperature and moisture content. Other factors that the science will need to encompass include heat transfer mechanisms, air and water permeability through structure and fabric, relative humidity to avoid condensation forming and an awareness of the factors which cause corrosion of metal and timber decay. However, on an international scale, such guidance may vary from one country to the other depending on the predominant forms of construction in a particular country and the climatic conditions. For instance, what is appropriate construction for a desert climate, e.g., United Arab Emirates, may be very different from a building in an arctic climate such as Finland. Clearly there will be different forces and climatic conditions imposed in different areas of the world with buildings in areas prone to earthquakes, flooding, extremes of temperature and heavy winds requiring special design and construction technologies to mitigate these factors. Consideration of such factors will affect the ways and means by which buildings are procured. In addition, it is important to consider at the design stages of buildings how they are going to be inspected and maintained post completion. For instance, if a building has a high degree of glass curtain walling, it is imperative to consider how the glazing system is going to be regularly cleaned. In the same way where mechanical and electrical plants are located on roof areas, it is crucial to plan for a safe access system and procedure to be designed into the build that will allow qualified engineers to safely and easily access plants and from time to time to replace certain component parts. Where building elements are inaccessible this can be problematic for inspecting their condition and maintenance planning coupled with arranging access for any remedial works. This can sometimes cause untold difficulties in faults emerging that escalate into more serious defects because of inspection restrictions. An example could be a roof space in an existing dwelling that has restricted or no access. As this confined and inaccessible loft space cannot be inspected there may be a long history of water penetration through the roof covering that has gone unnoticed and undetected for many years. This may have caused ancillary defects related to dampness in the timber structure of the roof, leading to widespread dry rot infestation throughout. The remedial works of initially rectifying the initial roof problem may be insignificant to the vast amount of expense of dry rot eradication works and timber replacement. It is a prerequisite of being a building designer, for all the aforementioned reasons, that they fully understand their client's future maintenance regimes and requirements at an early stage and allow for all the necessary provisions and equipment to facilitate their operations. In this regard they should consider worst-case scenarios where a fault or defect could arise that will require access for inspection and to carry out the remedial works.

In addition, forward planning and consideration should be had to those elements of buildings which are potentially prone to failure in building locations where they can be easily inspected and maintained. An example of this could be not to site a cold-water tank or related pipework above a ceiling to a data communications room, as the consequential damage to the technology and the effects on business continuity in the event of leaking water could be immense. Furthermore, what may represent an initial additional capital cost in the build may actually save more across the life cycle of the building in not having to expend on temporary measures. One example could be the installation of a gantry to access a roof atrium and thereby avoid the need for temporary scaffolding or the use of scissor lifts to access high areas of the space. In the UK the Construction Design and Management (CDM) Regulations 2015 impose health and safety legal requirements on the whole design team and clients to allow, in the design, for safely accessing buildings with the focus on reducing accidents. On very tall buildings it is becoming more popular in recent years to install self-cleaning glass, which, albeit is more expensive than standard glazing, eliminates the need for manual cleaning, therein saving future maintenance costs and avoiding the potential hazards of individuals working at height. The balance between initial capital costs and incorporating measures such as these that can reduce future maintenance costs will be discussed later in the book under life cycle analysis.

4.8 Summary

The pathology of building maintenance is related to the school of applied sciences and revolves around the investigation, inspection and reporting of building defects with principal aims and objectives around the need to comprehend the processes of building degradation. Such investigation work should identify the source or sources of building problems through diagnosis of the underlying technical issues and then determine the extent and specification of the required remedial works. In the UK building pathologists are normally building surveyors but if a particular building defect is related to major complicated technical problems, then specialist engineers may be appointed. When building surveyors are determining remedial works, following diagnosis of defects, it is important for them to focus on the root problem (cause) of the problem rather than simply address the symptoms of the defect. Certain building failures result from poor design and quality of build during the construction stage or refurbishment whereas others may simply arise due to natural forces associated with weathering and normal use. The condition of a building is not always correlated with its age, on the basis that some newer buildings can have more defects and signs of wear and tear than older buildings, as a result of poor design, workmanship and quality of build. Another factor relating to building condition and whether they are preserved or deteriorate will depend on the degree of maintenance they receive over their respective life cycles.

The diagnosis of building defects is not always straightforward as there may be more than one force at play in terms of multiple sources of different problems. In such cases this can sometimes complicate normal lines of investigation and enquiry and present difficulties in determining the cause and associated remedial works. When considering the typology of defects in buildings one should distinguish between different types of problems and these can be categorised by type and cause into i) settlement and subsidence, ii) condensation, dampness and water penetration, iii) loss of bond and adhesion, iv) decay and corrosion and v) movement and cracking. It is important to understand the nature and effect of agents that can lead to building defects and these include electromagnetic, mechanical, chemical, building users, thermal and biological forces and effects. The most common defects arise from poorly manufactured materials, components and elements alongside poor workmanship in the construction process and it is therefore important wherever possible to only use materials and building components that are adequately accredited for the standards or quality. Another common cause of premature failure and defects is a general understanding and lack of maintenance of buildings. Accordingly planned preventative maintenance systems should be instigated to address maintenance issues before they become more serious defects. There are still those who will only perform works on a reactive basis. Furthermore, design team consultants should be diligent in their designs and specifications to ensure buildings are as maintenance free as possible and allow for access for inspections and maintenance works.

The repair or replace question frequently arises when considering remedial works and building components and elements and how they interact and work with each other. It should be considered whether a patch repair will be sufficient over time or merely storing up a bigger problem for a later time. Sometimes remedial works do not solely look to repair or renew a particular element back to its former state but to improve and adopt the design to make it function more effectively. Other considerations for the repair/renewal decision-making process could be grant funding for replacement works, disruption during the works, sustainability, life cycle costing and clearly affordability for building owners. Notwithstanding this premise, the options that surveyors and designers recommend should be arrived at through carefully considered mechanisms and consultation with their clients. Normally such recommendations would be accompanied by technical reports which clearly set out the different options, and the benefits, disadvantages and implications of each one. In a scenario where the repair or renewal costs are disproportionately high to the value of a building, clients may decide to demolish and redevelop, change the use of the building or simply sell the building and relocate elsewhere.

When considering remedial works to address defects and maintenance issues it is important to factor in the degree of complexity and wide-ranging specialist nature of the repair contract. For this reason, it is essential to only appoint suitably qualified and appropriately selected main contractors, subcontractors and specialist contractors who are suitably qualified for the nature of work they are required to carry out. When the agreed remedial works are agreed upon between the client and their supervising officer/surveyor overseeing the work, a brief outlining the works should then be prepared with budget costings and a proposed plan of implementation with timelines. Remedial works will be closely managed by the supervising officer, normally the surveyor who is overseeing the works, and this will involve frequent inspections of the works and certification at various stages of the project. It may be necessary to have reviews of the work if there are unforeseen areas of work that cannot be adequately specified owing to access difficulties and/or opening up works via destructive intrusive investigation. Sometimes trials and tests of areas of repair will be undertaken to evidence that a particular repair method is working correctly before being applied over a much larger area. During the remedial works it is essential to have levels of supervision to reflect the nature and complexity of the works. On specialist works a competent and specialist supervisor may have to be commissioned to undertake the inspections and to test the areas of repair where appropriate or deemed necessary. On completion of the remedial works, periodic inspections should be carried out to ensure that the repair system is performing adequately and possibly to detect any early signs of failure that can be addressed before further problems emerge.

Notwithstanding the array of information that is available for diagnosis and addressing building defects there are still wide-ranging degrees of building defects that do frequently occur. In some more severe cases such defects are left to develop unchecked without the necessary remedial works which subsequently causes more damage, e.g., dampness and timber decay. In a similar way to the health sector, best practice in surveying should continue to be disseminated around prevention of defects as well as cure. Certainly, avoidance of defects to mitigating measures and strategies are normally preferable and this can be achieved through regular building surveys as part of planned preventative maintenance policies and systems.

References

Higham, A., Challender, J. and Watts, G. (2021). *Introduction to built asset management*. Blackwell-Wiley.

Thomas, M. (2013). *Managing building pathology and maintenance construction managers' library*. Leonardo da Vinci.

Chapter 5

Maintenance and conservation of historic buildings

5.1 Introduction to chapter

This chapter will start by exploring the complexities, challenges and dilemmas associated with historic buildings. It will examine the problems and issues that building owners face in terms of developing robust maintenance management systems and policies. Following on from this, the role and importance of formulating and utilising maintenance plans to assist building owners and their advisors with maintenance management will be described.

It will consider the statutory protection that is afforded to various types of heritage buildings from the perspective of restoration, refurbishment, alterations and maintenance works. In this pursuit it will distinguish between properties located in conservation areas, listed buildings and scheduled monuments. For each of these three categories it will explain the level of statutory protection afforded and give examples of how this may affect the types of repair and alterations that may be deemed acceptable. In the case of listed buildings, it will identify the different grades of listing and the level of historical and architectural interest for each. In addition, it will articulate the types of proposed work on protected buildings that may require planning permission. In this regard it will reference statutory protection afforded by both the Planning (Listed Buildings and Conservation Areas) Act 1990 and the Town and Country Planning Act 1990. In the case of conservation areas, it will discuss what types of alterations and extensions which may come under permitted development and those which fall outside. Examples of listed buildings in the UK will be referenced in the chapter alongside photographs of each for context.

It will discuss the role of Heritage England and local authority conservation officers in the planning application process, and the level of information that may be required for applications. Enforcement action for failure to obtain the necessary consents or for breaches in complying with planning conditions will be examined alongside statutory periods for action to be instigated by local authorities.

Conservation and heritage protection of buildings and the types of issues that could arise will be covered, using case studies for two different projects in which the lead author was personally involved, namely the former Stockport Union Workhouse and Gorton Monastery. For each case study the nature of the projects will be described alongside a narrative to explain the roles of Historic England and the local authority conservation officer. The potential conflict between conservation and meeting current standards e.g. building regulations on one of the case study projects will be articulated and how this was finally overcome.

Finally, the chapter will examine appropriate building technologies and materials for restoration works using real-life issues faced on a case study restoration project undertaken by the authors. One would like to thank his son, Bobby Challender, for his contributions in carrying out these restoration works with him and overcoming some of the challenges and complications faced.

DOI: 10.1201/9781003358800-5

5.2 Complexities associated with historical buildings

5.2.1 The importance of regular inspections and building maintenance reports

According to Historic England, regular inspection and a robust maintenance regime should be applied to all heritage buildings, especially those that are in conservation areas and/or are listed.

Having maintenance policies and systems that are robust should aid the conservation of historic buildings and ensure that defects such as water penetration and rot do not unduly take hold and cause disproportional damage to the property structures and fabric. Such maintenance policies, strategies and systems should not simply rely on doing frequent and regular building inspections but having effective procedures for pre-empting future problems and instigation of prompt measures to remedy problems and defects. In this way regular effective maintenance should avoid health and safety implications to building users and members of the public, brought about by problems of disrepair and potentially dangerous structures. Notwithstanding this premise, Historic England reports that some practitioners regard such repetitive and endless regular inspections as unnecessary and mundane. Others, however, appreciate their importance in identifying emergency repairs that could manifest themselves in the need for significant remedial works if left unnoticed and therein not recorded. One example could be water penetration through a pitched roof. This may not be visible externally or even internally, but regular inspections of the roof void should reveal this defect so that remedial works can be undertaken quickly to prevent further water penetration. Although the repair of the problem may be only minor such as replacing lead flashing or damaged/displaced roof tiling the consequential damage if left unchecked could be disproportionate to the initial problem. An example could be water penetration through a roof covering which over a prolonged period of time could lead to dampness and timber decay of roof timbers. It is not unknown for timber decay of this kind to spread throughout the roof void unnoticed leading in extreme cases to complete roof replacement, structure and covering.

Similar examples of other damp problems could occur in walls or floors. Water ingress for instance through walls could be a result of open joints in walling brought about by worn or missing pointing or where gaps have developed around window and door junctions. In a similar scenario to the aforementioned roof repairs these could also give rise to disproportionate damage and consequently require urgent remedial works. Another example could be brickwork cracking as a result of building subsidence and differential movement. Such cracking may not be automatically noticed by building end users but should be identified and recorded at the time of building inspections. It could be that the cracking could be 'caught early' as maybe the symptoms of a leaking underground water main and drainage pipe. Again, early intervention could cure such underground leakages and prevent further movement of the structure thereby lessening the risk of costly and disruptive wall replacement works. This could be particularly devastating for listed buildings as some of the original structure may be beyond repair and must be replaced. For all the aforementioned reasons maintenance management, underpinned by regular inspections, is regarded as the pinnacle of building conservation.

5.2.2 Maintenance plans

Robust maintenance management should involve the formulation and utilisation of 'maintenance plans', and these were previously referred to in Chapters 2 and 3. These consider individual building requirements and accordingly should be tailored to properties depending on their heritage value and condition. For instance, an Ancient Monument or Grade I Listed Building holding significant heritage, architectural and cultural value in a potentially vulnerable state of repair and condition would present a very different scenario from a non-listed building, of little heritage value in good condition.

5.3 Historical protection given to buildings, structures and monuments

This section of the chapter will consider the statutory protection that is afforded to buildings and structures which have historical and architectural worth and significance. In this sense it will discuss conservation areas, listed buildings, scheduled monuments and regulatory restrictions that apply for each.

5.3.1 Conservation areas

Conservation areas are designated places of historical or architectural interest whereby there is a desire to enhance and preserve the appearance and character of such places. The designation of conservation areas is normally undertaken by local authorities. In this way they look to designate any areas within their catchment boundaries as having special historical and architectural interest and significance, and where the appearance and character of the area is worth protecting and preserving. This is in line with Section 72 of the Planning (Listed Buildings and Conservation Areas) Act 1990. According to Historic England there are approximately 10,000 conservation areas in England.

One can identify which areas have been designated conservation areas by contacting the planning authority within the local authority. They should be able to advise if a particular location falls within a conservation area and how far the boundary extends. Furthermore, they should be able to advise on when the conservation area was established, the level of legal protection that it has and the reasons for its creation.

Planning restrictions in conservation areas are normally concerned with the implications that any developments or alterations have on the external appearance of properties. The principle of conservation areas, as a general rule, is that permission must be sought and granted for any alterations to the exterior of buildings. This may include extensions or simply a change of material or element on the roofs and walls.

Notwithstanding this restriction, Article 4 Directions of the Town and Country Planning Act 1990 stipulates that work can sometimes be undertaken to a single-occupation dwelling house within a conservation area without the need to apply for planning permission. This is referred to as permitted development and applies to small extensions or alterations at the rear of domestic properties. Careful reference will need to be made by building owners to what is permitted prior to them embarking on any works, to ensure their proposals do not fall outside permitted development.

As a guide under Article 4 Directions planning permission may be required for any of the requires areas of development contained in Figure 5.1, and especially if these are visible on the front of properties.

In conservation areas there is sometimes confusion on whether window replacement requires planning permission. The general rule is that building owners, in a single-occupancy house, can change their windows without permission if they are a similar design and material to the

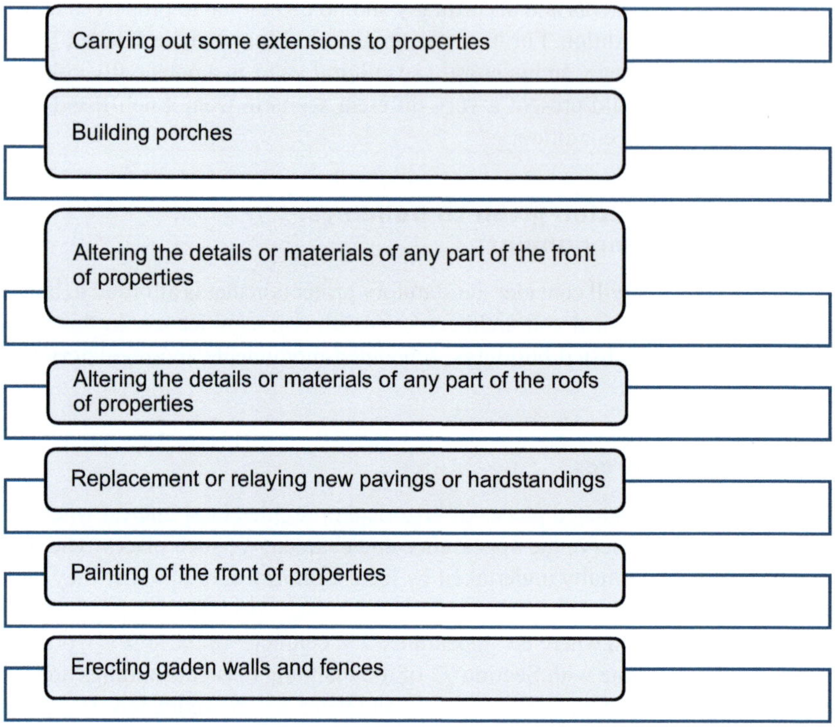

Figure 5.1 Works which may require planning permission under Article 4 Directions

existing. However, replacement of timber windows with alternative designs and materials such as uPVC and aluminium does not form part of permitted development and will sometimes be deemed unacceptable. There may be instances where a choice of alternative material for windows may be acceptable if the replacement windows mirror the size, form and design of the originals, but normally will be subject to planning applications. In addition, planning approval will normally be required for any proposed demolition works including buildings, walls, gates and railings. This demolition proviso extends to trees, and it can become a criminal offence, punishable by fines or even in some extreme cases imprisonment, to fell, prune or damage in any way a tree without giving notice and obtaining consent for the works. Normally this would be an online application to the planning authority justifying the reasons for the works, e.g., diseased trees with risk of collapse.

There is normally a four-year statutory period where the local authority can instigate enforcement action against building owners who do not have the required planning permission for their properties. Enforcement action of this kind may include the local authority requiring work to be removed and buildings reinstated to their original form prior to the alterations taking place.

5.3.2 *Listed buildings*

In the UK a building, monument or structure can become listed when it is deemed to have special historic and architectural interest and considered to have national importance and be worthy of protection. In such cases they are included on one of four statutory lists which are maintained by four different agencies depending on location and these are included in Figure 5.2.

Examples of listed buildings in the UK include the following buildings listed in Figure 5.3 and with photographs in Figure 5.4.

There are three categories of grading that apply to listed buildings in England and Wales and as set down by the Historic Buildings and Monuments Commission. These are set down in Figure 5.5.

- Historic England in England
- Cadw in Wales
- Environmental Scotland in Scotland
- Northern Ireland Environment Agency in Northern Ireland

Figure 5.2 The four different agencies responsible for listed buildings, structures and monuments

- Montacute House, Somerset
- Quarr Abey, Isle of Wight
- The Lloyds Building, City of London
- Battersea Power Station, South West London
- Chatsworth House, Derbyshire
- Plas Newydd, Angelsey
- The Outlook Tower, Edinburgh

Figure 5.3 Examples of listed buildings in the UK

Figure 5.4 Photograph of Battersea Power Station

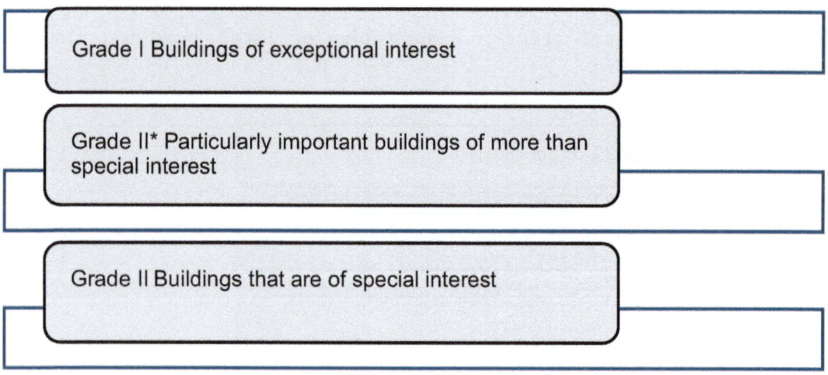

Figure 5.5 Categories of listed buildings in England and Wales

Grade I, pertaining to buildings of exceptional interest, represent approximately 2.5% of all listed buildings so are particularly rare. Grade II* buildings, referring to properties that are of particular importance, are slightly more common and represent approximately 5.8% of all properties listed. Grade II buildings, deemed to have special interest, represent the remainder and are therefore much more common than Grade I or Grade II* properties.

Listed buildings tend to have more protection than those purely located in conservation areas. Whilst conservation areas are focused on external alterations to properties, buildings designated as listed cannot have their exterior or interior altered without permission. There are exemptions for secular controls on listed buildings for some types of buildings including places of worship. However, even in these cases the religious organisations have to operate their own robust permissions controls which are equivalent to approval and management processes for listed buildings.

There are numerous interventions to listed buildings that applicants are unlikely to gain consent for. These include removing building features such as decorative stonework, moulded plaster ceilings, fireplaces, window mullions and wall panelling which are all considered to constitute special architectural features. Building owners are legally obliged in some cases to maintain and repair their listed buildings and may face prosecution in criminal courts if they allow them to fall into disrepair. The same outcome could also apply if building owners carry out any unauthorised alterations.

As part of the listed building approval process, it will be necessary for building owners and their appointed design consultants to justify the selection of repairs or alterations proposed in their applications. This will normally involve an explanation of specific materials and techniques which will be used and the reasons and justifications pertaining to their proposals. As a general rule, conservation officers will normally want to see like for like repairs and replacements using the same original materials and methods of construction. An example could be where stone facades are being restored and any new stone facings should be the same type of stone and appearance to match the original. Similarly, the raking out of mortar joints in this example should be undertaken by hand and not by using any mechanical grinding equipment which could damage the stonework. Furthermore, the choice of mortar may have to mirror the original and be of a lime base rather than a cement alternative.

5.3.3 *Other historical protection given to buildings, structures and monuments; scheduled monuments*

According to Historic England, scheduling represents the earliest form of providing protection to buildings and monuments. Although the first scheduling did not begin until 1913, its roots date back to the 1882 Ancient Monuments Protection Act, which created state protection for prehistoric monuments.

Although in England, Historic England maintains its own programmes for scheduling, anybody can nominate a site for scheduling, with the final decision being made by the government's Secretary of State for Culture, Media and Sport. Scheduling involves the identification and selection of archaeological sites which are deemed to have national importance. There is a mandate that these will be passed down through different generations in much the same condition and state as they were originally, although some changes may be considered in special circumstances. These may include instances where scheduled ancient buildings have become structurally unstable and at risk of collapse. In these circumstances new technologies to support the affected dangerous structure may be permitted but will need to be robustly justified in any applications submitted. All applications related to carrying out any works on a scheduled monument must be made to the Secretary of State for Culture, Media and Sport. Scheduled monument consent must be received before any works commence and any considerations and all conditions of the approval must be strictly adhered to during the works. This is designed to protect scheduled monuments from unlawful works or becoming damaged through poor design and construction management proposals.

Scheduled monuments are not exclusively ancient and there are numerous different categories on the list. These include monasteries, castles, medieval sites and abandoned farmsteads and in England there are no less than 20,000 scheduled monuments listed.

5.4 Conservation and heritage protection of buildings using case studies

Buildings and sites of heritage value represent an irreplaceable resource and government agencies alongside responsible building owners have a legal and moral duty to adopt measures designed to conserve them. Sometimes where the original use of a heritage building has become obsolete, strategies should be adopted to create sustainable and viable uses for these buildings, consistent with their conservation. This could involve creative reuse solutions which may draw on economic, cultural, social and environmental benefits that conserving and preserving buildings of special architectural and historical significance can bring. The following two case studies provide real-life examples where creative reuse of obsolete, derelict and vulnerable buildings through major restoration and change of use have taken place.

5.4.1 Case study 1: a vulnerable listed building; former Stockport Union Workhouse restoration, Stockport, Cheshire

Client Project Sponsor and Project Director: Jason Challender, Stockport College
Project Manager: Hawre Baban
Architects: Austin Smith Lord
Main contractor: Naylor and Walkden Building Contractors Ltd.
Contract value: £5m

This case study, relating to the former Stockport Union Workhouse in Stockport, presents a good example of restoration for creative reuse and which the author was personally responsible for overseeing. The author at this time was employed by Stockport College which acquired the building from the Stockport Primary Care Trust. Although the building was steeped in history as a Victorian workhouse it had been for many years operated as a primary care trust hospital, namely St. Thomas Hospital. The building was regarded as being both architecturally and historically significant since it was originally the Stockport Union workhouse which opened on Christmas Day in 1845. For this reason, it was designated a Grade II listed building which clearly provides a legal framework for its conservation and preservation.

On acquisition by the college from the Stockport Primary Care Trust in 2006, it was clear that the building was in a dire condition especially those areas with floors above ground level that were not used by the hospital, and which laid derelict, mothballed and unmaintained for decades. Figure 5.6 illustrates some of the damage to the structure and fabric of the building at this time. Many building inspections and surveys had been instigated by the author through specialist consultants. These revealed that the building had been suffering from several major defects owing to a lack of maintenance and had fallen into severe disrepair over an extremely long period of time. It was clear that the building was unsafe in many areas with rotten timber floors, extensive dampness and water penetration through the roof. This made the building particularly difficult to survey owing to health and safety issues. For this reason, a robust risk assessment had to be undertaken with mitigations and conditions which had to be strictly adhered to for any persons accessing the building.

Figure 5.6 Damage and defects to the former Stockport Union Workhouse prior to restoration works

In consideration of the scale and value of the project which was approximately £5m, it was vitally important to obtain listed building consent prior to any commencement of the works. For three years prior to the restoration works commencing, this involved extensive dialogue between the author and the college's design team. In addition, there was considerable liaison with English Heritage (now Historic England) and the conservation officer from the local authority, namely the Stockport Borough Council. This involved many meetings to update these external agencies on progress with the design and to discuss and agree on what the scope of the project may involve. It was also useful at this stage to assess what the college could use the building for, and the degree of any building interventions required to make this possible. It was decided to use the building predominantly as a higher education facility for the college, which involved conversion of most of the internal spaces into classroom accommodation.

Numerous specialist surveys were carried out to identify and assess those elements of the building's structure and fabric particularly sensitive from a historical and architectural perspective. These included undertaking a historical and architectural report on the building, recording elements which were regarded as being particularly important for conservation and preservation reasons. This survey was known as a 'Survey of Significance', and involved having to research desktop studies and through archives about how some parts of the building were originally constructed. In addition, it provided useful context on how the original Victorian building functioned and was utilised as a workhouse. These surveys also assisted the design team, where deemed appropriate, to develop plans to return some parts of the building back to its original form and appearance.

This period of discussion and dialogue with English Heritage and the local authority conservation officer took approximately three years as certain initial plans presented were deemed inappropriate for a building with listed status. Consequently, it was decided to limit any intervention works including demolition or additions to an absolute minimum to preserve the original layout and form of the building as a memory of what the building comprised in Victorian times. Detailed drawings for all the proposals had to be prepared and presented at various times to gain feedback and consensus during the design development process. In addition, sometimes it was necessary to provide samples where old materials were proposed to be replaced by new ones. Examples of this included having to produce sample replacement roof slates and handmade bricks where it was important to match them with the existing ones. In the case of brickwork, it was also required to produce a small sample area of restored brickwork on the building incorporating selected new bricks and replacement lime mortar pointing. The amount of time, cost and effort which was expended during this period was considerable and involved an approval process of three years. This is regarded as not representing an unusually long period when dealing with listed buildings. However, the duration of this period, leading up to listed building approval, delayed the college in progressing the commencement date of the works. Consequently, building owners and developers should not underestimate the period necessary to conduct this process and build into their project programmes realistic time durations for the listed building approval process.

The first phase of this extensive restoration, redevelopment and refurbishment project on the main workhouse building commenced on site in 2009 with the author as Project Director of the project. This phase was designed as an enabling project to carry out all essential repairs and put the building into a fully wind-resistant and watertight condition and structurally stable state, and thereby reverse the effects of many years of disrepair. Repairs, restoration and refurbishment work included the external works in Figure 5.7 and the internal works in Figure 5.8.

Figure 5.7 External repairs, restoration and refurbishment work

Figure 5.8 Internal repairs, restoration and refurbishment work

During the contract period it was necessary to expose certain parts of the structure and fabric of the building to decide on the scope of repair or renewal most appropriate. The survey of significance previously referred to assisted in this process. It allowed the client, design team and conservation officer to agree what building elements and components needed to be maintained in their present form rather than being repaired or removed entirely. Furthermore, it influenced the degree and scale of those elements which were required to be repaired rather than renewed. An example related to the original timber window frames which had to be preserved although heavily rotten. Undertaking repairs would not normally have been regarded as an economically viable option for any other building other than one which was listed owing to the condition of the windows. Consequently, extensive repairs had to be undertaken to splice in new timbers to replace those rotten ones. In addition, some window frames had to be completely carefully removed and rebuilt in factories before being refixed in situ, which was a time consuming and costly undertaking. Other examples included retaining some of the original roof slates that could be reused for the new roof covering. Although most slates removed were unusable owing to their condition some of them were carefully put aside to bolster the numbers of reclaimed slates being brought in from elsewhere. Figure 5.9 includes photographs of the restoration works progressing during the construction period.

Figure 5.9 Restoration works progressing during the construction period

One of the main challenges with repair/renewal decision-making revolved around the internal plaster wall finishes. Most of these above ground level comprised the original Victorian finishes and paintwork. The conservation officer from the council was very keen to preserve as much of the original plasterwork as possible. However, on closer inspection all the plaster within the building was deemed to be so badly damaged and detached from the walls through dampness that it had to be removed and replaced for new. Samples of plaster and original paint finishes had to be tested prior to removal for the presence of lead or anthrax. It is not unusual for both contaminants to be contained in such materials used in Victorian times. These debates on what could be sensibly saved represent the type of conversations that should be had on dealing with any restoration project, with the aim of preserving as many original elements of a buildings structure and fabric as possible. Notwithstanding this premise, preservation is not solely linked to buildings themselves but should consider any original fixtures, fittings and artefacts which can be retrieved. On this project, other than the original bell and clock which were overhauled, there was little in the way of fixture, fittings or equipment present when the building was acquired. There were however many different artefacts dating back to its checkered history as a Victorian workhouse uncovered during the works. These included inmates' shoes, cutlery, drinking mugs and smoking pipes uncovered when floors were exposed. These artefacts were carefully removed with dialogue with the curator of the local museum in Stockport. They were then safely stored and exhibited for the public to see in the Stockport Story Museum.

One of the other challenges associated with this project was the conflict between conservation and preservation requirements and those relating to other requirements. One example of this related to the walls and windows of the building. In the case of the walls the building control officer of the local authority, who was overseeing the building control application, was requiring the existing external wall to be insulated. This was related to increasing the U-value of the walling and therein complying with the building regulations on energy conservation and heat loss. Both the conservation officer and English Heritage objected to the internal lining of external walls to provide the degree of insulation required. They protested and argued that this would materially affect the form of the building fabric and compromise the heritage value of the building. As an alternative proposal, they requested that all exposed brickwork as shown in Figure 5.10 should be retained as an internal finish. A similar issue occurred with the windows where the building control officer was requiring double glazing or secondary glazing to be installed, again to comply with regulations relating to energy conservation and heat loss through the building envelope. This provoked a conversation regarding a material change in the form of the original windows. The conservation officer protested that this change would severely alter the construction of the windows to the extent that any heritage value would be lost or at least heavily compromised. These issues emanating from different perspectives and legal standpoints i.e. building regulations vs conservation and preservation demonstrate that opposing arguments and diametrically different opinions can become apparent. In this particular case the two opposing influences were both generated within the same council, namely the Stockport Borough Council. The solution to end any potential stalemate on this project was to apply for a relaxation on building regulation for compliance with the minimum energy efficiency levels on the grounds of the listed status of the property.

Figure 5.10 Exposed brickwork as an internal finish

The building restoration works were carried out over a contract period of 18 months and successfully completed in December 2010. The building was brought back to a state of excellent repair and condition and one which would have resembled its original form. It went on to receive commendations from many different professional bodies including the Royal Institute of British Architects and the Victorian Society. Unfortunately, the building did not go on to be converted into an educational facility for the college owing to unexpected funding challenges and was subsequently sold for residential development after the author left the college in March 2011.

5.4.2 Case study 2: A vulnerable listed building; Gorton Monastery, Gorton Manchester

Client: The Monastery of St. Francis and Gorton Trust
Architects: Austin Smith Lord
Value: First phase approximately £20m

An Ancient Monument or Grade I Listed Building represents significant heritage, architectural and cultural value as referred to earlier in the chapter. They can become very vulnerable when they are allowed to fall into disrepair. An example of this dilemma related to a Grade I listed building which prior to restoration commencing in 2006 featured in the world's most endangered historical buildings. It had been left in a terrible state of repair and condition following being bought by an American developer in 1989 and then subsequently abandoned and left exposed to the elements, theft and vandalism for many years after.

The author came across this vulnerable building in December 1999, when he inspected it with the director and founder of a charitable trust, tasked with saving the building from the severe deterioration inflicted upon it. On inspecting the building with the director from the Monastery of St. Francis and Gorton Trust the author was alarmed to witness the scale of disrepair. The roof had partially collapsed and water penetration through the roof covering was causing a huge amount of dampness to transgress through the structure and fabric of the building. It was clear to the author on this visit that time was not on the side of the building. Avoiding any prolonged delay in addressing numerous major defects and stabilising the building from irretrievable damage requiring its demolition was therefore considered imperative.

For context a short history of the monastery will be provided to assist readers in undertaking the historical significance of the building.

The monastery was commonly regarded as Edward Welby Pugin's greatest architectural masterpiece and was originally known as the Church and Friary of St Francis, in West Gorton and home to Franciscan friars. The building was completed in 1865 and constructed with a striking façade with a central crucifix and heavy buttresses. It incorporated stained glass windows that allowed light to penetrate through to the high altar which represented the building's focal point. The building was designed to be an imposing piece of architecture with cathedral-like proportions incorporating a 30m high nave. Figure 5.11 contains a photograph taken in the 1950s showing the front façade in its former glory.

Figure 5.11 Front façade of Gorton Monastery taken in 1950s

The church and its friary flourished for almost 100 years but with a fall in the congregation this impacted finances through church collections. This adversity was coupled with mass unemployment in the surrounding area of Gorton caused through closure of engineering works and further exacerbated by slum clearances which led to an exodus of the local community into other areas of the city of Manchester. The building closed in 1989 and was sold to a developer with an intended use for luxury apartments. Unfortunately, this development never happened, and the buildings and site were left vulnerable and exposed with little or no security arrangements. Consequently, vandals attacked the unprotected site and ransacked or damaged many sculptures and other artifacts. Over many years the building was not maintained, damp ingress through the building structure and fabric caused part of the roof and nave to collapse in and walls began to perish (see Figure 5.12).

Figure 5.12 Resultant damage to monastery caused through disrepair

The Monastery of St Francis and Gorton Trust was set up in 1996 in an attempt to save the church and adjoining friary. This trust oversaw a team of local community volunteers to attempt to reverse the deterioration and preserve the monastery and adjoining site through restoration. In 1997 the building was placed on the World Monuments Watch List of 100 Most Endangered Sites. This provided the impetus for a global call on immediate intervention and corrective action for this important heritage site. This eventually proved fruitful in 2005 when tireless lobbying and campaigning by the Trust provided a successful outcome. The building had been awarded both ERDF (European Regional Development Funding) and Heritage Lottery Funding, which represented one of the largest community-led heritage restoration projects of its kind in the UK.

The extensive restoration programme on the building was successfully completed in 2007 and the venue now operates for hire and as a heritage visitor attraction (see Figure 5.13). All visitor proceeds are allocated to the continued maintenance and ongoing preservation of this building and its site. The success of this project was predicated on its creative reuse as otherwise funding may not have been available to finance the expensive restoration works. Furthermore, its story provides a constructive example and catalyst for all other endangered buildings that with the right motivation and funding available can still be saved.

Figure 5.13 Gorton Monastery once restoration works completed

5.5 Examination of appropriate building technologies and materials for restoration works using a case study

This section of the chapter will draw on sources from Historic England and reference a restoration project on a protected Victorian house, carried out by the author in Ramsbottom as a case study.

Restoration works on all classifications of historical and protected buildings should wherever possible involve using authentic natural materials, coupled with forging the balance between traditional construction techniques and modern advancements. Key construction trades for restoration in this regard include stone and brick masonry, carpentry and joinery, roofing and decoration. In selecting authentic materials and construction methods, this assists in retaining the character of historic buildings and can help support traditional craft, skills and industries. Some skills such as thatching are regarded as becoming a 'dying art' and this can affect the availability of tradesmen that undertake such specialist work.

Materials used for restoration projects do not always have to be on a like-for-like basis for those building elements being repaired. For instance, it may be appropriate to use alternative materials if this allows more of the original fabric of buildings to be conserved than traditional materials. An example could be to introduce steel supporting structures into traditional masonry-constructed properties, to support unstable walls and floors. Notwithstanding this premise, if alternative materials are being considered for historic buildings they must be proven to be tried and tested and not experimented with for the first time. They should also be justified for appropriateness for restoration works and where possible complement the original elements of existing buildings.

Whichever materials are selected they must be highly compatible and sympathetic to original materials used within the structure and fabric of buildings. It is also important to consider the visual appearance of the materials to be used and whether these are in keeping with the style, age and design of buildings. Furthermore, one needs to examine the physical qualities of some materials given that some could have adverse effects on future deterioration. Durability and longevity of repair materials used in the restoration process should be considered to avoid those which could have implications on future maintenance. Whilst projects must consider budgetary boundaries it could represent false economy in the long run to opt for cheaper inferior materials that could have lesser compatibility and longevity.

Historic England holds a register with specialist contractors competently qualified with the necessary expertise to undertake the most complicated restoration project. They also have a suppliers list that can provide a wide range of different materials used commonly in restoration programmes. These include lime for mortar, handmade bricks, clay roofing tiles and hair for plaster. According to Historic England, the selection process for the most appropriate materials to use for restoration works is an area that requires expertise, analysis and sometimes investigation works. Such investigations may involve intrusive exposure works by opening up hidden parts of buildings. An example of intrusive investigations may include carefully taking up an existing relatively modern floor finish to reveal an original floor finish below. On a restoration project carried out by the author on a Victorian property in Ramsbottom, the heavily damp and deteriorated ply sheeting and carpeting to the floor was carefully taken up. The sheeting and the carpet had created a barrier for any natural ventilation to the floor structure and prevented the floor from 'breathing'. This process of removal revealed the original stone floor as can be seen in Figure 5.14. In this case the stone floor was repointed with lime mortar where mortar joints had perished. A proprietary treatment was then applied to it to clean the stone and prevent future staining. This preservation treatment was designed to allow the floor to 'breathe' and in this way avoid dampness being trapped within the floor.

On the same project, existing hard cement mortar applied to some walls which was trapping water in walls leading to dampness and efflorescence. The author decided to remove this render and expose the stonework structure of the walls. The stone was carefully cleaned to remove all signs of remnants of cement and then damaged mortar joints were carefully raked out by hand. Following this the walls were repointed with a lime mortar which offered flexibility and breathability. As illustrated in Figure 5.15 the walls became a very attractive feature, exposing their original structure. In addition, the walls were able to dry out, free from trapped water, and therein breathe normally from air movements around them. Where there is a conservation technique that can prevent future deterioration from the effects of defects such as dampness then this should be considered for any restoration project.

In the restoration of old buildings, it is quite common to find that at some point in the past original features have been removed and inappropriately substituted for modern alternatives. On the aforementioned project in Ramsbottom the original fireplaces had been removed and replaced with modern fire surrounds in the 1980s. The author took the decision to source two original timber Victorian fire surrounds, one carved timber and the other cast iron, which can be seen in Figure 5.16. Whilst these two additions were clearly not original to the house itself, they are nonetheless presented considered as a worthy addition in providing a similar age and design of fireplace to the ones that would have been originally installed. In this sense they were considered appropriate and sympathetic to this historic building.

Figure 5.14 Original stone floor exposed and restored in a Victorian property

Figure 5.15 Exposed and repointed stone walls on a Victorian property

Figure 5.16 Fire surrounds installations as part of restoration works

In addition to the fireplaces the original timber doors had been removed in the past and replaced with modern-day alternatives. It was felt that the character of the Victorian property was affected by such changes. Accordingly, the author took the decision to source some reclaimed pitched pine doors which would have been salvaged from a building of a similar age. These are illustrated in Figure 5.17.

Figure 5.17 Salvaged Victorian doors as part of restoration works

In terms of modern additions to historic buildings these can vary in appropriateness and quality. On the same project in Ramsbottom there had been some very poor additions and accretions in the form of a glass porch, extension of window openings and replacement uPVC windows. These alterations were considered by the author as unwelcome and compromised the historical nature of Victorian architecture. Accordingly, the porch was removed to expose the original external stone door opening and new timber corded vertical sliding sash windows, similar to those used originally (see Figure 5.18).

Figure 5.18 Front elevation of property showing porch removed, window opening reinstated to their original form and new window frames installed

5.6 Summary of chapter

In this chapter the following areas have been covered:

- Complexities associated with historical buildings
- The importance of regular inspections and building maintenance reports
- Historical protection given to buildings, structures and monuments and how this applies to conservation areas, listed building and scheduled monuments
- Conservation and heritage protection of buildings using case studies from the Stockport Union Workhouse and Gorton Monastery
- Examination of appropriate building technologies and materials for restoration works using a third case study on a dwelling in Ramsbottom

Restoration works on all classifications of historical and protected buildings should wherever possible involve using authentic natural materials, coupled with forging the balance between traditional construction techniques and modern advancements. Materials used for restoration projects do not always have to be on a like-for-like basis for those building elements being repaired. However, if alternative materials are being considered for historic buildings they must be proven to be tried and tested.

Historic England holds a register with specialist contractors competently qualified with the necessary expertise to undertake the most complicated restoration project. They also have a suppliers list that can provide a wide range of materials used commonly in restoration programmes. Whichever materials are selected they must be highly compatible and sympathetic to original materials used within the structure and fabric of buildings.

Regular inspection and a robust maintenance regime should be applied to all heritage buildings, especially those that are in conservation areas and/or are listed. This should involve building owners having robust maintenance policies and systems in place coupled with the formulation and utilisation of 'maintenance plans'. These should consider individual building requirements and accordingly should be tailored to properties depending on their heritage value and condition.

Conservation areas are designated places of historical or architectural interest whereby there is a desire to enhance and preserve the appearance and character of such places. One can identify which areas have been designated conservation areas by contacting the planning authority within the local authority. Planning restrictions in conservation areas are normally concerned with the implications that any developments or alterations have on the external appearance of properties. The principle of conservation areas, as a rule, is that permission must be sought and granted for any alterations to the exterior of buildings, unless works come within the parameters of 'permitted development'.

In the UK a building, monument or structure can become listed when it is deemed to have special historic and architectural interest and considered to have national importance and worthy of protection. There are three categories of listed building, and these relate to Grade I (buildings of exceptional interest), Grade II * (buildings of more than special interest) and Grade II (buildings of special interest). Whilst conservation areas are focused on external alterations to properties, buildings designated as listed cannot have their exterior or interior altered without permission. Examples and photographs of UK listed buildings have been given in the chapter. Building owners are legally obliged in some cases to maintain and repair their listed buildings and may face prosecution in criminal courts if they allow them to fall into disrepair. There are numerous interventions to listed buildings that applicants are unlikely to gain consent for and these include the removal or alterations to historic or architectural features.

Case studies have been described in this chapter to articulate some of the many issues and challenges faced when embarking on restoration projects. These related to vulnerable listed buildings in Greater Manchester which the author was personally involved in. The first case study related to the former Stockport Union workhouse, Cheshire under the project direction of the author and the second Gorton Monastery, East Manchester. In addition, a third case study relating to a restoration project carried out by the author examined appropriate building technologies and materials for restoration works.

Chapter 6

Common internal maintenance issues, defects and remedial works

6.1 Introduction to chapter

This chapter of the book will identify, describe and analyse the problems and challenges pertaining to internal defects which commonly occur in all domestic and non-domestic buildings. It should be read in conjunction with the next chapter, Chapter 7, which will identify, describe and analyse external defects, as these can be closely linked to and give rise to internal building problems. An example of this could be damage to roof coverings, identified externally, which causes water penetration internally. In this sense the importance of linking external and internal defects will be reinforced in the chapter.

The chapter will start by explaining the different types of building and the conventional or nonconventional nature of them, and the importance of considering their construction, use, condition and age. These factors will help surveyors when undertaking building inspections to understand some of the possible defects they may encounter with a particular building type.

The categories of the different internal defects will be discussed in detail which will include the following:

- Rising dampness
- Condensation
- Water penetration through buildings
- Timber decay. Wood-boring insect
- Presence of asbestos and other hazardous materials
- Wall, floor and roof defects
- Defective electrical, gas and water installations

For each one of these, there will be an explanation on how to identify them and the underlying causes of these defects. This will be followed by a comprehensive description of how to properly investigate them, leading to their correct diagnosis. Remedial works to address each defect will be articulated and, in some cases, the alternative repair works which could be adopted considering the different building types and scenarios e.g. listed buildings. With some defects such as dampness within buildings it is common for surveyors to misdiagnose the exact causes. For example, sometimes symptoms of rising dampness, water penetration and condensation, as three alternative sources of dampness can be similar. To address this dilemma, there will be precautionary advice within the chapter on the ways and means to avoid the pitfalls of misdiagnosing the underlying causes of building problems. In this regard, there will be an emphasis

DOI: 10.1201/9781003358800-6

on the importance of reaching the correct diagnosis to avoid costly and unnecessary remedial works associated with misdiagnosis.

To explain the conditions and building faults which can cause some of the defects, figures will be included to clearly illustrate these. In addition, in some cases, examples of each one will be presented to assist readers in understanding the practical implications and challenges they can create. Furthermore, to illustrate the internal defects commonly found in buildings, the chapter will contain photographs showing the possible visible symptoms. Again, this is designed to assist readers in identifying the telltale signs when conducting building surveys. Finally, for some of the defects, photographs showing remedial works to overcome them will be included in the chapter.

6.2 Common internal maintenance issues, defects and remedial works

When considering the types of internal maintenance issues, one needs to closely look at the construction of the buildings and focus particular emphasis on any elements of them which could give rise to defects and remedial works. For instance, if it is an older building, elements of it may not be constructed from conventional materials. Examples could include historical buildings where timber frames may have been utilised as the main structure for load-bearing walls. Particular emphasis on checking and potentially probing the timber for timber decay may prove a useful and necessary undertaking. In some parts of the South West of the UK and throughout developing nations, the use of cobb may have been used for walls and care and attention will be required to assess any signs of dampness and structural movement associated with them. For domestic properties roof coverings may again be nonconventional and examples could be the use of thatch and other natural coverings such as cedar shingles. Conversely some modern technically advanced roof coverings, such as zinc, could be classified as nonconventional and as such may present their own problems. The construction type of most buildings will also vary according to their intended use and age. For instance, newer large industrial buildings will rarely be constructed from traditional methods including load-bearing internal walls. Alternatively, these industrial buildings are more likely consist of steel or concrete framed structures, which are relatively quick to erect, and which give clear internal space free from load-bearing walls. Having clear internal environments free from obstructions can be especially useful for industrial operations such as manufacturing and to make them fit for purpose in terms of functional performance. That is not to say that industrial buildings have fewer maintenance issues than domestic buildings, it is just the nature of the defects will vary between these two building types. Industrial buildings such as distribution centres may have metal cladding which exhibit visible signs of corrosion and in some extreme cases may be constructed from potentially hazardous materials such as asbestos cement. The recent case of the Grenfell Tower disaster should also highlight the need to inspect buildings of all types for possible use of claddings that has been deemed unsafe and potentially hazardous. Where potentially hazardous building elements are suspected to be present, further detailed surveys or specialist inspections should be undertaken.

Whilst the categories of internal defects can be extensive, Figure 6.1 illustrates some of the common defects which can be present in all different types of buildings from domestic dwellings to commercial properties. These will be discussed separately in this chapter from the causes, telltale indicators and potential remedial works to address the defects.

are damp smells in the affected rooms caused by growing fungi on the damp areas of wall. One other noticeable sign of rising damp could be surface deposits of hydroscopic salts, otherwise known as efflorescence. If one or more of these indicators are present at lower levels in ground floor rooms, it is likely that this will be rising dampness. Notwithstanding this premise, rising dampness can be sometimes misdiagnosed for something else.

6.3.3 Misdiagnosis of rising dampness

Rising dampness can be commonly misdiagnosed by those who are not sufficiently experienced or qualified. It is not usual to the untrained eye to see some of the symptoms referred to in Figure 6.2 and assume that it can only be rising dampness. For instance, some building occupants may see dark patches and tidemarks on their walls and wrongly assume they are experiencing rising damp. Accordingly, it is important before formulating a diagnosis that the following questions in Figure 6.3 be answered.

If there is any uncertainty about any of the questions in Figure 6.3 then it may be that the problem is not in fact rising dampness but something else entirely. For instance, dampness synonymous with rising damp transcending vertically through external walls could be caused by water penetration transcending horizontally. An example of this could be where there are wide cracks or where mortar joints are open in external walls. Both these defects could facilitate wind-blown rain to traverse through walls causing internal dampness. Alternatively, water

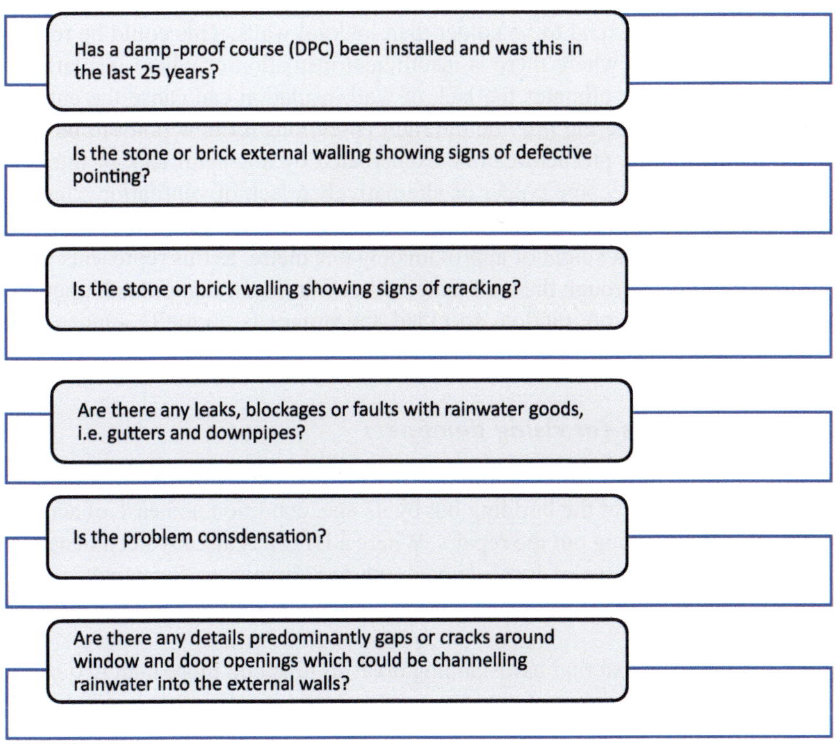

Figure 6.3 Questions that need to be answered before diagnosing rising dampness

penetration through walls or roofs may be emanating from blockages or damage to gutters and downpipes. Furthermore, gaps around door and window openings in external walls can cause dampness on internal wall surface. For all these reasons misdiagnosis of rising damp is common which can result in unnecessary and costly remedial works which will not cure the problems and issues being experienced. It is therefore crucial wherever possible to commission the advice of a qualified specialist, preferably a chartered building surveyor or preservation specialist. This should ensure an accurate defect diagnosis and the correct recommendations for remedial works being made. This advice should involve, prior to any treatment, undertaking a thorough, detailed damp survey of the area of walling that has been affected. This should allow surveyors to determine the cause of the damp and thereafter formulate legitimate recommendations for appropriate remedial works. Sometimes this may call for invasive explorative inspection works to open up and expose small areas of walling not visible from the surface. This may be appropriate in checking the condition of existing damp-proof courses (DPCs) where it is suspected that they may be damaged. This opening-up work could not only assist with diagnosis of the problem but help specify the most appropriate and suitable remedial work. The type and extent of any repairs will depend on the type of property being inspected as some types of remedial works may be more conducive for some buildings than others. For instance, the age and historical nature of a particular building could present an incumbrance for undertaking certain repairs if the building is listed and thereby subject to formal listed building consent.

Rising dampness is sometimes misdiagnosed where dampness caused through condensation is the real cause. Condensation will be referred to later in this chapter, but it is normally caused when moisture in the air reaches 'dew point' on the internal surface of walls, predominantly external walls which tend to be colder than internal walls. This could be resulting from defective wall construction where there is insufficient insulation or where insulation has been wrongly installed. In colder climates the lack of wall insulation can cause the entire fabric of walls to become cold and therein provide the right conditions for dew point to be reached and condensation to occur. This problem can be exacerbated by a general lack of internal heating causing the internal air to become colder or alternatively a lack of ventilation causing internal air to retain more humidity. As a general rule rising dampness normally affects the area at the base of external walls up to a height of approximately one metre, as this represents the limit that capillary action can work through the wall materials. Accordingly, dampness higher up the wall surfaces and displaying a more random speckled appearance is normally condensation rather than rising dampness.

6.3.4 Remedial works for rising dampness

The remedial works to treat rising dampness are sometimes complicated and will be affected not just by the construction of the building but by its age, condition and ease of access for both exposing the wall and carrying out the repairs. Where a DPC has already been installed it could involve addressing the sources of local damage where a breach or 'bridging' of the DPC is occurring. This may include the following works:

- Relaying and lowering external hardstanding areas to finish no more than two brick courses below the level of the damp-proof course
- Repairing broken and damaged rainwater downpipes and gutters spraying water over lower levels of brickwork above DPC level
- Removing defective deposits and inappropriate insulation from wall cavities

If the source of the problem through investigative works has concluded that the damage to the damp-proof course is limited and damp bridges through it are minor, then isolated repairs maybe present an affordable option. With traditional bituminous felt or plastic laid damp-proof courses this may involve removing small areas of walling and reinstalling small sections of the physical barrier where the material has been breached. Notwithstanding the aforementioned premise where it is deemed that there is no damp-proof course present or where the DPC is severely damaged or been installed incorrectly a new DPC installation may be required.

The traditional way of installing a new DPC within a property was in the past predicated on hacking off approximately 1 m of internal plasterwork to expose underlying brick or blockwork, before installing a chemical (usually silicon) injection DPC through predrilled holes in the external and internal leaves. The objective of this system is to provide a continuous horizontal band of impervious material which is waterproof and as such can act as a weatherproof barrier. This barrier is designed to block the path of moisture from the ground which is attempting to rise up masonry walls through capillary action. On completion, the walls should then be internally replastered with a 'renovating plaster' which would not react with the DPC chemicals, and thus prevent further salt migration, and then painted. In many older properties where solid walls consist of thick stonework with numerous voids, a chemical injection DPC cannot always be effective. In the past other damp proof course systems have been deployed and the electro-osmotic system is one example. This system relies on installing strips and anodes sometimes drilled into the walls incorporating high electrical conducive metal, normally titanium coated with a platinum group element. These are designed to generate an electrical voltage within the wall which can repel water and therein acts as a barrier to damp rising. In this way it can provide a cost-effective and reliable solution which is resistance to chemical and physical attack. This system may be more conducive where walls are very old and solidly built as it works for most wall types irrespective of their make-up, condition and age. It also represents an environmentally greener and safer product as it is completely chemical free with no associated health risks.

Some causes of rising dampness have been predicated by very hard cement based internal wall finishes which have effectively trapped dampness rising from the ground and prevented walls from breathing and drying out. Treatments which have involved removal of such finishes and reinstatement with alternative, 'breathable' finishes such as lime-based mortars have allowed walls to reverse the effects of trapped moisture and effectively dry out.

6.4 Condensation

The dictionary definition of condensation is the conversion of a vapour or a gas to a liquid (Oxford Languages). For buildings it normally occurs when water collects as droplets on a cold surface when air containing a high moisture content meets it. It has sometimes been described as the reverse action of evaporation or vaporisation when vapor is produced from liquid water when heated.

When the temperature of the air falls on contact with this cold surface it is less capable of retaining its moisture content. At this point there is an increase in what is commonly referred to as *relative humidity*. When this relative humidity reaches 100% this is referred to as the 'saturation point' or 'dew point'. If the air temperature continues to fall after this point moisture will begin to convert to liquid as condensation. Typically, this can happen in buildings when moist warm air comes into contact with surfaces which are much cooler and below the dew point. Window surfaces are a good example of such surfaces where condemnation is prevalent for this reason.

6.4.1 *Explanation of what constitutes condensation and its causes*

Surface condensation, as previously referred to, will normally occur in buildings where moisture within the air in the form of water vapour meets surfaces which are colder and below dew point temperatures. For this reason, cold internal surfaces of external walls, where there is insufficient insulative qualities, can be particularly vulnerable. Condensation can be exacerbated when other defects in buildings provide the conditions for damp air to be generated. This may result from buildings which are not fully wind and watertight. For example, water penetration through walls and roofs could be prevalent owing to breaches of roof coverings and penetrations through walls. The damp ingress resulting from such water penetration can generate high levels of humidity for internal air which on contact with cold surfaces condenses to form condensation. Condensation does not always have to occur on the surface of walls as moisture can be formed in what is referred to as 'interstitial condensation', where it is created within the inner layers of the fabric of building elements, mostly walling. This interstitial condensation results from the diffusion of air from the relatively warmer internal environment of buildings to the colder parts of the building fabric. Here it reaches dew point within the fabric of building elements rather than on internal surfaces.

6.4.2 *Investigation and diagnosis of condensation*

There are numerous visible signs and problems caused by condensation and some of these are included in Figure 6.4

The visible signs of condensation could include evidence of speckled black mould growth across the internal surfaces of walls, and are particularly prevalent in the corners of external walls where surface temperatures can be colder. This is illustrated in Figure 6.5.

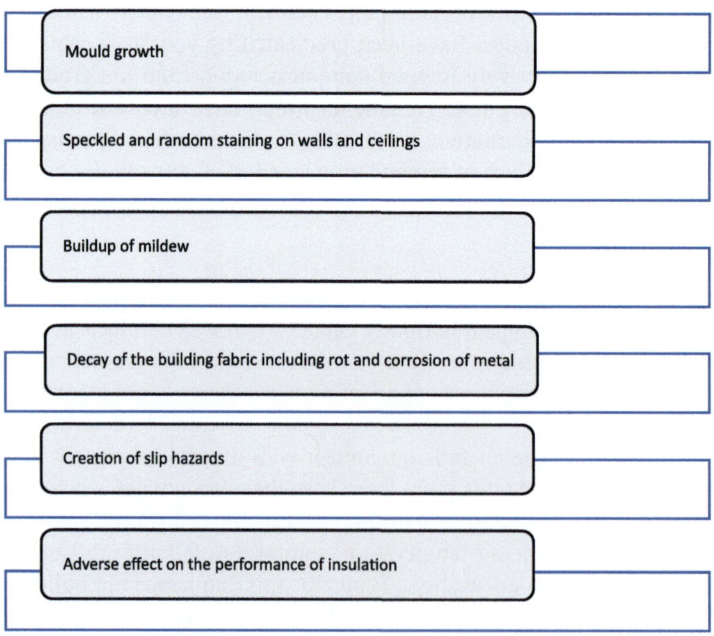

Figure 6.4 Visible signs and problems caused by condensation

Figure 6.5 Photograph showing black mould growth resulting from condensation

In certain instances, this can be harmful to health as people who have been exposed to this for a prolonged period of time may suffer from respiratory problems and develop allergies. Damp staining on walls and ceiling and sometimes floors tend to be more random and widespread that dampness caused through water penetration and rising damp which is more concentrated in specific areas. Condensation can cause damage to equipment, particularly electronics and other electrical appliances. The dampness caused by condensation can have a deteriorative effect on building components and elements. Examples of this would be timber decay and corrosion of nonferrous metal throughout the building.

6.4.3 *Misdiagnosis of condensation*

In the same way rising dampness can be mistaken for condensation, as referred to earlier in the chapter, condensation can also in some cases be wrongly misdiagnosed. For instance, the actual damp problem may have emanated from other defects such as water penetration through walls, floors and roofs or be rising up the lower section of external walls as rising dampness. It is therefore imperative to look at all the telltale signs as articulated in this chapter to avoid misdiagnosis leading to unnecessary and costly repairs for building owners. In addition, a misdiagnosis by a building professional could in certain circumstances lead to a compensation claim against them for professional negligence. For these reasons, it is best not to jump to initial conclusions without sufficient evidence but alternatively adopt a robust building pathology standpoint to assess the visible evidence and where unsure seek a second opinion or recommend further investigation/inspection works. This process of diagnosis could be compared with the process that doctors and clinical pathologists go through in diagnosing medical conditions and diseases in their patients.

6.4.4 *Remedial measures and remedial works for condensation*

Some of the measures that can be adopted to control condensation are included in Figure 6.6.

In some cases, expensive remedial works may be avoided when considering how to reduce condensation. An example of this would be to restrict and reduce the sources producing moisture within a given building. For instance, if there are old gas and oil heaters that are flueless, then these should be replaced not just to reduce the moisture they produce but to comply with

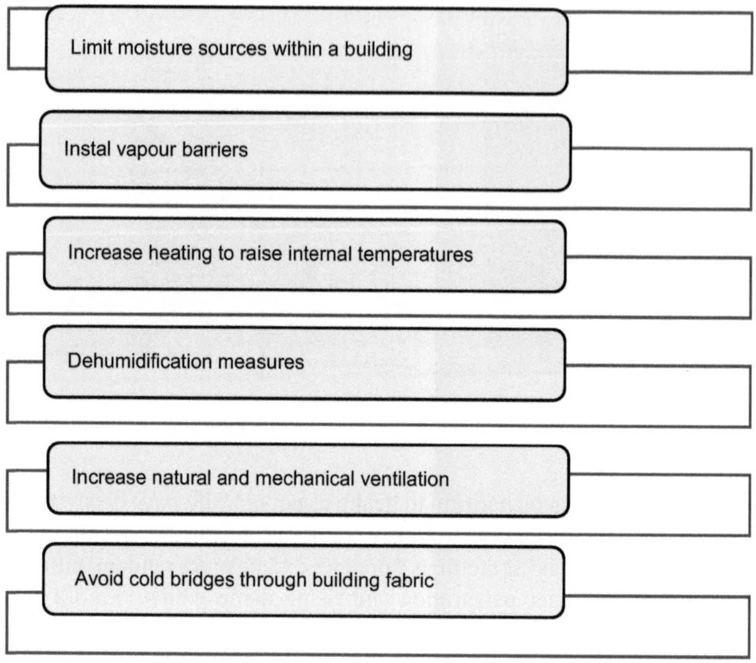

Figure 6.6 Possible measures to control condensation

building regulations. One should also look to provide ventilated spaces for those spaces where heavier build-up of moisture takes places. Examples of such spaces would include kitchen, bathroom, utility and drying rooms. Ventilation in these types of spaces can be provided naturally via opening windows or alternatively by mechanical means through extraction fans linked to the outside. Mechanical extraction fans are particularly useful when linked to humidity stats, rather than just light switches, which activate them when a certain level of room humidity is reached. If these types of rooms, clearly prone to high levels of humidity, are currently unventilated then remedial measures to introduce ventilation may include installing new vents or opening windows. Other relatively simple ways to avoid condensation is to increase internal air temperatures through heating, especially in cold months of the year. Increasing air temperatures will not only dry buildings affected but increase air temperatures to prevent further condensation occurring.

In mostly commercial buildings humidity can be controlled by the installation of full air-conditioning systems which control the flow, humidity and temperature of the air flowing into the internal spaces. Retrofitting ventilation may also be important in 'cold roofs' where insulation is placed below the structure of the roofs rather than above them. In these types of roofs, the outside structure and finish are exposed to cold temperatures which can exacerbate condensation occurring if there are not sufficient design measures to ventilate the roof structures. In such cases 'cross' ventilation through ventilation openings in the roof should be created to avoid buildup of cold damp air which can easily condensate and cause dampness.

Some buildings may have been initially constructed with what is referred to as 'cold bridges'. Cold bridging occurs in cases where a direct connection exists between the outside and the inside of the building fabric. This problem may occur in more than one element of the building where such elements are thermally more conductive than the remainder of the building. This causes cold spots on the inside of buildings where condensation can form. Cold bridging is more common with buildings which are older and may not have been constructed in compliance with modern day building regulations and standards. Examples of cold bridges in older buildings could be synonymous with those having poor insulation, constructed from a single skin envelope or having windows and doors fitted with single glazing. Whilst less common cold bridging can occur on modern buildings when compared with their older counterparts, it can still exist where they have been poorly designed or where construction workmanship has been substandard. This could occur where certain structural elements pass through any insulated building fabric. An example could be balconies where steel supporting members pass through walls from outside to inside and frequently have not been insulated.

Remedial works to address cold elements of the building fabric such as walls can be complex but could include the installation of proprietary insulation systems to the external face of walls followed by external cladding or rendering. This effectively insulates the whole wall fabric and prevents the temperature of the internal surface dropping below dew point. Such measures would not always be appropriate especially for historic buildings with period facades. A common mistake to try to address the temperature of the internal face is to install insulation on the inside. This essentially can compound the problem and can encourage interstitial condensation to develop as the temperature of the inner wall fabric drops. Other ways to address condensation, as previously referred to, could be to increase ventilation of the internal spaces to lower the moisture content of the internal air. This is particularly applicable to areas that naturally produce high moisture levels with the air such as bathrooms and kitchens. Additional ventilation can be achieved through air bricks or by mechanical fans through extraction. It is also possible to increase internal heating to avoid the air reducing to below dew point on contact with internal surfaces. However, this can only really be effective if cold bridging to walls is not a serious problem. In some timber-framed buildings, like those commonly built in Scandinavia, problems could occur with condensation within the fabric of the timber-framed wall and roof structures if there are no means to prevent moist internal areas passing through the wall structure. Good building design would normally allow for vapour 'check' membranes to be installed on the inside face of such walling systems to prevent such moisture passage but if these have been breached though gaps and holes then remedial repairs should be undertaken.

6.5 Water penetration through buildings

Water penetration otherwise known as water ingress occurs when water transgresses laterally from the outside to the inside of a building through its fabric. This occurrence can happen in a variety of ways and for different reasons. Water penetration through walls can also be referred to as penetrating damp owing to the way the water makes its way through walls and into properties.

6.5.1 *Explanation of what constitutes water penetration and its causes*

There are many different ways that water can penetrate into buildings and Figure 6.7 sets down a number of different sources and causes in this regard.

Perhaps the most common source of water penetration occurs when roofs are not fully wind- and watertight. On older buildings this may be owing to the condition of the roofs. Old flat roofs are particularly culpable to defects especially when they have suffered from a lack of

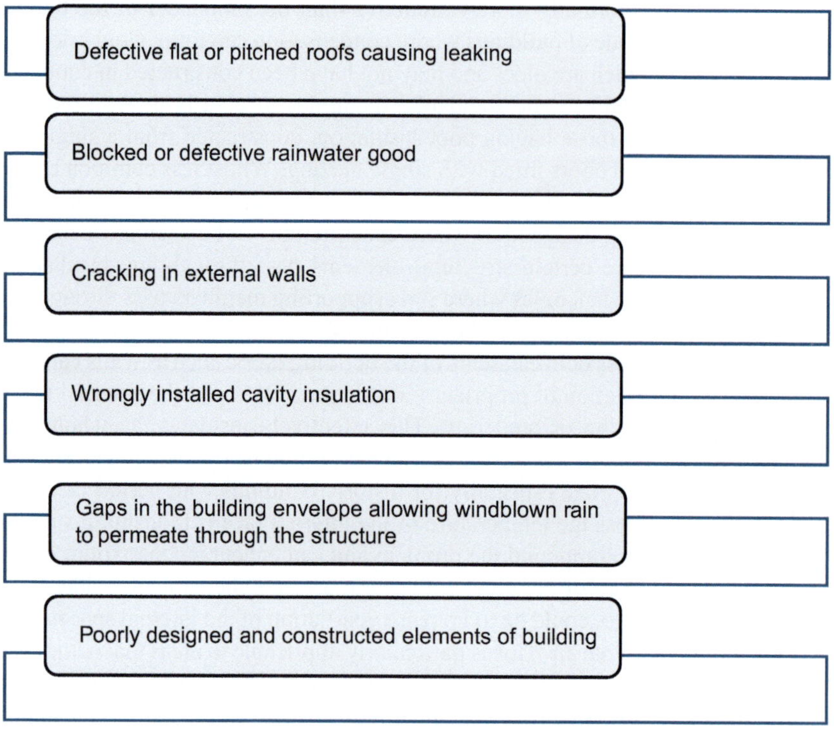

Figure 6.7 The different causes and sources of water penetration into buildings

maintenance in the past. Flat roof materials such as bituminous felt and asphalt can if maintained correctly have prolonged life spans but when defects occur, they should be repaired without delay. Defects on flat roofs will be covered in the next chapter but can include blistering, and cracking of the flat roof material giving rise to water ingress. In the same way, pitched roofs are not immune from water penetration problems owing to the exposed nature of these structures and especially when susceptible to damage from heavy winds, rain or snow. The types of defects and remedial works for roofs will again be covered in the next chapter dealing with examples of external building defects.

Wrongly installed cavity fill insulation can also have an adverse effect in causing water penetration. This is particularly prevalent on southwest-facing walls, exposed to driving rain which could bridge wall cavities, designed to provide barriers to lateral movement of water. Bridging of wall cavities could in a similar way occur if cavities are blocked with construction debris. It is not uncommon for cavity wall ties during construction to be deposited with mortar dropping and rubble. This prevents cavities being able to drain water ingress to the outside of the walls through small apertures referred to as weep holes.

Cracking in external walls can provide a pathway for wind driven rain to be blown into the wall fabric to the inside. This is particularly common in older buildings that have solid wall construction and therein do not benefit from having cavities as drainage barriers. In addition, older properties may be in generally poorer condition than their modern counterparts. In such cases, owing to their age, they may be more prone to suffering from a lack of appropriate maintenance which again increases the risks of water penetration.

6.5.2 *Investigation and diagnosis of water penetration*

Dampness can frequently be clearly visibly from the inside of buildings. Notwithstanding this premise, the diagnosis of the cause of the problem may be problematic as symptoms in certain cases may resemble those associated with rising damp and condensation (previously referred to earlier in the chapter). In general terms one should examine the location and configuration of any noticeable damp patches. If the dampness is at a lower level on the ground floor, as previously mentioned, it may be resulting from rising damp and not penetrating damp. Likewise, if the dampness is spread across the whole area of the wall where speckles of mould growth have developed this could be resulting from condensation. The main clue to the source of internal dampness through water penetration is to inspect the building closely from the outside and this area will be covered more broadly in the next chapter. If the external walling viewed from the outside shows signs of dampness, mould growth and possibly efflorescence in the vicinity of the internal dampness then it may be possible to conclude that the two are related in terms of the underlying cause. If there are signs of visible defects in these areas such as broken or blocked rainwater gutters and downpipes, cracking to walling, poor pointing or gaps between wall make up and window and door opening these could all present telltale signs. Furthermore, in the case of rendered buildings, paintwork and the render itself may be flaking which should present an obvious telltale sign of water penetration. Figure 6.8 illustrates a photograph of such damage.

Sometimes diagnosing the cause of any dampness is problematic and may require specialist surveys and equipment which could include thermal imaging cameras or electrical resistance meters (commonly known as Protometers). This is particularly a problem with large flat roofs where water could be permutating through the covering in one area but becoming visible a large distance away owing to the horizontal flow of water. Surveying roofs over supermarket car park roofs is one example where specialist thermal imaging may be required to accurately locate the cause of damp penetration, and therein carry out repairs.

Other causes of water penetration could be to examine roof structures and finishes externally and internally to identify any tears in flat roof coverings, poor abutment details, missing flashings or missing or cracked roof tiles. Once again these could present telltale signs that these structures are not wind- and watertight and consequently allowing the passage of water.

Figure 6.8 Paint flaking from render indicating possible water penetration

6.5.3 Remedial measures and remedial works for water penetration

Remedial measures associated with water penetration especially through roofs and walls will be mostly covered in the next chapter dealing with external building defects. However, water penetration could be coming up from solid floors especially where properties are older and may not have been constructed in compliance with modern-day building regulations. For instance, some older properties may have solid floors constructed from stone without any damp-proof membrane to prevent groundwater rising up through the floor through hydrostatic pressure. In such cases remedial works can be complicated and expensive especially where floors must be excavated, and new floors constructed. This may not always be necessary providing that heating and ventilation of the internal environment is maintained, and floors are allowed to 'breathe' with the natural internal environment. Hard mortar pointing between floor slabs and poor attempts at surface sealing old stone floors have in the past have had the effect of trapping water in the floors and not allowing floors to properly breathe and dry out. In these cases, this could lead to damp building up below the floor surface which would eventually percolate into the internal space.

6.6 Timber decay

Timber represents one of the most common and frequently used building materials throughout the world and is particularly prevalent in historic budlings. It has aesthetic, practical and structural properties which make it a popular choice as a construction material. As a natural product it is a renewable resource and represents an energy-efficient and environmentally friendly resource. In the wrong conditions, however, timber can decay as a biodegradable material and be digested by fungi or beetle attack. This can occur when environmental conditions linked to temperature, humidity and a lack of ventilation provide the right conditions for timber decay to prosper. Timber decay normally occurs in conditions where the moisture content in timber reaches high levels, normally more than 20%. The buildup of such high levels of moisture can emanate from water being deposited on the timber directly from leakages or alternatively the timber coming into contact with damp walls, roofs or floors that are suffering from dampness. For this reason, it is important to control the environmental conditions in buildings to avoid creating the conditions where timber decay can prosper. In this section the forms of timber decay will be restricted to dry rot, wet rot and beetle attack.

6.6.1 Dry Rot

Dry rot is a more commonly known name for serpula lacrymans. It is a spore-based fungus that destroys timber and can spread widely once developed. This type of wood decay has the ability to quickly transition and spread throughout buildings. This makes it particularly harmful and can cause widespread structural damage owing to the fact that it digests those parts which give the wood its integrity and strength. Another factor allowing it to cause widespread damage rests in its ability to travel through other materials other than timber e.g. brickwork and blockwork. The depletion of wood that can be brought about by dry rot can leave timber in a brittle brown-coloured state, rendering it hazardous and unsafe in most circumstances.

Identifying and therein diagnosing dry rot can be difficult in certain circumstances as it traditionally develops in areas which are damp but not exposed to light. This means that it prospers and can grow in hidden cavities and void areas which are not initially visible without opening up explorative works. Notwithstanding this premise, once developed it can become visible from the surface of internal finishes in the form of snow like strands, known as mycelium and sometimes, when very mature a fruiting body (spororophore), reminiscent of a large orange coloured mushroom, can grow. In less severe cases it can sometimes be mistaken for wet rot. Its particularly pungent smell and visible signs of cuboidal cracking at the surface of affected timber surfaces are normally a telltale sign of its effects. Figure 6.9, 6.10 and 6.11 show photographs of dry rot that has developed within buildings.

Some of these indicators of dry rot are included in Figure 6.12.

Figure 6.9 Photograph to show the signs of a dry rot infestation causing white mycelium strands to develop

Figure 6.10 Photograph to show the signs of a dry rot infestation causing fruiting body strands to develop

Figure 6.11 Photograph to show the signs of a dry rot infestation causing cuboidal cracking to develop

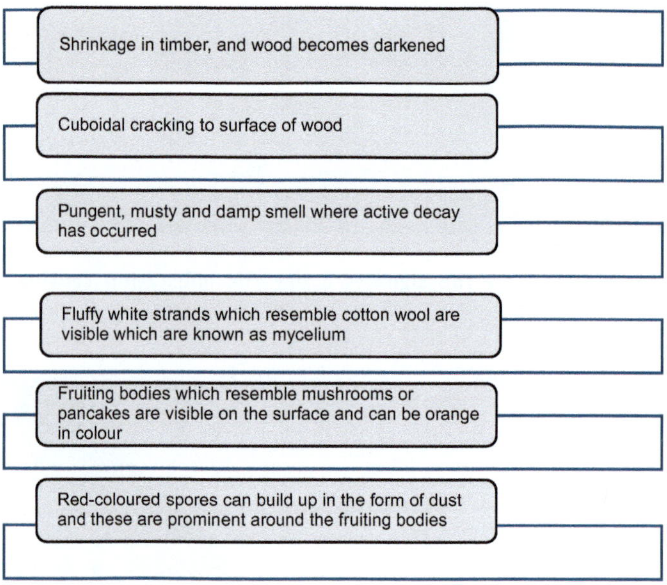

Figure 6.12 Visual indicators to identify dry rot

Although the moisture conditions in timber which can give rise to dry rot are approximately 20%, the optimum conditions for growth occur at between 30–40%, with spore germination occurring in timber occurring when moisture content reaches 30%. In terms of temperature the optimum requirements for dry rot are between 23–25 degrees C. Temperatures above these levels will preclude dry rot from forming and it will be killed at temperatures in excess of 40 degrees C.

The main difference between wet rot and dry rot is that the latter is much more aggressive in the rate it can develop, spread and the damage it can inflict on buildings. This makes dry rot a much riskier prospect than its counterpart in terms of resultant damage. In addition, whilst wet rot will normally remain within the vicinity of the moisture source which caused it, dry rot can spread much wider and far beyond the source of the initial dampness. From the authors' direct experiences, they have experienced extreme cases where dry rot has spread through many stories of buildings rendering them unsafe and dangerous structures requiring complete demolition. This is particularly the case with empty and dilapidated older buildings where unchecked the fungus can have both the right conditions and the time to spread. Another difference between the two types of rot is that dry rot can spread through materials other than timber such as brickwork whereas wet rot cannot.

In terms of treating dry rot, it is important to remove the source of dampness that has ultimately led to its existence. This can be leaking pipework and damp penetration through roofs or walls. Eliminating the source of moisture should form the core of any dry rot eradication strategy as it affects damp timbers that have moisture contents greater than 20%.

Timber can become damp for a number of other reasons which could include leakage from baths and showers in bathrooms or sometimes excessive condensation. The dampness can also come from outside the building and examples include leaking roofs, rising dampness or dampness penetrating through walls. Whatever the source of the dampness, if it is rectified and the timber allowed to properly dry out, the dry rot will eventually be controlled. However, it is not always possible or practical to be sure that the timbers will remain dry in the long term. Therefore, it is important that secondary measures are taken to defend against re-infection. Any affected timbers should be removed and replaced with pre-treated timber. Any remaining timbers at risk of being affected by the dry rot should be treated with an effective fungicidal spray. In previous years the advice was to remove not only affected timbers but those unaffected timbers within a proximity of 1 m from affected timbers but in modern days owing to the technological advancements in fungicidal treatments this is no longer recommended. In addition, there are timber resin repairs which may be able to be carried out to damaged timbers rather than complete replacement. This is particularly beneficial for historical buildings e.g. listed buildings when removing and replacing affected timbers may not be appropriate. Or permitted. Where the dry rot has passed through masonry, it should be isolated using physical containment and/or masonry sterilisation. Certain preservatives are particularly suitable fungicides for the treatment of dry rot, as they are able to spread much more deeply into the timber than conventional preservatives. This gives them an extensive performance advantage, as no wood preservative can start working until it comes into contact with the fungi that it is designed to defend against.

6.6.2 Wet rot

Wet rot is a cellular fungus which can break down timber and deplete its strength over prolonged periods of time. It tends to be more common than dry rot. It belongs to a group of wood destroying fungi known as basidiomycetes which attack the lignin or cellulose in wood and therein renders it weak and brittle. Wet rot can be in the form of white and brown rot. The most common form of wet rot is coniophora puteana, otherwise known as cellar fungus, but other types can include phellinus spp. and fibroporia vaillantii.

Rather like dry rot, wet rot occurs when there is excessive moisture in timbers caused by those factors previously referred to that can create dampness. It does not tend to spread as quickly as dry rot but can be damaging nonetheless especially when it attacks structural timbers. When there is dampness in external walls, timber joist ends can become prone to attack by wet rot, and this can sometimes lead to whole sections of floor requiring renewal. A similar scenario can occur with roof rafters that are exposed to rainwater penetration. Several factors which create excessive moisture can also include plumbing leakages, rising damp and water penetration through roof and walls. This is the main reason why it is important, in preventing wet rot, to conduct regular routine building inspections, particularly focusing on those elements which can generate leaks, such as plumbing pipes. The signs of wet rot can be a discoloured sponge-like consistency and surface cracking of timber as shown in Figure 6.13. It is mainly local to the source of dampness which gave rise to its existence and surveyors use prodding techniques using a knife or screwdriver to check for weaknesses. It is particularly common on old timber window frames that are untreated, and that have been affected by rain penetration through damaged paintwork. Corners of windows and door frames where joints occur are particularly vulnerable where water ingress has occurred over long periods of time.

In some cases, rather like dry rot, wet rot can develop in concealed spaces of a building owing to hidden dampness in certain areas. This makes areas behind floor skirtings, under floorboards, damp basements and below fittings and fixtures particularly vulnerable. Other areas can also be at risk where prone to leaking such as below hand-wash basins, toilets, baths and washing machines.

Wet rot in its early stages can develop spiderlike strands known as hypha which can grow into fern configurations if unchecked, as illustrated in Figure 6.14.

In a similar way to dry rot, wet rot can at later stages develop a white coating and eventually mushroom-shaped fruiting bodies. These are known as mycelium and can be identified not only visibly but by a musty pungent odour. Figure 6.15 represents some of the ways and means to identify wet rot.

Figure 6.13 Photograph of wet rot fungal attack showing surface cracking and discoloured spongelike effect

Figure 6.14 Photograph showing early stages of wet rot which can develop spiderlike strands known as hypha

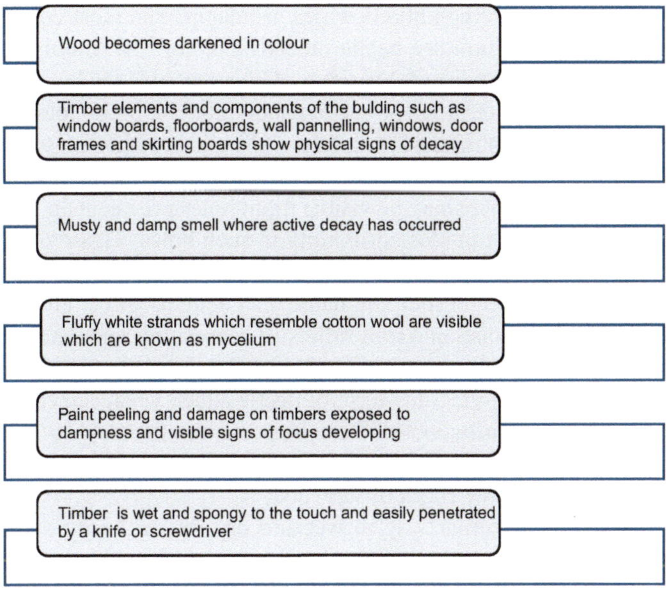

Figure 6.15 Some of the ways and means to identify wet rot

Wet rot normally occurs when there are persistent damp conditions pertaining to moisture contents of between 50% and 60% which is clearly much higher than dry rot, although these will vary between the many different species of wet rot fungi.

In seeking ways to alleviate the wet rot, one should first address the source of the dampness which was the initial cause. This may call for roof repairs or repairs to external walls to prevent further dampness. The affected areas of timber should be cut out and replaced and on window and door frames this could involve splicing new timbers into the old. This is particularly common with listed buildings where complete window replacement may not be possible owing to conservation requirement. For all other buildings a replacement of windows and doors may prove more economical and appropriate especially if they are designed and built to an improved environmental standard with double or triple glazing. If the wet rot has affected structural timbers such as roof rafters and floor joist the best solution may be to replace whole sections of timber to ensure structural integrity rather than risk patch repairs. It is sensible where timbers are built into structures which can become damp, to treat them with preservative treatment or use pressure-treated timbers. One example could be the timbers that make up external decking that are prone to exposure from the weather.

6.6.3 Wood-boring insect

A wood-boring insect is a term which covers a multitude of different beetle species where their larvae eat and destroy timber. Their presence is frequently associated with buildings which have poor ventilation and high levels of humidity. This makes them particularly prevalent in voids between under floor spaces which provide the perfect conditions for hibernation of their larvae, and which then go on to develop into woodworm beetles. Beetle attack is more commonly referred to as 'woodworm' and can occur when wood boring beetles lay their eggs within timbers, and which go on to produce wood-boring larvae. Wood-boring beetles, like dry and wet rot, are particularly suited to conditions which are damp. The most common type of beetle in the UK is the common furniture beetle otherwise known as Anobium punctatum, and the deathwatch (Xestobium rufovillosum) varieties. Others include the house longhorn and the wood-boring weevil. The most destructive beetle tends to be the house longhorn although this species does tend to be prevalent in the South East of England. It can cause particular damage as it can burrow through the core of timbers rendering them structurally unstable. One can normally identify beetle attack as there may be visible flight holes prevalent on timber surfaces and possible evidence of timber dust in close proximity to such holes. Figure 6.16 shows a photograph of timber affected by beetle attack.

The common furniture beetle, despite the name, can actually affect all timber elements of buildings such as floorboards, joists and roof rafters. It causes tiny round holes, less than 2 mm in diameter, exit holes and bore dust to be left behind. These represent telltale signs which surveyors need to look for in identifying this form of beetle attack when carrying out their inspections. They tend to only infest softwoods and can be present in the majority of buildings across the UK. Unlike other beetle such as the longhorn, it should not inflict structural damage as in most cases it consumes the outside edges of sapwood.

Wood-boring insects, in a similar way to wet and dry rot, will only attack timbers when dampness conditions within properties are at a level at which they can thrive. Rather like preventative measures for wood-rotting fungi, preventative measures should include regular checks on buildings, particularly those hidden areas like loft spaces and under floor crawl voids where they can prosper. Wood-boring insects tend to prefer certain types of timber such as sapwood but occasionally heartwood can be susceptible also in certain cases.

The deathwatch beetle, after the common furniture beetle, is the next most common beetle found in the UK. However, the main difference from its counterpart is that it is renowned for being one of the most prevalent for affecting properties as can infest hardwood. This can cause structural damage to these timbers where the moisture content is above approximately 15%. In a similar way to the common furniture beetle, it can thrive where dampness is present from dampness and water leaks. The house longhorn beetle is renowned as the beetle which can cause

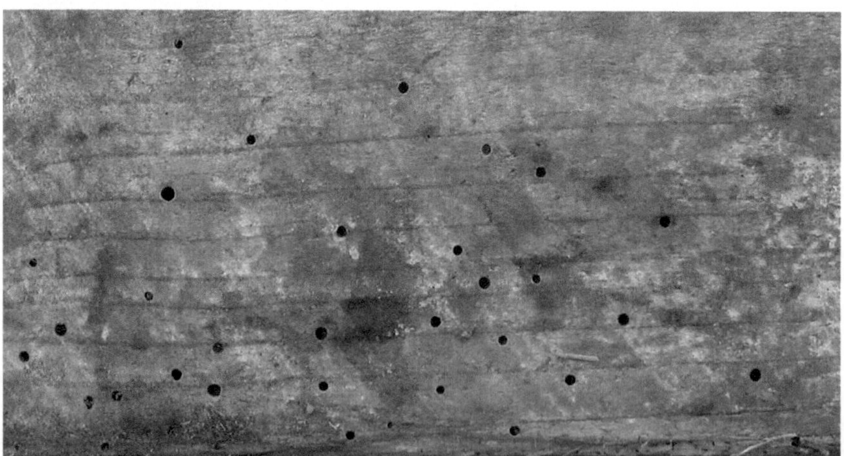

Figure 6.16 Photograph of timber affected by beetle attack

the most damage to properties although it only tends to be present in certain areas of the UK around the Surrey area. It can attack the whole core of structural timbers and for this reason can be devastating sometimes rendering properties structurally unsound. Identifying and diagnosing the presence of wood-boring insects can present challenges and if the signs are not spotted at the early stages, they can remain hidden up to five years dependent on the beetle species.

In terms of remedial works for a wood-boring insect infestation, like wet and dry rot, the first measure would be to address the damp problems within a property. Wood-boring insects do not tend to function on dry timbers and those with less than 15% moisture content so bringing the levels below this threshold is imperative. This can be achieved by stopping the causes of dampness and heating and ventilating the affected areas.

If the damage afflicted to timbers is considered significant then those affected timbers especially if structural members may be beyond repair and will require replacement. If the damage is less severe, then it may be possible to adopt a programme of specialist treatment which includes targeting the insects located within the timbers. This treatment should kill the woodworm on exit through the wood. In most cases this could include the application of appropriate and specialist preservatives to every area of the timber affected for maximum effect. Common furniture beetle attacks can be treated with a permethrin-based concentrate and be either brushed or sprayed into the affected timbers. Clearly where beetle attack is more advanced, and this has affected the structural integrity of timbers, then timber replacement will normally be required coupled with preservative treatment as a further precautionary measure to avoid any re-infestation.

6.7 Wall, floor and roof defects

Visible defects to walls, floors and roofs which become visible internally are in the majority of cases linked to imperfections present in the structure and fabric of buildings. Such defects may have resulted from penetrations and damage to these structures caused by inadequate design or poor workmanship during the initial construction. In addition, movement cracking, poor maintenance, damage from external forces including wind, snow and rain may present other contributory factors. Although these defects can be very visible internally, they are in most cases caused or at least linked with what is happening externally. For surveyors undertaking building inspections it is therefore imperative that they undertake a close overall inspection of properties to forge associations or links with defects that become apparent internally and externally. For example, a case of internal dampness to ceilings on the top level of a block of flats may be coming from a flat or pitched roof defect that can only be visible externally. This hopefully demonstrates the point that one cannot simply report on internal and external defects separately as in most cases they are the same defect that has manifesting itself externally and internally. External defects will be covered in the next chapter.

6.7.1 *Walls*

Small hairline cracks can develop in wall plaster especially following renovation works or in recently completed new builds. This is usually through shrinkage of the newly applied plaster finish and is completely normal in most cases. In such cases these cracks are purely cosmetic and can be repaired with an appropriate filler prior to redecoration. In older properties, especially Victorian properties in the UK, internal walls may have been constructed from timber laths and over time the poor adhesive properties of the material may cause plaster to crack which can result in an uneven surface across wall areas. From a conservation perspective especially in the case of listed buildings, repairs may be viable by cutting out and removing lumps of plaster and then carefully screwing back the laths to the timber studwork prior to patch plastering the affected area. However, where damage is significant and lath work rotten, consideration should be given to over boarding with plasterboard or in some cases removal and replacement with plasterboard. This treatment may not, however, be acceptable if there are conservation restrictions on the property e.g. a listed building. Complete removal of plasterboard may be more disruptive especially in occupied premises owing to the demolition works which can generate debris for removal and a significant amount of dust.

Wall finishes may be showing signs of dampness and discolouration in patches and there may be a smell of water-soaked plaster. In such cases wallpaper may be peeling and painted wall finishes flaking and blistering. These problems were referred to earlier in this chapter and the source of dampness whether it be condensation, rising damp or penetrating dampness should be alleviated prior to drying out and repairs undertaken to wall finishes.

In some cases, plaster cracking can be more severe than the aforementioned shrinkage cracking. Instead of hairline randomly configured crazing or cracking from natural shrinkage there may be cases that may signify structural cracking of the internal or external walling. This can manifest itself as vertical, horizontal or diagonal cracking on the internal surfaces with relatively wide-open joints. Any cracks wider than approximately 4 mm and which show up in this regard should therefore be properly investigated. Diagonal cracking may signify differential movement of the building structure which could in turn have resulted in subsidence. Figure 6.17 shows an illustration of the appearance of such cracking. Vertical cracking especially those that emanate from corners of external walls could in some cases indicate that one side of an external wall is moving away from another, or that one of the walls are bulging. In both these cases this could pose a serious structural problem for the building. Again, it is important to reiterate that external inspection of walls in the vicinity of where these internal cracks are visible present telltale signs and suggestions of the possible cause.

Wall cracking if left unchecked could exacerbate over time causing cracks to become more extensive and wider. This is the reason it is imperative to have a suitable qualified professional carry out exploration/investigation works to identify both the source of the problem and recommend the necessary and appropriate remedial works. In severe cases of structural movement, the advice of a qualified structural engineer rather than a chartered building surveyor may be deemed necessary. It is not unusual for structural engineers to recommend fixing telltale crack monitors over such cracks. These are illustrated in Figure 6.18 and will record if the crack is getting wider which will in normal circumstances confirm the wall is still moving and urgent

Figure 6.17 Differential movement cracking to walling viewed internally

Figure 6.18 Telltale crack monitors

structural remedial works are required. In some cases where these devices do not record further movement it may be deemed that cracks are 'historical' and where the source of any previous problems have been resolved and only cosmetic repairs to plaster surfaces are required. Examples such as this could include where drainage or water main leakages underground have for a relatively short period caused localised subsistence minor wall cracking, but movement has stopped following plumbing and drainage repairs.

6.7.2 *Ceilings*

In older properties, especially Victorian properties in the UK, original ceilings were constructed by lath and plaster. Over time plaster can become dislodged from the laths and this can cause cracking of the plaster ceilings. In more serious cases plaster can become broken and dislodged, exposing the laths above. In both these cases if the lath remains intact and the damage to plaster is not too severe then patch repairs to the plaster finish can be undertaken relatively easily and inexpensively. However, when lath work becomes rotten or dislodged from its fixings to ceiling or roof joints this can cause the ceilings to sag and possibly crack. In such cases the lath and plaster ceiling has become unstable and will need to be entirely removed and replaced with a conventional plaster board ceiling, prior to finishing with gypsum plaster skimming and redecoration.

In some ceilings, especially to the underside of roofs, cracking could be present, and one example is shown in Figure 6.19. In this particular case the ceiling is fixed to the underside of a pitched roof constructed from trussed rafters. The roof owing to the fact that it not been fully

Figure 6.19 Photograph showing horizonal cracking to a ceiling

anchored to the external walling was lifting slightly in heavy winds. The cracking to the ceiling carried down to the connecting wall is a telltale sign that some structural remedial works to prevent further roof movement is required.

6.7.3 Floors

When inspecting floors due consideration should be had for signs of cracking, sagging or sloping. Cracking would normally be associated with solid ground floors and could be caused through structural movement, possibly subsidence of buildings where parts of a building's structure are moving differently when compared to other parts. Such movement can result in sloping of solid floors where one corner or side of a building may have suffered from local subsidence. Where solid ground floors show extensive random cracking and possible upward bulging this may signify upward ground forces are causing the concrete floors to rise up from hydrostatic pressure exerted by groundwater below. This could have resulted from poorly designed or constructed floor slabs where the thickness and structural stability are inadequate. In such cases remedial works might have to include complete excavation of the slabs and replacement with a properly constructed new floors. The same remedial works may be necessary if solid floors are sagging and showing signs of major cracking and settlement. This could signify that the slab or the soil beneath the slab is unstable. In this latter case prior to reinstatement of the concrete slab any loose soil should be replaced with a well-compressed buildup of graded hardcore to avoid such problems occurring again.

Defects are not limited to solid floors and suspended timber floors can exhibit signs of sagging. This may have been caused due to inadequately design and constructed floors where floor joist may have deflected over time, possibly under heavy imposed loading. In addition, timber floor joist ends are prone to timber decay where they are built into walls which have been damp over prolonged period of time. Remedial works may entail replacement of joist ends with metal joist hangers but in some cases could warrant complete joist replacement.

6.7.4 *Basements*

Walls and floors in basements, especially in older buildings e.g. Victorian cellars, are particularly prone to dampness. Their original design may not have factored in any waterproofing treatments as they were possibly designed only as non-habitable spaces for storage purposes. The absence of any waterproofing causes damp penetration through basement floors that do not have damp proof membranes and basement walls. Accordingly with no waterproofing barriers to floors and walls coupled with an absence of ventilation it is not surprising that such spaces can become very damp and in extreme cases this can lead to ponding of groundwater on basement floors. Figure 6.20 shows some photographs of cellars to illustrate these types of endemic problem.

Where basements are to become habitable spaces then a proprietary 'tanking' solution may be required to render the spaces watertight. Such installations on existing basements may involve

Figure 6.20 Photographs of cellars to illustrate dampness

Figure 6.20 (Continued)

a system with 'envelopes', the internal face of the walls and floor with an impermeable membrane. This system allows penetrating dampness from groundwater to flow through the walls and down the concealed face of the waterproofing membrane and then drain away through proprietary drainage channels linked to mains drainage. This system can be compared to the construction of swimming pools in the sense that any penetrations or imperfections can give way to leakage through the membrane. It is therefore imperative that lapping and sealing of joints alongside special proprietary waterproof fixings are deployed. Depending on the level of the basements in relation to invert levels of mains drainage, mechanical pumps may be required as opposed to relying on gravity solutions for the passage of water. Figure 6.21 shows photographs

Figure 6.21 Proprietary waterproof tanking solution to an existing basement

of this type of waterproofing system. This is a specialist piece of work, and this should only be undertaken by an experienced and qualified waterproofing contractor. Normally installations such as these come with long guarantees against design and workmanship. In such cases special care and attention should be given to passage of services through the tanking mechanism and for the introduction of adequate ventilation and heating to avoid the effects of condensation and further dampness.

6.7.5 *Roofs*

Roofs are clearly a major building element of any building and water penetration caused by imperfections and defects of their roof coverings will culminate in damage to internal elements and finishes. Leakages through roof coverings can cause ceilings to sag or collapse or cause other possibly more serious problems such as timber decay as referred to earlier in the chapter. Leakages may on further investigation be found to be coming from plumbing in loft spaces such as water tangs or heating pipework. This can cause similar damage to defective roof coverings and remedial works to address the leakages should be carried out as soon as possible to avoid further damage to ceilings, walls and floors.

6.8 Presence of asbestos and other hazardous materials

Asbestos is a naturally occurring fibrous material which has properties which are resistant to corrosion chemical attack and heat. For this reason, it was used extensively over past decades for such things as insulation and fire protection to building elements. It was a relatively inexpensive material and strong which made it a popular choice as an integral component of buildings constructed before the year 2000. It is estimated that the health risk for individuals who have been exposed to it has been responsible for 5,000 deaths a year in the UK from asbestos cancers like mesothelioma and lung cancer (Health and Safety Executive). Asbestos is part of a group of six naturally occurring minerals composed over very then microscopic fibres and these include Actinolite, Amosite, Anthophyllite, Chrysotile, Crocidolite and Tremolite.

In buildings, Chrysotile, also known as white asbestos, was the most commonly used. Asbestos material can be categorised as friable (easily crumbled by hand) or non-friable. As a general rule friable asbestos products and materials pose the greatest danger to health as asbestos fibres are more prone to becoming airborne and breathed in by individuals. The fibres are so microscopic that once breathed in can block alveoli in the lungs where they can, over many years, lead to asbestos-related cancer. The presence of asbestos can sometimes be difficult to detect and for this reason all business premises in the UK are required by law to keep an asbestos register of their premises to avoid any asbestos being disturbed which is when it can be most dangerous to health. One of the main reasons detection of asbestos can be difficult is that it is sometimes hidden behind wall, floor and ceiling finishes or inside cavities and fire doors. For this reason, any demolition or drilling of finishes should always be undertaken carefully and with reference to the asbestos register. Furthermore, asbestos material may have been used in electrical installations as shown in Figure 6.22; it was present in an old fuse carrier. There are numerous ways and means to identify if a product or material contains asbestos and these include examining the age of the product and looking for warning labels. However, the only accurate way to establish that a product or material contains asbestos is to conduct a professional survey with sampling and laboratory testing.

The other factor which makes asbestos difficult to identify is that it was used in many different building products over many years which include insulation, vinyl floors tiles, cement and finishes. Products used in the 1970s and 1980s such as Artex for finishing walls and ceilings have been associated with having asbestos content and a list of other asbestos products are included in Figure 6.23.

For non-domestic building owners there is a legal duty for building owners to identify all asbestos-related materials and products within a building that could become disturbed and damaged. Such a legal duty emanates from the Control of Asbestos Regulations 2012. This would

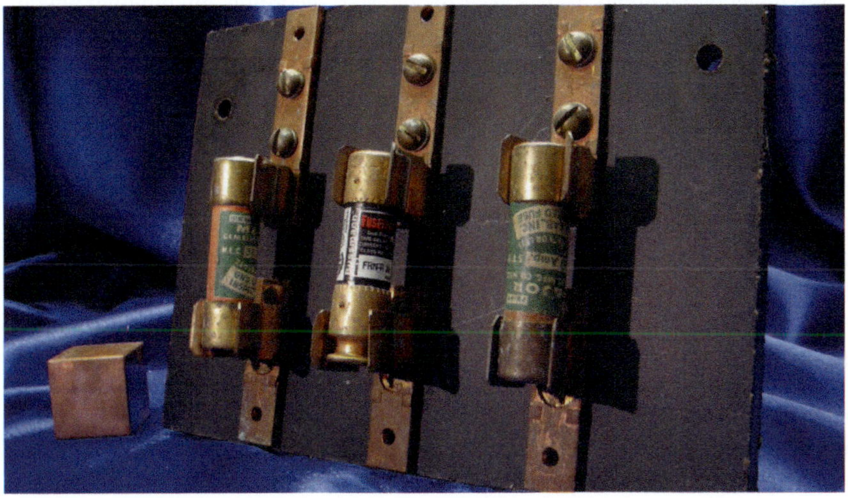

Figure 6.22 Example of asbestos found in electrical apparatus

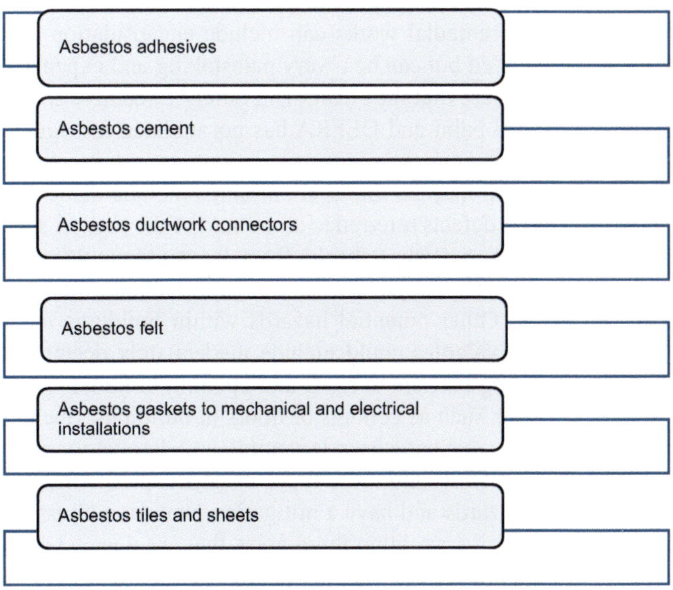

Figure 6.23 List of other asbestos products

normally, as referred to previously, require a record to be produced by a qualified specialist of the type and extent of asbestos in a particular locality, known as an asbestos register. Such specialists may be required to collect samples of materials they suspect to contain asbestos for testing in a laboratory. It may be feasible to have asbestos removed by accredited specialist contractors, but this may not be possible for a multitude of reasons including cost and disruption. If asbestos cannot be removed and must be left in situ, it must be managed under the Control of Asbestos Regulations 2012. For compliance management provisions under the regulations will require maintaining an up-to-date asbestos register, recording any building alterations and closely managing all building works including routine maintenance and cleaning.

Lead is regarded as a hazardous material which can be present in old buildings in the form of water and drainage pipes which can normally be easily identified from their appearance. In such cases all installations and pipework should be removed and replaced with modern components. Where lead is contained in paint finishes, again frequently used in older buildings, this can be more problematic to diagnose. Lead-based paints were banned in the UK in the 1990s but clearly are still present in many different properties built or renovated before this time. Minute lead particles can be spread into the internal air if disturbed during paint preparation, by sanding or heat removal during renovation works. Heat removal of paint can create dangerous vapours which can spread through internal environments. Similarly, air borne particles created by sanding can contaminate other building elements including finishes, fixtures and fittings that come into contact with it, causing a whole property to become affected in worst case scenarios. Human exposure to lead once ingested into the lungs can have serious health effects and lead to many medical complications including high blood pressure, aching joints and cardiovascular problems. Furthermore, lead inhalation in the past has been linked with complications during pregnancy including miscarriages. Lead contamination can be only investigated and diagnosed competently by specialists as it will not be visible and meters and sampling may be required to confirm diagnosis in this regard. Chemical paste may prove a better alternative to sanding or heat guns as it will tend to contain lead deposits rather than release them into the atmosphere. Notwithstanding this premise, remedial works can include encapsulation or safe removal by specialist should be recommended but can be a very painstaking and expensive operation. Perhaps the most worrying aspects are that the Health and Safety Executive still do not have clear guidelines on dealing with lead paint and DEFRA has not updated its regulations on lead for many years.

Other hazardous issues which may be found in buildings include damp and mould growth. These can be caused from those defects referred to at the start of the chapter such as rising damp, water penetration and condensation. Risks to health from damp and mould growth could include individuals developing symptoms synonymous with fungal infections, alongside affecting their social and mental well-being. Other potential hazards within buildings include those which could result in trips or falls. Examples could include inadequately designed and constructed staircases, insufficient or missing barriers to roofs and opening in floors. Where there is a risk of collapse of building elements such as ceilings or floors in buildings. Such risks may present greater threats in older buildings that have been left empty and derelict for many years. Surveyors should ensure that when entering into these types of unoccupied buildings they undertake a risk assessment to identify hazards and have a mitigation plan for addressing all risks during their inspections. This may involve avoiding these areas that are deemed unsafe to enter into. Other hazards could include problems associated with defective electrical, water and gas installations where heating may be insufficient or problems with the quality of drinking. Other dangers associated with such service installations will be referred to in the next section.

6.9 Defective electrical, gas and water installations

Mechanical and electric installations, otherwise known as building services, can pose a real and imminent danger for building occupants or alternatively in the case of unoccupied buildings, surveyors and other visitors that enter them. This section of the chapter will look specifically at hazards and safety aspects associated with electrical, gas and water installations.

6.9.1 Electrical installations

According to the Health and Safety Executive, electricity can cause death, severe injury to people or damage to property. The main hazards associated with electrical installations in buildings are contained in Figure 6.24.

In addition to the main hazards in Figure 6.24 an electrical shock from a defective electrical installation can result in other injuries from falling especially when working at height. Notwithstanding this premise there are simple precautions which can be adopted when coming into contact with electrical installations. For instance, an assessment should be undertaken to identify any electrical hazards. The assessment should articulate the severity of each and every risk and the mitigations and precautions to reduce and control risk. In addition, building owners should ensure that all electrical installations and pieces of equipment are safe and properly maintained to prevent any dangers to the health and safety of the building occupants. Any repair or improvement works to electrical installations deemed necessary should be undertaken by suitably

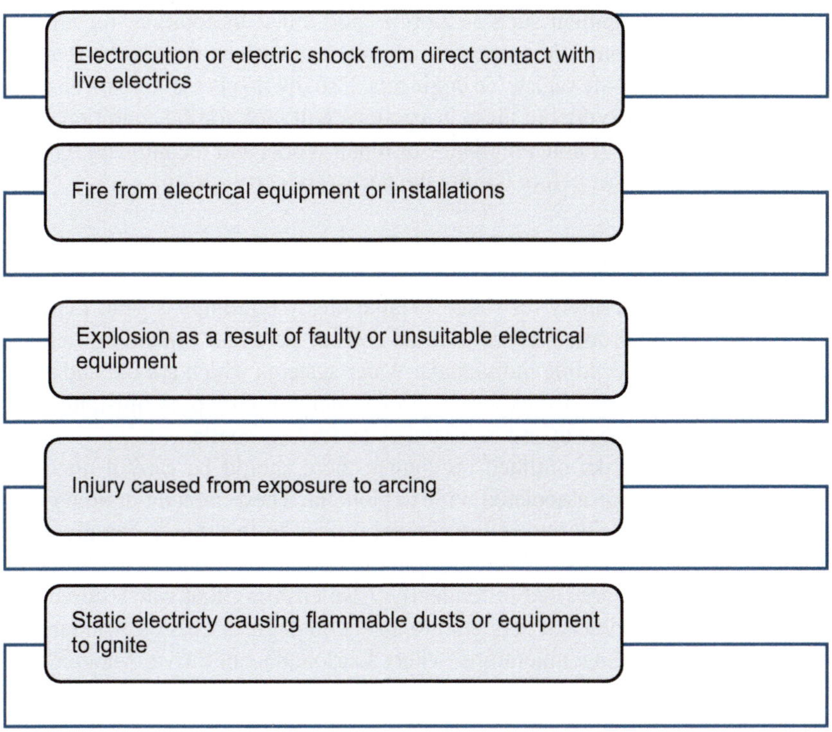

Figure 6.24 The main hazards associated with electrical installations in buildings

qualified persons commonly referred to as competent persons. All fixed wiring installations should be tested and inspected regularly which will include electrical sockets, light switches, wired-in equipment and all electrical wiring and apparatus including distribution board. This should ascertain the condition of the installations and whether they meet current regulations, and outline any repair or improvements works deemed necessary.

An Electrical Safety Certificate, sometimes referred to as an Electrical Installation Condition Report (EICR) or a Fixed Wiring Certificate, is required under law in the UK for all commercial property. For domestic properties it is also a requirement for landlords when renting flats or houses to tenants. This is to ensure the health and safety of building occupants are free from any dangers associated with noncompliance, defects and faults. For all new electrical installations and material alterations to existing installations Part P of the Building Regulations will apply.

6.9.2 *Gas installations*

Most commercial and domestic buildings, certainly in the UK, despite calls from the government to phase out gas installations, still rely on gas as a valuable resource for heating or cooking. For this reason, gas services and equipment must be installed and maintained correctly by law to prevent hazards associated with gas or carbon monoxide release, possibly leading to asphyxiation, fire and explosion.

By law in the UK all gas services and equipment can only be installed by a competent and qualified engineer registered under the Gas Safe certified scheme. This covers gas boilers and heaters including mobile heaters. It also includes cooking equipment including hobs and ovens and other refrigeration equipment such as mobile food units. In addition, for all commercial properties and for rented domestic properties, all gas installations need to be regularly maintained and serviced by Gas Safe registered engineers. Usually this is required on an annual level.

Competent engineers carrying out these inspections will look for the condition of gas installations for any telltale signs of noncompliance or repair works and recommend remedial works to be undertaken in such cases before issuing the necessary certification.

6.9.3 *Water installations*

The main risk to health and safety for water installations in buildings is from Legionella. This occurs when microscopic droplets of water in an aerosol state and containing Legionella bacteria are breathed in by susceptible individuals. Water systems which are old and poorly maintained present an increased risk for exposure to Legionella. Figure 6.25 illustrates some of the main issues which can give rise to and be associated with Legionella.

When considering the risks outlined previously there should be careful measures implemented to reduce the hazards associated with Legionella. These include draining down water systems before demolition or during refurbishment works. In this way it can prevent the stagnation of water. Other measures may be to regularly 'flush out' water outlets such as shower heads, especially those that are used infrequently. Also ensuring that water stored as part of a hot water system is maintained at a temperature above 60 degrees Celsius or in tanks below 20 degrees Celsius can avoid the temperatures where Legionella can thrive. Another safeguard is to kill Legionella and any other microorganisms in the water using an appropriate biocide or alternative water treatment processes.

Another alternative hazard for water installations in properties is to maintain clean and safe drinking water free from contaminants for occupants and visitors. Drinking water fed directly

Figure 6.25 Water hazard through ceiling resulting from bathroom appliance leakage above space

from mains installation is less susceptible to contamination but building owners should still be vigilant if the clearness of drinking water becomes visible. Any discolouration could indicate that the water main is damaged and contaminated by possibly groundwater, and clearly plumbing repairs should be instigated in those circumstances. The main risks come from drinking water systems fed from storage tanks which are more prominent in high rise buildings. Overhead tanks must be designed to be completely sealed and therein are always free from contamination in water storage areas. Regular testing and servicing of the equipment as part of these types of water installation should be carried out regularly by qualified, experienced water engineers.

In addition to the problems related to water hazards there is the obvious problems associated with leakage of water pipes. Even the smallest amount of leakage over time can cause extensive damage as shown in Figure 6.25 where a water supply pipe to an upstairs bathroom hand-wash basin came loose to cause major damage to a ceiling below. The extent of the damage involved the majority of the ceiling requiring replacement which was a disruptive and expensive undertaking.

6.10 Summary of chapter

The categories of the internal different defects referred to in this chapter are:

- Rising dampness
- Condensation
- Water penetration through buildings
- Timber decay. Wood-boring insect
- Presence of asbestos and other hazardous materials
- Wall, floor and roof defects
- Defective electrical, gas and water installations

When considering the types of internal maintenance issues, one needs to closely look at the construction of the buildings and focus particular emphasis on any elements of them which could give rise to defects and remedial works. For instance, if it is an older building, elements of it may not be constructed from conventional materials.

Rising dampness can be described as prevalent where there is either no damp-proof course (DPC) installed in any wall structure or an existing DPC has become damaged. Damp-proof courses can over time become damaged or compromised allowing moisture 'bridging' to take place and rise up the wall by capillary action. Dampness can cause internal damage to finishes including plasterwork and can lead to timber decay of floors, architraves and skirtings if it is prevalent for long periods of time. The noticeable signs of rising dampness can be water staining, possibly tide marks on wall finishes and mould growth usually no more than one metre from ground level. There are many different visual indicators that a building is suffering from the effects of rising dampness, and these are normally present on walls no more than 1 m above ground floor level. These include dark staining, efflorescence and paint flaking and blistering. In addition to these visible signs there may also be a damp odour in rooms affected. Rising damp can sometimes be misdiagnosed as condensation or water penetration and it is therefore imperative for surveyors to carefully consider all the symptoms and not initially jump to conclusions. Remedial works to address rising dampness may vary and can include isolated repairs to damp-proof course (DPC), or complete reinstallation of a new DPC system.

Condensation can happen in buildings when moist warm air comes into contact with surfaces which are much cooler and below the dew point. Window surfaces are a good example of such surfaces where condemnation is prevalent for this reason. Surface condensation will normally occur in buildings where moisture within the air in the form of water vapour meets surfaces which are colder and below dew point temperatures. For this reason, cold internal surfaces of external walls, where there is insufficient insulative qualities, can be particularly vulnerable. There are numerous visible signs and problems caused by condensation, and these could include random staining on walls and ceilings and buildup of mildew on surfaces. Other visible signs of condensation could include evidence of speckled black mould growth across the internal surfaces of walls, and particularly prevalent in the corners of external walls where surface temperatures can be colder. However interstitial condensation can occur within the fabric of buildings which may be hidden and therein more difficult to diagnose. Some of the measures that can be adopted to control condensation can include ways and means to reduce the humidity whilst increasing air temperature within rooms. This can be achieved through increased ventilation and other dehumidifying measures whilst increasing heating. If the source of the humidity is cold bridging or dampness, then measures to address these defects should be instigated. Remedial works to address cold elements of the building fabric such as walls can be complex but could include the installation of proprietary insulation systems to the external face of walls followed by external cladding or rendering.

Water penetration, otherwise known as water ingress, occurs when water transgresses laterally from the outside to the inside of a building through its fabric. Perhaps the most common source of water penetration occurs when roofs are not fully wind- and watertight. This occurrence can happen in a variety of ways and for different reasons. These include defective flat or pitched roofs causing leaking, wrongly installed cavity insulation, blocked or damaged rainwater goods or poorly installed cavity insulation. However, in a lot of cases it can occur from a general lack of maintenance of roof finishes and external wall e.g. damaged or missing pointing or render finishes. Sometimes diagnosing the cause of any dampness is problematic and may require specialist surveys and equipment which could include thermal imaging cameras or electrical resistance meters. Remedial measures and remedial works for water penetration can vary but will normally involve overcoming external defects associated with leakages through walls and roofs e.g. missing or damaged roof coverings.

Timber decay can include dry rot, wet rot or wood-boring beetle attack. Dry rot can quickly transition and spread throughout buildings and through other surfaces other than timber e.g. brickwork. This makes it particularly harmful and can cause widespread structural damage because it digests those parts which give the wood its integrity and strength. The optimum conditions for growth occur at between 30–40%, which is normally associated with dampness through the building structure or from internal leakage e.g. baths and other water installations. Identifying and therein diagnosing dry rot can be difficult in certain circumstances as it traditionally develops in areas which are damp but not exposed to light. Therefore, it can be hidden at early stages of its development although in later stages it can become visible as white mycelium strands or fruiting bodies resembling mushrooms. It can also become visible as cuboidal cracking in the damage it can inflict on timber. As well as these visible signs it can be identified from a musty pungent odour in those buildings affected.

Wet rot tends to be more common than dry rot. It can be in the form of white and brown rot. The signs of wet rot can be a discoloured spongelike consistency and surface cracking of timber. Corners of windows and door frames where joints occur are particularly vulnerable where water ingress has occurred over long periods of time. Timber affected by wet rot tends to become spongy to the touch and easily penetrated by a knife or screwdriver. Wet rot normally occurs when there are persistent damp conditions pertaining to moisture contents of between 50% and 60% which is clearly much higher than dry rot. In most cases, remedial works will involve cutting out affected areas and timber for replacement. However, in some cases where damage has significantly affected a building, e.g., wet rot in windows, it may render complete replacement of the affected component.

A wood-boring insect is a term which covers a multitude of different beetle species where their larvae eat and destroy timber. Beetle attack is more commonly referred to as 'woodworm'. Wood-boring beetles, like dry and wet rot, are particularly suited to conditions which are damp. The most common type of beetle in the UK is the common furniture beetle, although the deathwatch beetle and the house longhorn beetle can generally be more destructive. If the damage afflicted to timbers is considered significant then those affected timbers especially if structural members may be beyond repair and will require replacement. If the damage is less severe, then it may be possible to adopt a programme of specialist treatment which includes targeting the insects located within the timbers.

Wall, floor and roof defects are commonly linked to imperfections present in the structure and fabric of buildings. Such defects may have resulted from penetrations and damage to these structures caused by inadequate design or poor workmanship during the initial construction. They can be associated with movement cracking and poor maintenance, and damage from external forces including wind, snow and rain may present other contributory factors. Defects can be very visible internally; they are in most cases caused or at least linked with what is happening externally. Visible symptoms can include wall cracking to ceilings, walls and floors affected by movement of buildings, or in some cases bulging of walls and floors and sagging of ceilings. Walls and floors in basements, especially in older buildings e.g. Victorian cellars are particularly prone to dampness. Where basements are to become habitable spaces then a proprietary 'tanking' solution may be required to render the spaces watertight.

Asbestos is a naturally occurring fibrous material which has properties which are resistant to corrosion chemical attack and heat. Asbestos is part of a group of six naturally occurring minerals composed over very thin microscopic fibres and these include Actinolite, Amosite, Anthophyllite, Chrysotile, Crocidolite and Tremolite. It represents an inexpensive and strong material which made it a popular choice as an integral component of buildings constructed before the year 2000. It presents a major health risk for individuals who have been exposed to it. Notwithstanding this assertion, the presence of asbestos can sometimes be difficult to detect as it was used in many different building products over many years which include insulation, vinyl floors tiles, cement and finishes. All business premises in the UK are required by law to keep an asbestos register of their premises to avoid any asbestos being disturbed. In addition, it must be managed under the Control of Asbestos Regulations 2012.

Mechanical and electric installations, otherwise known as building services, can pose a real and imminent danger for building occupants or alternatively in the case of unoccupied buildings, surveyors and other visitors that enter into them. This chapter has identified hazards associated with electric, gas and water installations. Electrical hazards can include electric shocks, fire or explosion from unsuitable electrical installations and equipment. For this reason, an Electrical Safety Certificate, sometimes referred to as an Electrical Installation Condition Report (EICR) or a Fixed Wiring Certificate, is required under law in the UK for all commercial property. Gas services and equipment must be installed and maintained correctly by law to prevent hazards associated with gas or carbon monoxide release, possibly leading to asphyxiation, fire and explosion. By law in the UK all gas services and equipment can only be installed by a competent and qualified engineer registered under the Gas Safe certified scheme.

The main risk to health and safety for water installations in buildings is from Legionella. Water systems which are old and poorly maintained present an increased risk of exposure to Legionella and there should be careful measures implemented to reduce the hazards. These include draining down water systems before demolition or during refurbishment works. In this way it can prevent the stagnation of water. Another alternative hazard for water installations in properties is to maintain clean and safe drinking water free from contaminants for occupants and visitors.

Chapter 7

Common external maintenance issues, defects and remedial works

7.1 Introduction to chapter

This chapter of the book will identify, describe and analyse the problems and challenges pertaining to external defects which commonly occur in all domestic and non-domestic buildings. It should be read in conjunction with the previous chapter, Chapter 6, which has identified, described and analysed internal defects, as these can be closely linked to external building problems. An example of this could be damage noticed to external wall surfaces, possibly where rendered cracked or missing, which has given rise to damp penetration internally. In this sense the importance of linking external and internal defects is of paramount importance.

The many types of external defects pertaining to buildings are endless and therefore the defects referred to are by no means exhaustive. For this reason, this chapter will focus on those more commonly experienced. Accordingly, the categories of the different external defects which be discussed in detail which will include the following:

- Brickwork defects, pointing, cracking, dampness
- Differential settlement and subsidence
- Window and door defects
- Flat and pitched roof defects
- Defective rainwater goods
- Presence of hazardous and potentially flammable materials
- Defective external areas and hardstandings

There will be an explanation for each of the aforementioned defect categories, of how to identify them and the underlying causes of these defects. This will be followed by a comprehensive description of how to properly investigate them, leading to their correct diagnosis. Repair and remedial works to address each defect will be articulated and debated in some cases. In addition, where there are alternative repair works which could be adopted considering the different building types and scenarios, this will be discussed. For example, when dealing with heritage buildings a different way of approaching remedial works may have to be undertaken than a conventional modern dwelling. As previously referred to in Chapter 6, not all external defects are easily diagnosed. Some defects such as structural movement, possibly causing wall cracking externally, may be easily misdiagnosed in terms of the exact causes. For example, cracking to walls can be horizontal, vertical and diagonal and in each of these cases there could be a different cause such as subsidence, cavity toe corrosion or thermal movement. Accordingly, this chapter will stress the importance of surveyors reading the telltale signs associated with this type of problem to avoid the pitfalls of misdiagnosing the underlying causes of building problems. In

DOI: 10.1201/9781003358800-7

this regard, there will be an emphasis on the importance of reaching the correct diagnosis to avoid costly and unnecessary remedial works associated with misdiagnosis.

To explain the building issues which can give rise to some of the external defects, figures will be included to clearly depict possible sources. Furthermore, symptoms of each defect will be described to aid surveyors in correctly diagnosing building problems. In addition, examples of each defect will be included to assist readers in understanding the challenges and practical implications that they can create. To illustrate the external defects commonly found in buildings, the chapter will contain photographs showing the possible visible signs. Again, this is designed to assist readers in identifying the telltale signs of different building problems when conducting their surveys. Finally appropriate remedial works to overcome the various defects will be discussed and photographs of repair treatments will be included in some cases.

7.2 Common external maintenance issues, defects and remedial works

When considering the types of external maintenance issues, one needs to closely look at the construction, age and use of buildings. An older industrial unit for instance may exhibit very different signs of disrepair when compared to a modern dwelling. This could be synonymous with the different materials and construction techniques that were adopted or simply related to a general lack of maintenance in the case of older buildings. The chapter will stress the importance of assessing the appropriateness of different repair and remedial works to address the building defects and considering such factors as repair or renewal decisions, and any special status that buildings may possess as would be the case with listed buildings. In addition, where renewal decisions are taken, the chapter will discuss possible ways and means by which environmental sustainability should be considered. For instance, if an old roofing system is deemed to require replacement, should the new system be on a like-for-like basis? Alternatively, the roof renewal may provide the opportunity to introduce a new improved environmental alternative from responsibly sourced materials, providing better insulation and incorporating renewable energy sources.

The construction type of most buildings will also vary according to their intended use and age, as previously referred to in Chapter 6. For instance, modern city centre commercial buildings will rarely be constructed from traditional methods including loadbearing internal walls and pitched roofs. Alternatively, these types of building may be high rise and consist of steel or concrete framed structures, with curtain walling systems or stone or metal cladding, and flat roofs accommodating sophisticated air conditioning plants. Conversely most dwellings may be traditionally constructed from load-bearing brick and blockwork masonry cavity walls with pitched and tiled roofs. In these two very different scenarios the range of defects and areas of disrepair could vary tremendously. It is therefore important for building surveyors to be able to ascertain their construction type as this will ultimately affect the types of defects they may expect to find and strongly influence their recommendations for any remedial works. The recent case of the Grenfell Tower disaster should also highlight the need to inspect buildings of all types for possible use of aluminium composite cladding (ACM) that has been deemed unsafe and potentially hazardous. Where potentially hazardous building elements are suspected to be present, further detailed surveys or specialist inspections should be recommended and undertaken.

Whilst the categories of external defects can be extensive, Figure 7.1 illustrates some of the common defects which can be present in all different types of buildings from domestic dwellings to commercial properties. These will be discussed separately in this chapter from the causes, telltale indicators and potential remedial works to address the defects.

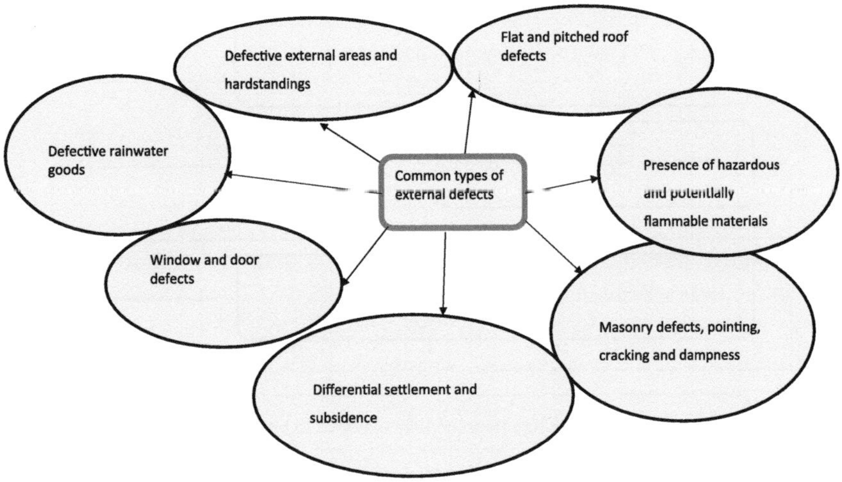

Figure 7.1 Some of the common defects which can be present in all different types of buildings

7.3 Masonry defects, pointing, cracking, dampness

7.3.1 *Masonry defects*

Masonry walling has been used for centuries and the use of brickwork as external walling is most common in the UK, particularly for domestic properties. The extent of its use and popularity as a construction material can relate to its longevity as if built correctly it can last for many years with relatively little in the way of maintenance. However, poor quality control during construction could affect external walls and mortar pointing between brick courses leading to defects. Such poor-quality control could have resulted from not having robust site management including adequate supervision of trades on construction sites. This deficiency could mean that the proper operating processes and procedures may have not been adopted, resulting in substandard workmanship. In addition, brickwork defects can result from the use of inferior materials or failure to follow specifications. Notwithstanding this premise, even when the quality of materials and workmanship is good, given external brickwork is exposed to the elements and extreme conditions, this can affect the best masonry walling over time.

One of the most common problems with brickwork emanates from the mortar pointing as bed joints between the course of bricks. Some of the common defects found in pointing to new brickwork are contained in Figure 7.2 and these will be discussed.

Where horizontal mortar joints, sometimes referred to as bed joints, have been constructed too wide between courses of brickwork. Mortar joints should typically be approximately 10 mm in depth but in cases where they may have been built 15 mm – 20 mm deep this could reduce the structural strength of a wall by as much as a third. In addition, mortar joints constructed in this way could cause brickwork to collapse during the construction process. This is particularly the case in bad weather where heavy winds may be experienced.

Other problems with new masonry pointing could be that mortar has been mixed incorrectly. Typically, mortar should contain approximately four parts of building sand to one part of cement, with a plasticiser added for increased workability during laying. If there is an excessive

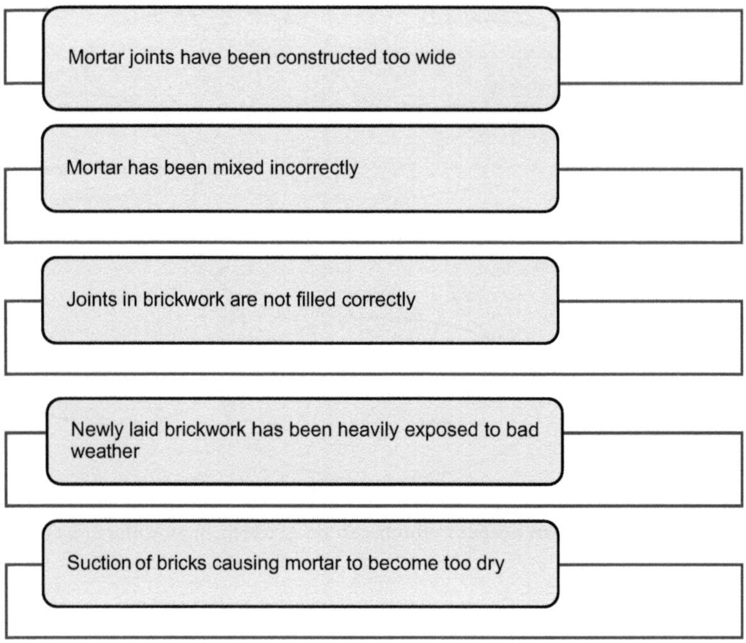

Figure 7.2 Common defects found in new brickwork

amount of sand in the mix, this can affect the strength of the mortar and can make it more porous to water penetration. It can also lead to cracking of the weakened joints mortar or a vulnerability for the mortar to become excessively eroded over time. Conversely, if there is excessive amount of cement then this can make the mortar too hard, inflexible and impervious. Having joints which are too hard and impervious can trap water in brickwork rather than allowing it to escape through the porosity of the mortar joints. This can lead to brickwork retaining water and in very cold conditions a risk of frost action. In addition, if the hardness and strength of mortar is excessive then this can have an adverse effect on the flexibility of external masonry walls to move under normal conditions. In some cases, this can cause cracking to brickwork that would otherwise rely on the joints for some degree of movement tolerance. When mortar joints become damaged for the reasons outlined previously any cracking or erosion can cause water penetration through the joints of the building causing further deterioration of the mortar through frost action and possibly leading to internal dampness. Mixing mortar also relies on the correct ration of water as any excessive water could compromise the sturdiness of the wall both during construction and thereafter. Conversely a mix which is too dry could be difficult for workability during construction and potentially leave voids which could channel water penetration during the life cycle of a building.

It is important when brickwork has been recently completed to protect it from exposure to extreme temperatures until the mortar has properly cured. Any exposure to extreme heat could cause the mortar to dry out too quickly. Such a loss of moisture from evaporation could affect the hydration process of mortar rendering it unable to reach its full strength. In extreme cold weather, the hydration process is again affected, and it is recommended for this reason that brickwork should not be laid in temperatures below six degrees C. In both cases it may be possible to undertake temporary works to cover and therein protect brickwork from the elements if temperatures are not too extreme.

Another quality control aspect during construction is to ensure that mortar joints have been filled correctly. As previously referred to, with excessively dry mortars, joints not filled correctly leave voids which could affect strength of brickwork and be prone to increased frost action or damp penetration. Mortar joints may also become excessively dry if adversely affected by the suction properties of very dry bricks, and therein become weakened and less stable. This can sometimes be the case in dry weather where bricks have been exposed to direct sunlight. In these cases, bricks should be sprayed with water to increase their moisture content. However, saturating the bricks prior to laying should be avoided as this could affect the bonding process of brickwork to mortar.

7.3.2 *Other brickwork defects*

In addition to the constriction problems referred to previously there are other defects which can affect brickwork which are contained in Figure 7.3. and will be discussed individually.

(1) Frost damage

Frost damage is more synonymous with the use of older bricks or alternatively bricks that became burnt during the manufacturing processes. This can leave them with small 'pockets' of space which can fill with water and become frozen in colder weather causing frost damage. Figure 7.4 illustrates the effect this can have on brickwork causing brickwork

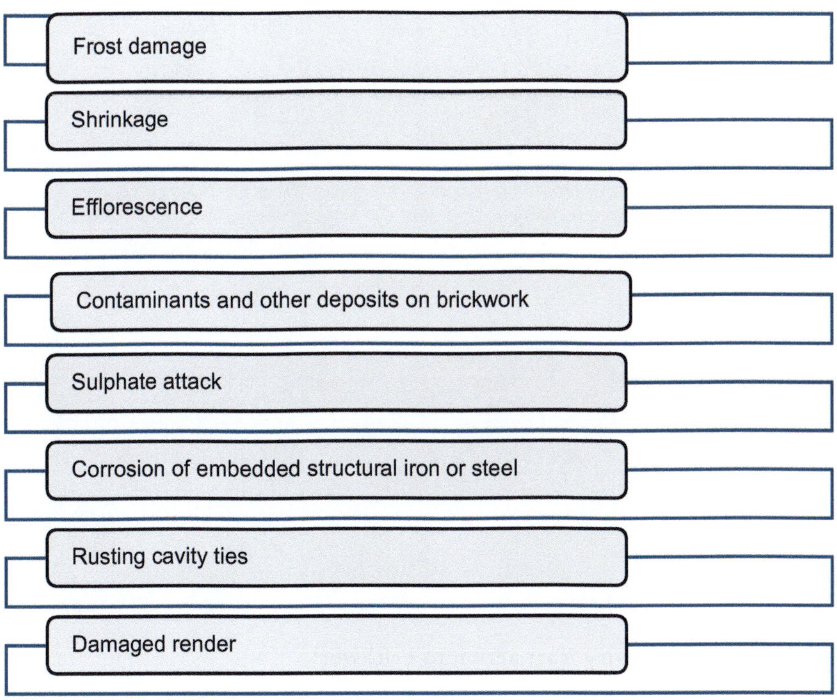

Figure 7.3 Other defects which could affect brickwork

Figure 7.4 Photograph showing frost action to brickwork

to spall and in some extreme cases the whole front of the bricks to fall away from the remainder. It should be noted that frost action is not exclusively linked to inadequacies in the brick firing processes but can affect brickwork where mortar joints have been excessively recessed or suffering from poor maintenance.

(2) Shrinkage of brickwork

As a porous construction material, bricks can swell when they absorb water and then shrink when the evaporation of water takes place. The movement will be greater the more porous the bricks are. Accordingly, handmade bricks, synonymous with older buildings, possibly heritage buildings, are more likely to be more porous than those mass produced in modern times and the greater the resultant shrinkage and expansion. Such forces can cause the mortar joints that are unable to accommodate such movement to crack. For this reason, in older properties particularly heritage buildings it is not uncommon for pointing to be lime rather than cement based. Although the curing times are greater, lime facilitates more flexibility than cement and therein is less likely to crack in such conditions. Sometimes hairline cracks can develop in mortar joints on drying due to shrinkage. In most cases this is perfectly normal as the structural strength and waterproofing qualities of brickwork may remain unaffected as can be illustrated in Figure 7.5. The ways to reduce shrinkage cracking could be to protect masonry from excessive moisture or by using a good quality brick and mortar.

(3) Efflorescence

Efflorescence can frequently affect brickwork and is caused by the crystallisation of excessive salts from the bricks that come into contact with water. It can be exacerbated by those exposed elevations of external walling prone to driving rain. Some would not describe efflorescence as a defect as it is purely cosmetic. It creates a white powdery appearance from those salt deposits which can be removed by repeated washing with a suitable cleaning detergent and brushing of the brick surface. Figure 7.6 shows an illustration of efflorescence on brickwork.

(4) Air pollution

Bricks especially in urban environments can become exposed to air pollution which can cause them to become dirty and blackened in extreme cases. Whilst this can be considered unsightly air pollution is widely regarded as not being damaging to brickwork and some

Figure 7.5 Hairline cracking in brick mortar joints

Figure 7.6 Effloresce on brickwork

would even decree that air borne deposits actually protect the surface of brickwork. Other deposits can build up over time and these include dust and dirt which can be easily cleaned down. In addition, mould growth can build up on walls which have been prone to dampness over prolonged periods of time.

(5) Sulphate attack

Sulphate attack can occur when the alumina contained in cement mortar or other chemicals in lime mortar adversely react with the sulphate salts contained in bricks. This can cause spalling and chipping of bricks and cracks to develop in brickwork joints and rendering. Sulphate attack is more prevalent where walls are exposed to large volume of dampness. This makes it synonymous with parapet walls, boundary walls and walls to manholes. Figure 7.7 shows photographs of the effects it can have on brickwork in such cases.

(6) Corrosion of embedded iron and steel

Iron and steel used in brickwork can become corroded over time as they are exposed to dampness. Iron or steel can be built into walls as structural reinforcement for them or where an adjoining structure such as a balcony requires support from them. If the metal is not treated the corrosion process can cause rusting to develop and outward expansion forces on the brickwork structures. This can lead to stress cracking of brickwork which left unchecked could cause water penetration. Where metal reinforcements are deemed required, it is good practice to encase them with a suitable covering to protect from the corrosion processes. Figure 7.8 illustrates corroding steel in a brickwork structure.

(7) Cavity tie corrosion

Cavity ties are for use when constructing traditional double skins walls and are designed to secure the external leaf of brickwork to the blockwork inner leaf. One of the most common failures of brickwork relates to cavity tie corrosion. The corrosion of the steel cavity ties causes them to rust, expand and cause horizontal cracking of mortar joints along brick courses which can cause water ingress. The main cause of the corrosion is poor quality ties or more likely mortar dropping on the ties during construction which react with the metal to exacerbate the corrosion process. Figure 7.9 shows an illustration of horizontal cracking of brickwork caused by cavity tie corrosion. In some cases, cavity ties can be carefully removed and new ones retrospectively installed but this is an expensive and time-consuming exercise.

Figure 7.7 Brickwork showing the effects of spalling and chipping of brickwork from sulphate attack

Figure 7.8 Corroding steel in a brickwork structure

Figure 7.9 Horizontal cracking of brickwork caused by cavity tie corrosion

Other failures associated with cavity ties include failure to properly bed them on the mortar joints or not installing the correct number of ties. This reduces their ability to tie the two leaves of wall securely together and prevent them moving apart, which can sometimes result in bulging of cavity walls. This is illustrated in Figure 7.10 and remedial works may have to include the rebuilding of the outer skin of brickwork.

(8) Dampness to external walls

External walls being exposed to the elements can be prone to water penetration and dampness. Rising dampness can affect external wall where damp-proof courses (DPCs) are missing or become damaged, or external hardstandings have been built too high in relation to the external walling, causing damp bridging. In addition, surveyors when carrying out their external inspections should observe any signs on walling that could give rise to water penetration. As referred to in the previous chapter these signs could include the following:

- Is the stone or brick external walling showing signs of defective pointing?
- Is the stone or brick walling showing signs of cracking?
- Are there any leaks, blockages or faults with rainwater goods, i.e. gutters and downpipes?
- Are there any details, predominantly gaps or cracks around window and door openings, which could be channelling rainwater into the external walls?

If any of these signs are present the presence of dark patches or mould growth may be noticeable. Figure 7.11 illustrates an example of the lower sections of a stone wall which has been affected by rising dampness causing mortar joints to become perished and recessed. Remedial works to address the cause of the water penetration should be undertaken as soon as possible as otherwise is likely to develop further and cause other complications such as internal dampness and time decay.

(9) Damage to render

Some external walls may have been constructed with sand cement render finishes. Other walls may have been retrospectively rendered to provide a weatherproof alternative to carrying out repairs to brickwork, where walls have been damaged by some of the defects referred to previously. Sometimes paint finishes to render can flake off as shown in Figure 7.12. This may be an indicator that the wall is becoming damp, and investigations should be carried out to assess the source of this. Over time render can become cracked and

Figure 7.10 Bulging of wall through cavity wall tie failure

damaged. This can be caused due to poor installation where a failure to adequately secure the render to the walls has occurred. In some cases, pieces of render can become loose and fall away exposing the brickwork or blockwork substrate as shown in Figure 7.13. This exposure can allow rainwater to percolate behind the render and cause damp ingress and further render damage through frost action. Remedial works may include patch repairs to render or in cases where render has been incorrectly installed a complete reinstatement of the render system.

Figure 7.11 Rising dampness on an external stone wall causing perished and missing mortar joints

Figure 7.12 Paint flaking off render, indicating possible dampness

Figure 7.13 Render damage exposing underlying brickwork

7.4 Differential settlement and subsidence

In the early years following construction all traditionally built properties will settle to some degree on their new sites. This is a natural process as the weight of buildings through their foundations exerts downward pressures on the ground below. The degree of settlement can be affected by ground conditions and the height and construction of buildings. Firmer soils such as clay will generally be more conducive to carrying heavier loading than silty soils. In normal cases a degree of settlement does not create problems when the whole building is settling at the same rate. What can cause difficulties is when differential settlement takes place. Differential settlement is when parts of a building, for instance one elevation, is settling at a different rate than anther. Structural tensile stresses in buildings are generated in these instances causing movement and resultant cracking. Differential settlement could occur where there are weak pockets of land under buildings, sometimes referred to as hot spots which could allow local downward movement in these areas. These weak points may have been in existence before and during construction. Site investigations should have been carried out prior to construction to try to identify any problem areas for remedial land treatment. In some cases, it has been known for the collapse of old sewers or culverts to have affected these local areas. Remedial land treatment may include digging them out and providing new made-up ground from graded hardcore aggregate as a firm substrate to build upon. However, in some cases the ground could have been affected at some later stage in a building's life cycle, long after the construction period. This could have been caused by a leaking underground water pipe installation or drainage below or in proximity to the affected areas. Investigations should be carried out to confirm any underground leakage and this would normally take the step of undertaking thermal imaging surveys or alternatively carefully excavating in the affected area. The most important aspect to stop the source of underground leakage to prevent any further movement. If the differential movement is minor, stopping the source of the problem may be all that is required other than patch repointing or rebuilding of any cracked brickwork.

Subsidence can in most cases be more serious than differential settlement and refers to the gradual caving in or sinking of an area of land (Oxford Languages). It may occur where land has been infilled in the past or where groundwater problems are causing land to be washed away. For buildings constructed on land, prone to subsidence, it can be devastating and may render them dangerous structures with demolition being the only alternative. Figure 7.14 shows a building clearly leaning to one side owing to subsidence.

Differential settlement and subsidence can have similar symptoms, and these can be illustrated in Figure 7.15

Walls can become heavily cracked as buildings struggle to cope with differential settlement and subsidence owing to the inflexible nature of traditional construction. Normally the configuration of cracks is diagonal as illustrated in Figure 7.16 and they can appear quite suddenly both internally and externally. This again reinforces the point that the internal and external inspections are mutually dependent and rely on each other for diagnosing some building defects.

The type of cracking depending on the severity of the movement will be much wider than hairline cracking. In some cases, the cracks can be wider at the top which normally is associated with 'rotational' forces where one wall is pulling away from another. Sometimes cracking in walls can be misdiagnosed for differential settlement and subsidence when they are being caused by something else. An example of this could be hairline vertical cracking in walls where walls have not been designed and constructed with suitable joints to allow thermal movement. Other examples could be where walls are becoming detached for others due to poor lateral support. This could be where floors are not adequately tied into the wall by floor joists and walls could be bulging or becoming detached from abutting walls. This is particularly common on front and gable elevations of Victorian terrace properties. Sometimes local failures of brickwork

Figure 7.14 Extreme case of a building suffering from subsidence

Figure 7.15 Symptoms of differential settlement and subsidence

Figure 7.16 Diagonal cracking in external walling indicating differential settlement or subsidence

Figure 7.17 Cracking of brickwork owing to defective lintel

can be easily noticeable where lintels carrying brickwork have been inadequately sized for the imposed loads they should have been designed to take. Figure 7.17 shows an example of how lintels can fail in such cases and diagonal cracking of brickwork above becomes evident.

Differential settlement and subsidence can cause floors to drop and become uneven. Telltale signs could be gaps opening up between skirtings around the perimeter of floors. Other symptoms may be that windows and doors do not open correctly and start to stick as buildings sink and walls become deformed and 'out of plumb'. Differential settlement can become most visible at weak spots in buildings, and this may be at an intersection of an extension and main building. Unless vertical movement joints have been installed to allow for movement, there is a tendency for local cracking to develop as the extensions try to pull away from the main building structure.

Cleary intervention works are required when signs of differential settlement and subsidence occur. Remedial works to address problems can vary according to the severity of the movement. The solution in minor cases of differential settlement may simply involve addressing the source of any underground water leak to stabilise the ground conditions. In more serious cases of subsidence, complete underpinning of buildings to provide them with new foundations may be required. Underpinning refers to the process of carefully digging below and underneath existing buildings and casting new concrete foundations at a greater depth than the existing and hopefully on firmer subsoil. This process has to be conducted in small segments of the wall at a given time to ensure the structural stability of a building. Figure 7.18 shows some underpinning as part of a basement conversion in Bolton.

Figure 7.18 Underpinning of external walling to form a new basement

Figure 7.19 Collapse of a property in London following poorly executed underpinning and temporary support works

In some cases, this underpinning process when not undertaken correctly can have catastrophic consequences. This occurred in Chelsea, London in November 2020 resulting in the complete collapse of two four-storey Georgian townhouse (see Figure 7.19).

There are proprietary treatment systems to inject material below buildings, at regular intervals to lift them back into place. This is an alternative treatment to underpinning or injection of cement grouting under pressure. This normally is a very specialist and expensive process involving the use of geo-polymer materials, pumped into the soil as a liquid and this expands below the ground. As the material expands it can infill voids in the ground and can dispel water. This can have the effect of stabilising and compacting the ground. The material will begin to create an expansive force onto building foundations and specialists using laser monitoring technologies can accurately track movement up to the point where the building that has subsided becomes level again. The same injection system can be used to raise floors.

7.5 Window and door defects

There are many different types of materials used in window and door manufacturing, and these include timber, uPVC, aluminium and composite. In addition, types of windows can vary widely, and examples include casement, sash, sliding and tilt and turn. The type of windows and doors and their materials are normally affected by the age of the property. For instance, in Victorian buildings original windows and doors would have been manufactured from timber and windows tend to be of the sash variety. The most common types of defects occur in timber windows and doors. Some of the typical problems which should be identified include those contained in Figure 7.20.

Timber decay especially on older properties is normally a sign that windows and doors have not been properly maintained for prolonged periods of time and this can be illustrated in Figure 7.21.

The most predominant type of timber decay for external windows and doors is wet rot. Visible signs to look out for include missing, blistered or damaged paint and darkening of timber especially on corners of frames. On prodding such area timber affected by wet rot is found to be soft and spongy in nature and easily penetrable with a knife or screwdriver. In some cases, it may be possible to splice in new timbers to frames to undertake repairs, and this may be more appropriate with protected buildings or where windows and doors have special architectural or historical significance. However, in most cases where wet rot has developed extensively it may be more appropriate to replace windows and doors completely with alternative materials which may reduce future maintenance e.g. aluminium or uPVC. Similar decisions to renew rather than repair may occur where warping of timbers window and door frames has occurred possibly as a result of manufacturing defects.

Glazing to both windows and doors can present issues. Sometimes in older buildings windows and doors have not been modernised to accommodate double glazing. Whilst this does not constitute a defect per se it could be recorded in a building survey report. Recommendations may be made for upgrading to double glazing to reduce condensation and heat loss and bolster the building's energy efficiency credentials. Moreover, where double glazing has been installed seals in the double-glazed units can sometimes fail causing interstitial condensation in the double-glazed cavity. In such cases the units should be replaced as any repairs are not normally considered economically viable.

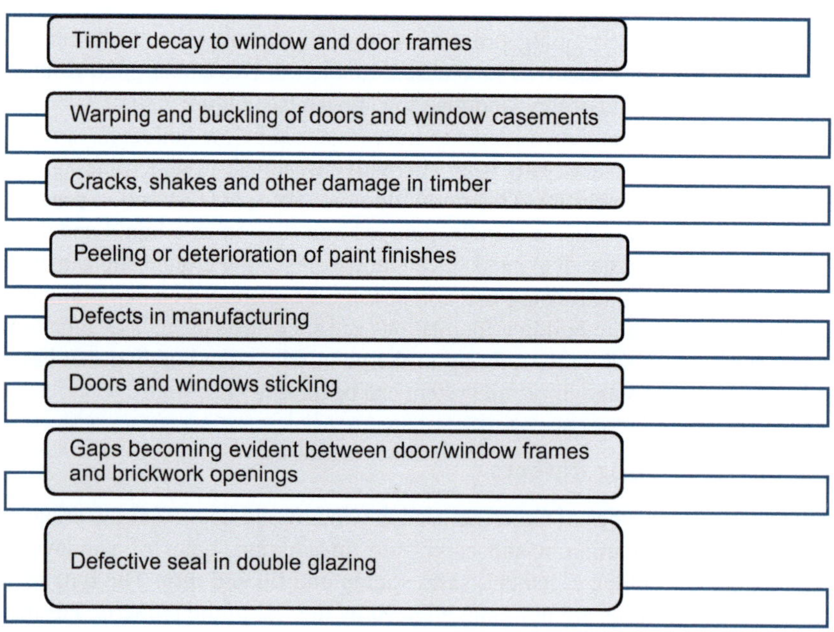

Figure 7.20 Typical defects to timber windows and doors

7.6 Flat and pitched roof defects

In this section of the chapter, flat roofs and pitched roofs will be dealt with separately owing to the difference in their structures and roof coverings. However, to some degree there can be commonality in the type of common defects which can affect them both. These are normally associated with water penetration through roof coverings that can cause some of the internal problems referred to in Chapter 6, including damage to ceiling and wall finishes and possibly timber decay.

7.6.1 *Flat roof defects*

There are many different forms of flat roof, and they can be categorised according to where insulation is installed. Where insulation is positioned within the roof structure these are generally known as 'cold' flat roofs. This contrasts with 'warm flat roofs' where insulation is placed on top of the roof structure. Cold flat roofs can be prone to condensation as the air temperature in voids within their predominantly timber roof structures can be low at certain times of the year. To overcome condensation, cross ventilation through the roof structures via vents should be introduced to reduce air moisture content and thereby avoid the dew point occurring. This is a lesser problem for warm flat roofs as the insulation above the roof covering protects the roof structure from lower temperature. Flat roofs can also be categorised according to the materials used for their membranes. Flat roof coverings used in the past vary and have traditionally included bituminous cement, rubber and asphalt. Proprietary system roofs in recent years have introduced alternative metal coverings including aluminium and zinc.

Some of the most common forms of defect in flat roofs are contained in Figure 7.21 and the circumstances that give rise to them and possible remedial works will be discussed in this section of the chapter.

(1) Inadequate design or poor workmanship and damage caused by the weather

Sometimes flat roofs suffer from inadequate design or poor construction workmanship. Common problems include providing insufficient roof pitches to allow rainwater to properly drain into outlets. Flat roofs, contrary to their name, should not be built completely flat but have a typical pitch of approximately 1 in 40. Flat roofs that are constructed with pitches less than 1 in 80 can hold water. For bituminous felt, rubber and asphalt roofs having water ponding on them can create temperature differences on roof surfaces in hot weather. This can cause differential movement stresses on the roof coverings especially those not protected by the sun through solar treatment. Such stresses can result in splitting, tearing, blistering of the covering which if left unattended will grow progressively worse. Similar symptoms can occur in cold weather when water on flat roof freezes and frost action occurs. Other design and construction issues may include poor roof detailing around the perimeter of flat roofs leaving them vulnerable to roof membranes becoming detached from their structure and therein introducing water penetration, or timber decay. The use of substandard roofing materials may also be a contributory factor for defects to emerge if a particular product does not have sufficient longevity owing to its quality. A flat roof should have a life expectancy of between 15–30 years typically. However, this duration is very much dependent on a number of factors including the quality of materials used, the degree of maintenance management, and owing to the exposed nature of roofs, the weather conditions they are exposed and subjected to.

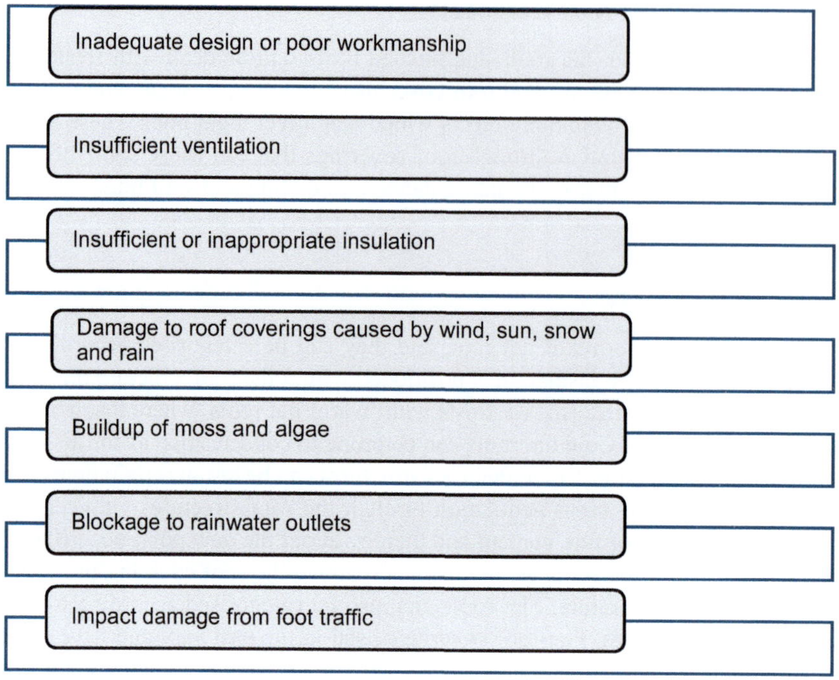

Figure 7.21 Most common categories of defect in flat roofs

(2) Insufficient or inappropriate insulation or ventilation

The installation of insulation can affect how roofs perform. In most cases warm flat roofs tend to perform better than cold flat roofs as they insulate and therein protect roof structure constructed from timber, concrete or metal from extremes. This prevents them from being exposed to extremities of temperature and therefore reduces the effects of excessive expansion and contraction from thermal movement. In addition, warm roofs are less vulnerable to interstitial condensation as referred to earlier which can create dampness within roof structures which could permeate through to internal ceilings. In order to reduce the risks of condensation within cold roofs especially, they should be designed to be ventilated to prevent buildup of stagnant, moist air.

(3) Solar radiation

Solar radiation can present a particular problem for flat roofs which can be very expose to the heat from the sun at certain time of the year. This can cause the surface of roof coverings such as bituminous cement and asphalt to expand, causing fissures and cracking to develop. Other times small pockets of trapped air below roof coverings may expand to cause blistering. To avoid the effects of solar radiation roof surface should be treated with an appropriate solar reflective covering to reduce the effects of both heat and UV radiation form the sun.

(4) Blockage to rainwater outlets and buildup of moss and algae

Blockages in rainwater outlets, from debris, dirt or vegetation can be a problem especially when roofs are hard to access for routine maintenance. Such blockages can result in a buildup of water on the roof surfaces, which can cause excessive amounts of moss and algae to develop, further affecting drainage of water. In circumstances where rainwater is prevented from being discharged away from the roof through drainage outlets and pipes, this can also present an inherent risk of water penetration through roof coverings. This emphasises the need for regular inspection of flat roofs as part of a robust maintenance management regime and where necessary undertake remedial works to repair roof imperfections which may become apparent.

(5) Impact damage from foot traffic

In more modern metal system roofs, such as aluminium and zinc, they can be very fragile. If not designed correctly they can become damaged by foot traffic when individuals access the roof for routine maintenance or for access to roof mounted plants. Damage can include deflection and warping of the sheet roofing membranes which could compromise their ability to prevent wind borne rain from penetrating through their roof structures. Accordingly, purpose-designed raised platforms should be incorporated above these types of roof coverings for maintenance access and to protect them from impact damage.

(6) Remedial works to address splits, cracks and blistering to roof coverings

Remedial works to address damaged roof coverings from splits, cracks and blistering caused by some of the aforementioned factors, can include carefully cutting out the areas of damage and reapplying a new patching membrane of the same material. How this membrane will be applied very much depends on the particular material. For instance, bituminous cement will be applied as a layer in sheet form whereas asphalt repairs tend to involve patch burning in of new asphalt material to replace the old.

7.6.2 *Pitch roof defects*

Pitched roofs for domestic properties are the most common form of roof in the UK and in most countries around the world. Gable pitched roofs are the most common type of pitched roof and these comprise two slopes. These slopes intersect at their highest point known as the ridge and drain rainwater to their lowest point into gutters and downpipes. However, there are many other forms of pitched roof normally differentiated by the number and configuration of their slopes. Figure 7.22 although not exhaustive contains some of these.

Roofs can also be differentiated in the way the roof structure has been designed and constructed. For gable roofs the most modern way to construct them is using trussed rafters which are fabricated in factories and simply lowered into place. These speeds up the installation process and accordingly is a quick, effective and relatively inexpensive method of construction. In older buildings roof structures may have been constructed from rafters supported along their spans by timber purlins which on gable roofs would be built into brick external walls. One of the advantages of this type of roof is that the roof void space is more accessible for access than those constructed from trussed rafters.

Generally speaking, the simpler the form of pitched roof, e.g., gable, the less likely that defects will occur. Where there are complicated types of roofs comprising many different roof slopes and abutment details, these tend to be more prone to defects occurring unless they are properly maintained. For instance, if there are several valley gutters where the different roof slopes drain into, the risk of blockages occurring is greater which could result in water penetration. Figure 7.23 illustrates some of the more common forms of defect that are common for pitched roofs and these will be discussed.

(1) Poorly designed and constructed roofs

Poorly designed and constructed roofs may present themselves as future defects waiting to happen. For instance, if they have been designed or wrongly constructed with pitches below the gradient recommended for a particular roof finish then it is not surprising that

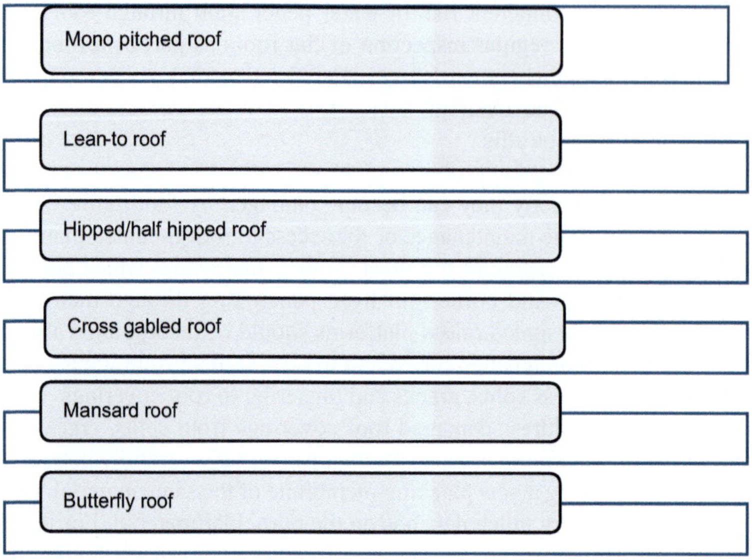

Figure 7.22 Types of pitched roof

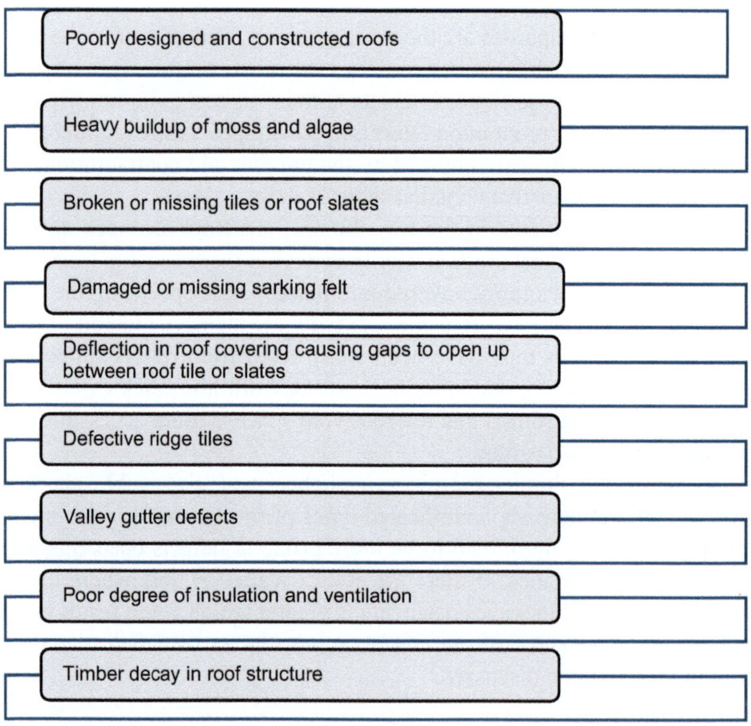

Figure 7.23 Different categories of roof defect

this can allow water penetration through the covering. The lead author recently inspected a roof constructed on a single store extension in Bolton with only an 11-degree pitch. Whilst the architectural design specified a 15-degree pitch the builder explained that it was not possible to create this pitch owing to the existing level of the first-floor windows on the rear elevation of the main house abutting the new extension. On investigation it was revealed that the type of roof tile installed had a minimum pitch requirement of 15 degrees. As the pitch was constructed four degrees short of this minimum threshold, severe water ingress was prevalent through the roof covering and into the extension space below. Remedial works included having to strip the roof of all its roof tiles and roof lights and install special roof tiles and rooflights which are designed to work at low pitches of 11 degrees. This was a very costly mistake for the builder and one that could have been easily avoided. Other design defects may include specifying the wrong material for roof tiles for a particular location or climate. For areas particularly exposed to heavy winds and gusts, possibly near coastlines, a heavier roof tile such as clay should be less likely to become uplifted than a lighter tile e.g. cedar.

(2) Heavy buildup of moss and algae

Heavy buildup of moss and lichen on roofs not only becomes unsightly but can cause growth between and under the roof tiles or slates and cause gaps to form. This can promote damp penetration in some cases as the vegetation in the gaps can provide the conduit for passage of water between tiles. In addition, heavy moss deposits can congregate in gutters which can cause them to block and deflect under the additional imposed weight.

(3) Broken or missing tiles or roof slates

Missing or broken roof tiles or slates are the most common form of pitched roof defects. They may have been dislodged or damaged by heavy winds passing over pitched roof slopes which can create an upward vacuum suction force with a tendency to lift tiles. Missing or damaged tiles and slates can create gaps in the roof coverings and therein allow wind driven rain to enter roof voids. Identifying damaged or missing tiles or slates could be through visual inspection of the roof slopes from street level externally and possibly with the assistance of binoculars. Figure 7.24 illustrates an example of a roof with damaged and missing slates.

It may also be able to spot signs in roof voids where daylight can be seen passing through the roof covering. Remedial works could include patch replacing damaged or missing tiles and securing with special clips. Unfortunately where roofs are damaged extensively reroofing the whole property may be the only realistic option available.

(4) Damaged sarking felt

Other signs whilst inspecting roof spaces may be the condition of sarking felt which is a layer of felt or plastic below the surface of the roof covering and acts as a secondary membrane in water penetrates through tiles or slates. If this is damaged or missing, then water penetration is more likely to enter the property and damage ceiling and wall finishes. Figure 7.25 shows an example of where sarking felt has been damaged.

(5) Deflection in roof coverings

Sometimes owing to the weight of the roof tiles, or where there is damage to the timber roof members, e.g., timber decay, roof structures can become prone to outward movement. This can result in deflection or sinking of the roof as the it becomes deflected. This type of movement can sometimes be easily noticeable, especially in older properties where their original heavy stone roof slabs are still in situ (see Figure 7.26). Noticeable signs are

Figure 7.24 Example of a roof with damaged and missing slates

Figure 7.25 Example of where sarking felt has been damaged

Figure 7.26 Deflection of a pitched stone roof

deflection in the ridge line or buckling of the roof at eaves level. This type of movement can compromise the performance of roofs by reducing their pitch in the deflected areas and allowing gaps to open up in the roof covering.

(6) Poor degree of insulation and ventilation

Pitched roofs are normally built as 'cold' roofs, where the insulation is installed below the roof covering. This means that the air in the roof void can become very cold especially in locations where very low temperatures are experienced at certain times of the year. It is important at these times to ensure that the roof spaces are adequately insulated as otherwise this will adversely affect the energy efficiency of the property in terms of heat loss. In addition to avoid the effects of condensation forming these roof voids should be adequately ventilated, and this is normally facilitated through vents at ridge level which promote cross ventilation.

(7) Defective ridge

Another common weakness point on pitched roofs occurs at their ridges. On traditionally constructed pitched roof interlocking ridge tiles, normally made of concrete or clay, are laid on a bed of sand cement mortar along the line of the ridge. In heavy winds they can become dislodged which tends to cause cracking of the mortar beds and pointing between them. Gaps can open up where mortar pointing becomes dislodged and provides a conduit for the passage of rainwater through the roof. Visual inspection externally will normally pinpoint damage at this level, and this is illustrated in Figure 7.27. Repointing or replacing damaged ridge tiles should be instigated. Owing to the height of most ridges access difficulties can be experienced. Accordingly, repairs should only be undertaken by qualified and experienced roofing contractors using an appropriate system of access scaffolding.

In addition to ridges similar defects can exist at roof eaves where stone copings are used to prevent the passage of rainwater through the roof covering. In cases where mortar joints between copings become moved and become damaged, water can traverse through the open joints and into the roof spaces below. Again, urgent repairs to reposition displaced coping stones and repoint all open joints between them should be instigated without delay.

(8) Valley gutter defects

Valley gutters occur at the intersection of two roof slopes. They can be formed from many different materials but most commonly from leadwork on traditional tiled and slated roofs. Both slopes will discharge rainwater into them making them particularly vulnerable

Figure 7.27 Cracking and missing mortar joints to ridge tiles

to water penetration when they become blocked by dirt, moss and leaves. Regular cleaning is therefore imperative to ensure that they do not block and discharge water efficiently.

(9) Timber decay in roof structures

Pitched roofs are clearly at risk of the alternative forms of defect which can cause water penetration through their coverings. If left unchecked this can create dampness conditions in roof voids for timber decay to develop. In the last chapter it was explained that timber decay through wet rot, dry rot or wood-boring beetle attack can occur when the moisture content of wood reaches certain levels. Prevention is better than cure and addressing the source of water penetration will hopefully prevent ant timber decay occurring. Where timber decay is prevalent then the most suitable form of timber repair, replacement or preservation works depending on the type of timber decay should be carried out. Please refer to Chapter 6 for remedial works associated with timber decay.

7.6.3 Defects at abutments and penetrations through flat and pitched roof coverings

Both flat and pitched roof coverings can become particularly vulnerable where they meet abutments or where there are penetrations through them for the passage of pipes or rooflights. It is important that all interfaces at abutments are correctly designed to prevent the passage of water. Flashings are designed to provide a waterproof seal at the interface where roofs meet abutments and where they become more susceptible to water penetration. Lead or rubber collars may be required where pipes pass through roof coverings, and these can again be more susceptible to leakages if damaged or not installed correctly as illustrated in Figure 7.28.

Figure 7.28 Damage to collar around soil and vent pipe pasting through slated roof covering

Broken flashings are one of the most common problems where flat and pitched roof adjoin abutments such as a gable walls, parapet walls, chimneys, perimeter upstands, rooflights and dormer windows. There are different types of flashing but the most popular tends to be leadwork coupled with lead soakers used on pitched roofs. Over time flashings can become damaged or displaced, as illustrated in Figure 7.29. At this point roofs become increasingly vulnerable to rainwater ingress where gaps may have opened up. If leadwork has simply become loose, then it may be possible to refix possibly by repointing them back into brickwork beds. However, they may require replacement if their condition and integrity is showing obvious signs of other damage such as tears, splits and cracks. At this time all other leadwork checked including soakers which sit below pitched roof tiles and slates.

Chimneys can present many problems for roofs and especially on older properties. Chimneys can sometimes be visibly leaning to one side which has had heavy exposure to prevailing winds. Other defects could include perished, cracked or displaced mortar joints, which will allow increased water penetration and inflict further damage by forces such as frost action. Figure 7.30 illustrates a chimney which has suffered from these types of defects.

Where chimney defects are relatively minor, repairs can be carried out which may include cleaning off vegetation growth and patch repointing. However, major damage will normally necessitate rebuilding or removing them altogether if redundant and extending pitch roof coverings over the chimney opening. Sometimes there can be missing or damaged chimney flaunching. Flaunchings are constructed from a strong mortar mix and are designed to hold chimney pots in place and prevent rainwater travelling vertically downwards through the structure of

Figure 7.29 Damaged and displaced roof flashings

Figure 7.30 Chimney suffering from brickwork damage and leaning

chimneys. Therefore, any damage to them can cause pots to become loose and allow damp ingress to internal walls and chimney breasts. Sometimes damaged flaunchings are difficult to see from external inspections viewed from street level. One telltale sign that there may be damage is when there is a heavy buildup of moss and algae on the chimney or where vegetation appears to be growing from the head of the chimney. This is illustrated in Figure 7.31. Where this is apparent damaged flaunchings should be repaired with new sand cement mortar as soon as possible.

7.7 Defective rainwater goods

Rainwater goods come in many different shapes, sizes and materials. On Victorian buildings cast iron was the most common material used for gutters and downpipes and these were prone to corrosion and other forms of damage including cracking. Most gutters and downpipes installed on modern houses now will tend to be uPVC. Some of the defects which can occur with all types rainwater goods are listed in Figure 7.32.

Blockages can frequently happen when vegetation such as leaves collects in gutters. This can not only block the gutters but rainwater outlets and downpipes also. For this reason, gutters and downpipe should be cleaned regularly to ensure that rainwater can be discharged safely

Figure 7.31 Chimney showing signs of moss, algae and vegetation growth

Figure 7.32 Common defects with rainwater gutters and downpipes

Figure 7.33 Defective gutters and downpipes

into underground drainage. Sometimes rainwater pipes can become cracked and split from the impact of falling ladders or tree branches. This can cause rainwater to be discharged down the face of external walling possibly leading to internal dampness. Dampness can also occur where gutters cannot release water when excessively deflected or where there are leakages at joints. Sometimes the cause of leakages is not the condition of the rainwater goods but that they have not been adequately sized to cope with the maximum volume of water that they can become exposed to under heavy rainfall. In these cases, they should be replaced with larger sectional gutters and downpipes. Some photographs of these types of problems are contained in Figure 7.33.

7.8 Presence of hazardous and potentially flammable materials

When conducting building inspections, surveyors should be looking for signs of any materials which could be hazardous or be deemed dangerous in terms of flammability. The presence of asbestos on older buildings is certainly one to be aware of and this was discussed in Chapter 6. Whilst it could be argued that the presence of asbestos is more likely to be internal than externally evident, there are still examples of where asbestos related materials have been used for roof and walling products and components. Asbestos cement has, for instance, in the past been used as roof sheeting for industrial buildings and was sometimes installed as an inner layer behind external wall cladding for fire protection. Prior to building surveys, desktop studies should be conducted with reference to asbestos registers to accurately locate any areas where asbestos may be present. Since the Grenfell Tower disaster in London in June 2017, causing 72 deaths, there has been a focused effort to identify the same or similar cladding that could prove flammable in the case of fire, to avoid a repetition of this tragedy. The type of cladding was classified as an aluminium composite cladding (ACM). A programme across the UK has been ongoing since this time to strip and remove such cladding. Building surveys reports should refer to the need to undertake further specialist inspections, possibly involving testing of materials if they suspect ACM is installed on any buildings they come into contact with.

7.9 Problems associated with reinforcement corrosion of concrete structures and facades

Whilst this chapter has been mainly focused on traditionally constructed buildings of a predominantly domestic nature, it is worth noting that a lot of commercial buildings and blocks of flats, still in existence in the UK, were constructed with exposed concrete frames or precast reinforced concrete façade panelling. This type of construction was a popular choice in the UK and Europe throughout the 1960s but design and construction defects have since this time become evident. Such defects and deficiencies have resulted from insufficient concrete encasement given to reinforcement cover during construction and is commonly referred to as 'concrete cancer'. Concrete cancer occurs when concrete structures, being porous, absorb water and air containing salts. The salts in certain circumstances absorb through the concrete and comes into contact with steel reinforcements. This can cause the steel reinforcements to corrode causing rust to develop on the steelwork. This rusting process causes the steel to expand which results in concrete becoming cracked and detached which can expose the steelwork and therein exacerbate the corrosion process. This deterioration can significantly weaken concrete and can become an obvious hazardous to the health and safety of individuals if concrete spalling from the external facades of buildings becomes evident. The typical signs of concrete cancer viewed externally can include those included in Figure 7.34 and illustrated in Figure 7.35.

Whilst the predominant cause of concrete cancer has been identified as insufficient concrete cover during construction, it can be exacerbated by the following aspects contained in Figure 7.36.

Concrete cancer, like cancer in humans, is best diagnosed earlier than later to limit its effects and the damage it can cause. Accordingly, if it is left to develop unchecked the costs of remedial works can be considerable as it can compromise the structural integrity of buildings, and in some cases may require them to be demolished. In terms of remedial works this section of the chapter will consider three alternative repair treatments namely:

- Polymer modified repair treatment
- Electrochemical treatment
- Replacement of affected concrete and steel

(1) Polymer modified repair treatment
The polymer modified repair can be utilised when there is insufficient concrete cover to steel reinforcements in some areas. The system relies on removal of the concrete in the

Figure 7.34 Typical signs of concrete cancer

Figure 7.35 Examples of concrete cancer

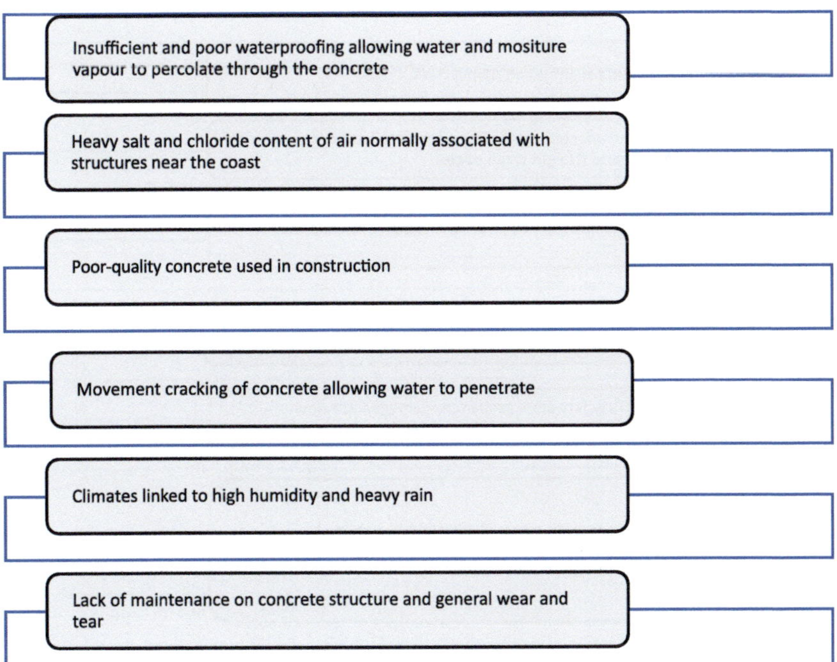

Figure 7.36 Aspects which can exacerbate concrete cancer

vicinity of the affected areas and cleaning down steel reinforcements which have become rusted. Following this, a specialist primer is applied to the exposed steelwork and thereafter covered with a polymer modified material. Thereafter an anti-carbonation sealant is normally applied across the entire surface of the concrete to create a robust and resilient barrier to water and moisture vapour.

(2) Electrochemical treatment

Electrochemical treatment is normally conducive to buildings located near coastlines and therein exposed to air borne salts. This system can reduce corrosion of the reinforcements by making their steelwork surface into a cathode of an electrical cell. It also relies on more corroded metal adjacent to the steel becoming a sacrificial anode. This is where a more easily corroded metal sacrifices itself by corroding before the steel reinforcement does. A waterproofing membrane layer should then be applied or installed where practical. This presents a physical barrier that prevents water from percolating through the structure.

(3) Replacement of affected concrete and steel

Replacement of affected concrete and steel as the name suggests involves removing those areas of concrete affected by concrete cancer and replacing them with new concrete and steel reinforcements. This can be a very intense and expensive treatment where there are extensive areas requiring attention and, in some cases, may be limited by city centre locations and access issues. A typical replacement system of this kind will normally involve the following steps in Figure 7.37.

It should be noted that repairing concrete cancer is a highly specialist job and one that needs to be carried out by certified and experienced contractors who can competently carry out all works, comply with all regulations and provide a guarantee for all works undertaken on completion.

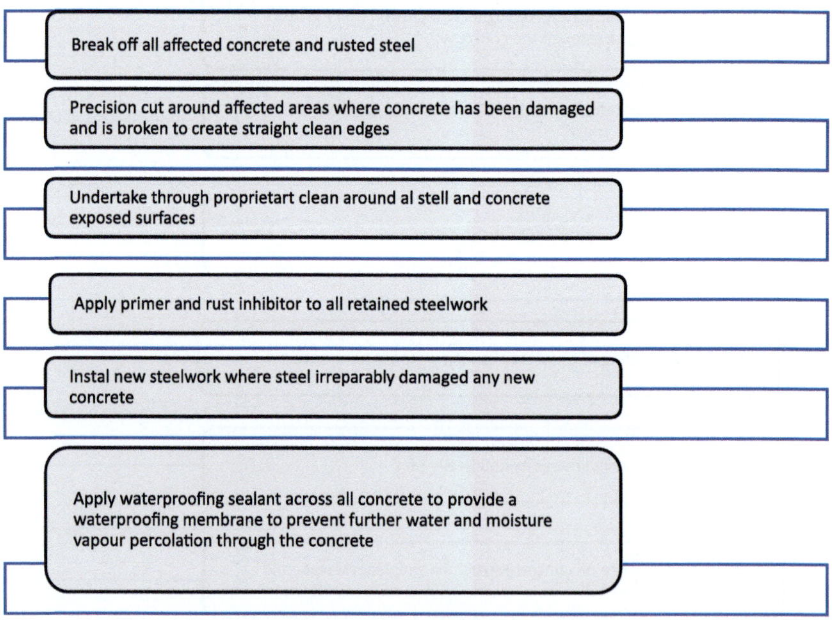

Figure 7.37 Normal steps to the replacement of concrete and steel affected by concrete cancer

7.10 Defective external areas and hardstandings

Defects in brick masonry have been previously referred to in this chapter. However, it is worth pointing out that boundary walls, retaining walls and walls forming manholes are particularly exposed to large volumes of water on all sides. This can exacerbate the brickwork defects referred to in 7.3.1 causing rapid deterioration of these structures if left unchecked and unmaintained. With boundary walls, it is important to have well-constructed coping stones which can provide a barrier to water traversing vertically through the walling from the top downwards. They should also be designed with a suitable projection or lip to shed water away from walls and not allow dripping of water down the walls which could be absorbed causing dampness. Figure 7.38 illustrates a boundary wall showing clear signs of brickwork disrepair caused by

Figure 7.38 Defective and damp boundary wall showing paint peeling

a general lack of maintenance, with open joints and the absence of any proper coping stones. Dampness of the brickwork as a result can clearly be seen resulting in masonry paint peeling and blistering.

A different example illustrates a boundary wall suffering from extreme dampness and where the face of the brickwork has spalled and become dislodged through frost action. In this example random stone has been installed to the head of the boundary in a poor attempt to install act as a coping stone. Water in this case will continue to percolate through the joints in the stone down through the wall and cause the wall to remain saturated.

Sometimes boundary walls can be retaining walls where there are changes in level, possibly on sloping sites, and they are structurally supporting higher areas of ground. They can suffer from all the same defects as previously described for boundary walls but can become unstable if these affect their structural ability to withstand the lateral pressures imposed on them from build up ground. These forces can also be exacerbated by the hydroscopic pressure when the ground they are supporting contains large quantity of water after heavy rainfall. Accordingly, these types of walls should be constructed with weep holes on their lower course to drain water away and therein reduce hydroscopic pressures. These can be fitted retrospectively as possible remedial works at a later stage. Sometimes the horizontal pressures exerted on masonry boundary wall cause them to lean over and crack. Repairs in such cases include repointing of brickwork and coping stones which may be deemed appropriate for minor damage. However, in some cases the possible addition of brickwork piers as supporting buttresses may be required to structurally stabilise walls. Where retaining walls have completely failed and are leaning significantly with large movement cracking as illustrated in Figure 7.39 the only measure will be to demolish and rebuild.

Hardstandings should be inspected as part of the external areas of buildings. These may comprise of solid paving areas such as concrete flagging or block paving. In these cases, common defects might include cracking, uneven surfaces and sinking of paving possibly brought about by vehicular traffic in the case of driveways. In such cases recommendations may be made to patch repair the worse areas or reinstate the whole area where subsidence has occurred. The condition of outbuildings including garages should be included in this part of the survey and any necessary remedial or demolition works.

External gullies and outlets serving paved areas should be visually inspected for obvious signs of damage and blockage. They are sometimes prone to blockages from leaves and debris and recommendations should be made for through cleaning possibly jet washing to allow water to drain normally. The removal of manhole covers, and inspection of underground drainage is not usually included in most building surveys as it is deemed beyond scope. However, in certain circumstances where problems with foul and rainwater drains, and sewers are suspected then separate drainage surveys can be conducted. Normally these would be undertaken by specialist drainage companies who have access to specialist equipment for working in confined spaces. It is normal for companies such as these to do camera surveys of the inside of drain runs which will identify any movement, collapse, fracture or cracking of the underground drainage system. In addition, the source of any blockages can be established in this way and tree roots growing into drainage could present just one example. Where drains have collapsed, camera surveys can pinpoint the exact location of failure to allow excavation and drain replacement to be carried out in the correct area without disturbing other areas of hardstanding. If damage to drainage is not too severe other remedial works may include proprietary drain lining works. This may be a suitable alternative where there is a need to reduce disruption to paved areas that may otherwise be affected.

Figure 7.39 Failure of a brickwork retaining wall causing leaning and cracking

7.11 Summary

The categories of the different external defects which have been discussed in this chapter include the following:

- Brickwork defects, pointing, cracking, dampness
- Differential settlement and subsidence
- Window and door defects
- Flat and pitched roof defects
- Defective rainwater goods
- Presence of hazardous and potentially flammable materials
- Defective external areas and hardstandings

When considering the types of external maintenance issues, one needs to closely look at the construction, age and use of buildings. An older industrial unit for instance may exhibit very different signs of disrepair when compared to a modern dwelling.

The use of brickwork and its popularity can relate to its longevity as, if built correctly, it can last for many years with relatively little in the way of maintenance. However, poor quality control during construction could affect external walls and mortar pointing between brick courses leading to defects. Examples given in this chapter have included mortar not being mixed with the correct ratios of sand and cement and new brickwork being exposed to extremes of weather during construction, without protection. Notwithstanding this premise even when the quality of materials and workmanship is good, given external brickwork is exposed to the elements and extreme conditions, this can affect the best masonry walling over time. Other problems for brickwork could include frost damage, shrinkage, efflorescence, contamination from air pollution, sulphate attack, rusting cavity ties and damaged render. Cavity tie corrosion can be a particular problem causing horizontal cracking of brickwork and bulging where the outer leaf of brickwork moves away from the inner leaf. In some cases, cavity ties can be carefully removed and new ones retrospectively installed but this is an expensive and time-consuming exercise.

Dampness can affect brick external walls, and this can be caused through water penetration or rising dampness. Both these sources of dampness can damage brickwork, causing joints to become cracked and perish.

Differential settlement and subsidence can cause major problems for buildings. Differential settlement occurs when parts of a building, for instance one elevation, is settling at a different rate than another. Differential settlement could occur where there are weak pockets of land under buildings, sometimes referred to as hot spots which could allow local downward movement in these areas. Structural tensile stresses in buildings are generated in these instances causing movement and resultant cracking. Subsidence can in most cases be more serious than differential settlement and refers to the gradual caving in or sinking of an area of land. It may occur where land has been infilled in the past or where groundwater problems are causing land to be washed away. For buildings constructed on land prone to subsidence it can be devastating and may render them dangerous structures with demolition the only alternative. Visible signs of differential settlement and subsidence could include cracking to external and internal wall, floors becoming unlevel, bulging walls and noticeable leaning of walls and properties. Remedial works can include underpinning to create new deeper foundations or proprietary injections system with cement grout or geo-polymer materials to stabilise and compact the ground.

Window and door defects could include timber decay, warping of frames, peeling or deterioration of paint finishes or defective double-glazing panels. The most predominant type of timber decay for external windows and doors is wet rot. Visible signs to look out for include missing, blistered or damaged paint and darkening of timber especially on corners of frames. On prodding such area timber affected by wet rot is found to be soft and spongy in nature and easily penetrable with a knife or screwdriver.

Flat roof defects can be synonymous with inadequate design and construction issues. These could include constructing flat roofs without adequate gradient to discharge rainwater insufficient. In addition, incorrectly installed insulation or a general lack of ventilation can cause problems associated with condensation in roof structures. Damage to roof coverings can emanate from the effects of wind, snow, sun and rain. Blockage to outlets can also result in ponding on roof surfaces which could lead to differential movement causing cacking and tearing of roof surfaces. There are many types of pitched roofs, but gabled roofs are the most common. They can be formed by trussed drafters or from traditional roof rafters and purlins. Defects can include broken or missing tiles and slates, damaged sarking felt, defective ridge tiles and timber decay in roof spaces. They can also be prone to a heavy buildup of vegetation growth particularly moss and algae. Flat and pitched roofs can be particularly vulnerable where they meet abutments such as chimneys, parapet walls or pipework. The intersections in each case must be carefully designed and constructed with flashings or collars, in the case of pipework, to protect the roof from water penetration at these vulnerable points. Defective rainwater goods serving roofs can be prone to blockages, cracking and splitting, deflection and leaking joints. Any leakage from rainwater goods could have an adverse effect on brickwork where walls become exposed to dampness.

When conducting building inspections, surveyors should be looking for signs of any materials which could be hazardous or be deemed dangerous in terms of flammability. The presence of asbestos on older buildings is certainly one to be aware of as this was used for many years in walling and roofing products. Aluminium composite cladding (ACM) which was a cause of the Grenfell Tower disaster should be identified wherever possible although this may be subject to separate specialist surveys.

Concrete cancer can be prevalent for a lot of older buildings in the UK and Europe as it was used for commercial buildings and tower blocks especially in the 1960s. Deterioration associated with this condition can significantly weaken concrete. Furthermore, it can become an obvious hazardous to health and safety of individuals if concrete spalling from the external facades of buildings becomes evident. It can be identified by cracking and displacement of concrete and can be exacerbated by climates linked to high humidity and heavy rain. Concrete cancer can be repaired using polymer modified or electrochemical treatment, but in most cases repair systems rely on the replacement of affected concrete and steel.

Inspection of external areas and hardstanding should be part of most building surveys. Boundary walls and retaining walls can be prone to excessive dampness which can lead to frost action damage to brick faces from frost action. Mortar joints can also become cracked and receded if dampness in external walling persists and repointing in such cases should be instigated. Where retaining walls can be exposed to lateral forces of higher ground which they are supporting and hydrostatic pressures from the underground water. This may cause them to move under these forces and lean outwards, become structurally unstable and in extreme cases require them to be completely rebuilt.

Chapter 8

Surveying existing buildings

8.1 Introduction to chapter

This chapter of the book is largely based on the previous work of the lead author which featured in Higham, A., Challender, J. and Watts, G (2021) *Introduction to Built Asset Management*. It will outline a background to conducting building surveys and identify the skills, steps and requirements for them to be completely successful. It will describe the methodology by which defects are correctly diagnosed through collection of evidence and observing telltale signs. Notwithstanding this premise, it will provide instances when this can sometimes be challenging, and misdiagnosis may remain a real factor. To assist readers, it will describe circumstances and practical examples where there may be more than one force of nature at play and how these can lead to different forms of building deterioration. Furthermore, it will provide examples where certain defects have been misdiagnosed for others and where symptoms are similar, e.g., alternative forms of structural cracking and dampness.

The chapter will also highlight the importance of client engagement both before and after building surveys and will describe the type and form of background information that may be useful to them. In this regard the usefulness of desktop studies and reviewing previous technical reports will be discussed. Following on from this, the importance of understanding the clients' requirements in the type and nature of information to be captured in the survey will be articulated and explained.

The importance and purpose of different types of building surveys will be discussed and why clients commission different surveys tailored to their specific requirements will be covered. In this regard the different types of survey will be articulated together with the benefits and possible restrictions of each. This will cover a description and explanation of each and will include mortgage valuation surveys, Homebuyers Level 1–3 surveys, schedules of condition and schedules of dilapidations. In terms of the latter two surveys reference will be made to leases and the application of each to serve tenants and landlords contractual obligations and responsibilities. In this regard the chapter will outline the procedure for possible breaches under the lease provisions in terms of adjudication and litigation. In addition, the range of different surveying equipment and tools will be identified. The use and application of each piece of equipment, to assist surveyors in their inspections will be covered, especially where access and visibility difficulties are experienced. Photographs of different building surveying equipment and apparatus will be provided.

The modus operandi by which surveyors will conduct their inspections will be discussed alongside factors such as health and safety and limitations which will need to be considered. In addition, the challenges and obstacles which sometimes confront building surveyors in the

DOI: 10.1201/9781003358800-8

course of their inspection will be identified especially where disruption to normal building operations remains an important factor. Examples of such factors will be given and an understanding of the difficulties that these constraints can impose on the building surveying process.

Following on from this, the importance of further investigation work will be outlined in circumstances that require opening up works or more specialist input will be examined. Finally, the various stages of managing the remedial work process will be described, alongside the importance of client consultation of the different options and preferred solutions. Examples of different scenarios leading up to the agreement of the agreed remedial works will be provided.

8.2 The various types of building surveys

8.2.1 Purpose of building surveys

Before commissioning building surveys, it is important to consider the purpose for which they are required, especially as they are designed as health checks for different types of buildings and should be tailored to the individual needs of clients. This is especially pertinent considering the cost of any type of survey and ensuring their remit covers the brief for which clients require in a particular instance. For instance, a house buyer may only require a relatively basic and non-detailed survey when buying a new house which automatically comes with a ten-year warranty. In the UK a Buildmark warranty is normally issued by the National House building Council for all new builds. This type of survey may satisfy the requirements of a mortgage provider or other third party financing the purchase, and this would normally be referred to as a mortgage valuation survey. Alternatively, if a buyer is purchasing an older property which shows evidence of disrepair, they may opt for a Homebuyers survey which will generally be more thorough and detailed than the mortgage valuation survey. These Homebuyers surveys can be Level 1, 2 or 3 depending on the depth of survey required, and a description of each will be given later in this chapter. Such surveys will also examine and analyse any defects and maintenance issues and make recommendations for addressing these in terms of suitable repair works. In some cases, an assessment of remedial costs associated with such defects and repairs may be included or recommendations for surveys to be conducted. This is particularly important as buyers may wish to know the full extent and cost of remedial works required prior to purchase to appraise whether the purchase is still viable for them. In addition, they may use the report detailing associated costs to renegotiate the purchase price of a property taking into account the report findings. Alternatively, if a prospective tenant is considering taking on a lease on a given property whereby they are responsible for future repairs they may wish to consider a schedule of condition to record the state of repair on taking over the lease. This could prevent them being responsible for costly repair works at the end of a lease on being served with a schedule of dilapidations.

In the UK most surveys are normally governed by the Royal Institution of Chartered Surveyors (RICS) as the professional body and many codes of conduct regulate for members as to how they should be conducted. Violation of such codes of conduct can result in members of the RICS facing disciplinary action from the professional body sometimes resulting in fines or in extreme cases their dismissal from the institution.

Although the following represents the main types of building surveys, and the vast majority carried out in the UK, it is not intended as an exhaustive list.

8.2.2 Mortgage valuation surveys

These types of survey are normally organised by the institution providing the mortgage for a particular property. In this sense it is important for prospective purchasers to note that these types of survey are carried out for the lender's purposes only and not for the benefit of the customer. For this reason, buyers should not rely on this type of survey when purchasing a property. However, this does not always mean that the buyer does not have to incur the cost of the survey as this is normally charged to them as a fee for transacting the mortgage. These surveys would confirm the value of the property, based on previous historical sales data in the area, to assure the lender via a loan-to-value (LTV) ratio that there is a suitable and acceptable level of security and comfort for transacting the loan. The LTV ratio represents the amount of borrowing against the value of the property and in some cases, owing to risk management on behalf of the lender, can affect the mortgage rate that a borrower is eligible for. These surveys are completed by surveyors who are qualified and normally members of the RICS. However, the reports they prepare normally rely on an automated valuation model (AVM) which calculates the value of the premises. This means that the surveyor does not always have to inspect the property to undertake the valuation, and in a lot of cases the survey reports can be prepared 'desktop'. This is especially the case as a consequence of these surveys not requiring any descriptions of the condition of the property. For this reason, mortgage valuation surveys can be prepared cheaply and quickly perhaps within a one-week period. The mortgage provider can then use the valuation findings to assist them in deciding whether a given property is eligible for a specific loan or not, depending on the security that the LTV ratio provides them with. In addition, mortgage valuations on those properties where a buyer has applied for a buy-to-let mortgage will normally have an estimation on how much rental the buyer can expect to receive. This rental value will be largely based on existing rents on similar properties in the local area. In cases where a given property receives a valuation less than expected, possibly due to market prices in the area, then this may result in the mortgage provider reducing the amount of loan they are prepared to offer to reduce their level of risk.

8.2.3 Homebuyer surveys

The main difference between Homebuyers surveys and mortgage valuation surveys is that the former will report on the condition of the property whereas as previously referred to the latter would only include the value of the property. It is therefore advisable for buyers to arrange these types of survey for themselves rather than just rely on a mortgage valuation survey to expose any building problems and therein lessen their risks. This is particularly the case when buying older properties and those constructed from nontraditional materials.

According to the RICS there are three types of Homebuyer surveys, categorised as follows:

(i) **RICS Home Survey Level 1**
This represents the survey with the least amount of information and for that reason it's the most basic. It will give a description of the building and its condition. In addition, it will identify defects which require urgent addressing alongside any significant potential risks and legal issues. Its limitations are that it will not identify other defects which are considered to be nonurgent.

(ii) **RICS Home Survey Level 2**
This is the most popular choice for home buyers especially for conventional buildings in reasonable condition and displaying no obvious defects or issues of disrepair. They tend to be significantly less expensive than Homebuyer Level 3 full structural surveys and

therefore suitable for clients purchasing a conventional property constructed from traditional building materials and in a generally good state of repair. The report should include details of the construction of the property, approximate age, information on the locality, environmental issues, alongside the condition of the various elements of the building. Such items may affect the value of the property and normally it is common practice for the report to articulate estimates relating to diminution of value against repairs. This may suit some prospective purchasers who only want to know the 'basics' or 'bottom line' to decide whether to progress the purchase without the need for more costly types of survey e.g. Homebuyers Level 3 surveys. Surveyors carrying out these surveys will normally inspect the walls, roof, floors, services and any other visible parts of the building and check for significant defective elements in this regard. They would not normally uncover and expose areas such as under floors to examine any suspected defects further but may recommend further investigations in their report. It may be helpful for surveyors carrying out such reports to be locally based. This could provide them with good local experience and knowledge of the area and any particular problems in an area. Examples of this could pertain to flooding, subsidence, pollution or any other environmental issues in the local area which clients would benefit from being made aware of.

These reports are normally presented in paper format which utilises a standard comprehensive system for condition rating. This system would categorise defects into three bands. Condition rating 1 would signify that no repairs are currently required, whereas condition rating 2 would articulate which defects need attention through repair or replacement, and which ones are considered urgent and serious. Alternatively, Condition rating 3 would include those defects which are considered significant and serious, and which require urgent investigation, repair or replacement. This categorisation provides an easy-to-follow report for clients who may not be familiar with building technology and thus prefer not to receive more technical reports which they may not be able to comprehend. A sample survey provided by the RICS is contained in Appendix 2.

(iii) **RICS Home Survey Level 3**

This represents the most detailed form of survey and was formerly known as full structural surveys and can be significantly more expensive than the Level 1 and Level 2 reports. These surveys would cover every basis including a thorough detailed description of the construction and every part of the building. They should include any known or suspected hazards that may be present, and these should be linked with any local knowledge in the case of floods, subsidence and environmental issues in the local area. They would normally report on each element of buildings separately and contain detailed information about their condition. Particular emphasis would be on those elements, such as the condition of timber possibly displaying signs of disrepair and possible further hidden defects that may require further inspection. Dampness through walls and roofs would certainly be reported and any associated problems which could culminate from this. An example could be examination of timber in roofs, walls and floors where water penetration has occurred to check for timber decay such as dry and wet rot. In certain cases, the report may recommend opening up works to expose elements of the building which are not currently visible for further surveys to be conducted perhaps by specialists. Examples of such recommendations could include drainage surveys, structural engineers' assessments and timber preservation reports. Owing to the more detailed type of survey, they are more suitable for those properties which are older, displaying signs of disrepair and defects or built from nonconventional, nontraditional materials.

8.2.4 Condition surveys

These types of survey are normally conducted by prospective tenants as part of a new lease they are progressing or have completed on. The surveys which are generated are referred to as Schedules of Condition. These schedules include details pertaining to the condition of a given property to be retained as a record for some time in the future and normally form part of an appendix to a lease. They are particularly helpful at a time approaching the end of a lease as a record of the condition of the property at the start of the lease. In this way they may lessen the liability of the tenant in terms of costly repair or renovation works at the end of the lease, on being issued with a schedule of dilapidations, if it can be shown that the property was not in a overall good condition at the start of the lease. Accordingly, they serve to limit the tenants' repair obligations and liabilities to the condition of the premises at the commencement of the lease.

Schedules of condition should be carried out by qualified professionals who are normally members of the RICS. They should comprise a detailed description of buildings and all elements within them including roof, walls, floors, finishes, fittings, equipment and building services. They should also incorporate photographs and sometimes drawings which help to build up a picture of the state of the property at the start of the lease. This is particularly important with a long lease where alterations, improvements and other changes may have taken place. The written evidence and photographs that are more important will provide a detailed account of the condition of the premises and any disrepair or defects in existence at the time of the survey.

A well-constructed schedule of condition should benefit both landlords and tenants as a term of reference at the end of leases, and thereby protect their respective interests. A lease may have changed hands over many years and without a record of the condition of the building at the start of the lease it can be very difficult to assess any damage or deterioration throughout the lease term. A sample sheet from a photographic schedule of condition is contained in Appendix 3.

8.2.5 Schedules of dilapidation surveys

A schedule of dilapidations contains a description and record of defects and disrepair pertaining to a property normally towards the end of a lease. In this regard defects and areas of disrepair represent breaches of the repairing covenants of a lease which could result in potentially expensive repair works for the tenants. They can also include redecoration, alterations and works required for statutory compliance of the demised premises. Normally the tenant has the opportunity to address these defects and repairs themselves, if the lease term has not expired, and at their expense. Alternatively, it is common for a financial settlement to be agreed with the landlord to enable the landlord to organise the works themselves, and especially the case if the lease has expired and the works have not been competed.

There are three types of schedules of dilapidations categorised as interim, terminal and final schedules.

(i) **Interim schedule**

An interim schedule is normally served within the lease term and outlines the breaches of the tenants' obligations in terms of maintenance responsibilities of the property. The servicing of the schedule normally acts as a catalyst for the tenant to bring the property back to the condition at the start of the lease and thereby preserve the property and safeguard against further disrepair and defects emerging. The schedule should not be costed but provide sufficient information on the areas to be addressed to allow the tenant to arrange the necessary remedial works. It normally follows that the landlord agent, normally a qualified RICS surveyor, will return to the premises after completion of the tenants works to ensure they have been correctly carried out.

(ii) **Terminal schedule**

A terminal schedule will normally be served upon the tenant within the last year of the lease but in certain instances can be served within a period up to three years before the expiry of the lease. Rather like the interim schedule it will contain information on breaches of tenants' obligations to repair and maintain the premises but more importantly it will contain costs for all elements of potential work.

(iii) **Final schedule**

A final schedule is normally served upon tenants after the lease has expired and describe the condition and overall state of the property as a final position in terms of potential breaches of the tenants' obligations to maintain and repair, especially as the tenants has no rights to return to the premises. This final schedule will normally be a terminal schedule updated to represent the current position of the landlord and all financial claim on the tenants. In this way it will set down the landlord position in terms of financial claim on the tenant. A *Statement of Quantified Demand* will normally accompany the final schedule, and this will detail the final claim based on the extent to which the landlord believes they have suffered. This will become an important document for both the landlord and the tenant if a disagreement arises and they cannot agree any part of the schedule. In such cases, it is normal practice for both the tenant and the landlord to employ the services of their own qualified Dilapidations Surveyor. These surveyors in turn will make their own assessments of the extent of repairs and formulate estimates of the cost of the required works. Disputes can be referred to adjudication through an expert practitioner or in extreme cases result in litigation through the courts.

8.3 A background to conducting building surveys

Carrying out building surveys calls for a comprehensive knowledge of building typologies, alongside an understanding of many different types of construction technologies alongside experience in the surveying process. Furthermore, they require the forensic identification, collection and recording of evidence to support a data-led approach to defect diagnosis and remedial works. In this pursuit it is vitally important for building surveyors to assess all the possible causes of defects and not simply jump to assumptions. For instance, there may be multiple causes of defects in some cases, and it would be easy to determine that dampness in external walling is being caused by a visible problem such as a leaking gutter but could transpire to be a more complicated issue including rising dampness and water penetration through a defective cavity tray above a window or door opening. In other cases, relatively straight forward building problems could be misconstrued for more complicated and expensive building defects. If a building defect diagnosis is incorrect it is almost certain that remedial works to address the problem will also be wrong, resulting in potentially unnecessary expensive repairs which will have little or no benefit in rectifying the problem. An example could be condensation on internal faces of external walls being diagnosed wrongly as water penetration through the walling structure. In this case the remedial works to combat what is thought to be water penetration could involve resealing around all windows and doors and repointing brickwork, when the correct remedial works to address condensation would be to introduce more internal ventilation and possibly improve the insulative properties of external walls. Other examples of misdiagnosis of defects could quite easily be made where shrinkage and/or crazing of external wall render or internal plaster finishes are mistaken for failures in the structural integrity of buildings associated with movement cracking from subsidence and buckling or deflection of external walls. Such cases in the past have resulted in legal claims from building owners or leaseholders for professional negligence and damages for monies expended. For this reason, all surveying companies should maintain a minimum level of professional indemnity cover to ensure they are fully covered and protected in these instances. Where there is a potential case of uncertainty as to the source or diagnosis of a building defect, then where appropriate the surveyor should consult with colleagues or specialists to gain a second opinion, in the same way as a medical practitioner would do before diagnosing the source of an illness or disease.

Before surveyors agree to take on a commission to undertake a building survey, they should fully understand what they believe their clients require and the purpose of the survey. There are many different types of surveys that can be carried out on buildings and each one will very much depend on the purpose of what is required from the survey. If clients are not clear which type of survey, they require then surveyors should outline their options, the respective approximate costs of each and make recommendations where necessary. For instance, condition surveys may be prepared at the start of leases as a record of the condition and state of repair of buildings and used as a reference therefore in landlord and tenant agreements. At the end of leases, a schedule of dilapidations survey will be conducted with reference to the condition survey which will outline the wear and tear, deterioration and defects which have emerged during the term of the lease and set down remedial works to address these. It is normal for schedules of dilapidations to include budget costs for the remedial works and the tenant would be required to pay for these works or instigate the works themselves. Landlord and tenant deliberations on the extend and cost of dilapidations works can become a source of debate and sometimes dispute between the parties and it is not uncommon for legal action to result from this.

8.4 The process of undertaking building surveys

When purchasing a property, it is normal for the purchaser to commission building surveys to assess the condition of buildings. The most proficient survey is normally what is referred to as 'full structural surveys'. These are normally detailed reports and give a thorough description of the condition of all building structures, elements and components. They will normally also give a summary of any necessary remedial works to address defects or items of disrepair. The costs for undertaking these remedial works will normally be a source of negotiation between the buyer and the seller to agree who will incur the costs and sometimes a reduction in the agree sale price will be agreed between the parties to take account of the necessary works. When surveying dwellings in the UK structural surveys are not the only type of survey that can be carried out owing to the substantial costs of undertaking these. It is probably more common for 'home buyers surveys' or 'valuation surveys' to be undertaken. Home Buyer surveys would normally be of a high level and a summarised version of a full structural survey and would report on the condition of each integral part of a property and any remedial works. A valuation survey would be a relatively short and condensed report on the overall condition of a property and then give a value on the property for mortgage purposes. Valuation surveys normally comprise the minimum requirements that banks and building societies require in order to grant a mortgage on a property. Prior to agreeing to undertake surveys and making arrangements for the inspection a fee quotation setting out the scope and extent of the survey with any limitations should be provided in writing to clients and adopting a format in line with the Royal Institution of Chartered

Surveyors format and inclusions. The fee quotations should clearly make reference to any services outside the remit of the survey and contain a list of exclusions accordingly. It should also set down the provisions with regard to employer and indemnity insurance held by the surveying company and a complaints procedure in the unfortunate event of a dispute. It is important for a contract to be entered into on the basis of this fee quotation prior to surveys being carried out. It is also important for surveyors to agree with their clients the extent of opening up and making good, access into and around the premises in order to conduct the inspections.

All surveys will be subjected to a minimum level of due diligence in terms of covering all the building elements and sections included in the survey report and in the UK, they normally follow the Royal Institution of Chartered Surveyors format for presenting such reports. Most building survey reports will exclude certain items which are considered as either too specialist for the type of survey being carried out and therein beyond the scope of the survey. There may be areas of the building that are regarded as being inaccessible, e.g., underground drainage and manholes and this is another reason which can lead to them being excluded. There will be certain instances where a particular issue requires the services of another potentially more specialist consultant. For instance, if dampness is detected in external cavity walling there may be recommendations related to further investigatory work using borescope equipment. The borescope would be able to view inside cavities of external walls and assess whether there is any 'bridging' between the outer and inner masonry leaves, possibly caused by mortar droplets on cavity walls which could be resulting in damp penetration across the cavity. A photograph of a borescope is included in Figure 8.1. In addition to the aforementioned, there may be specialist reports carried out on a particular element of concern on a building. For instance, if a building is showing signs of differentiation settlement or subsidence then a structural engineers report may be commissioned on that particular problem. Likewise, a specialist survey may be undertaken solely on a roof where water penetration through a roof covering is being experienced. Both these types of specialist

Figure 8.1 Photograph of a borescope for inspecting cavities with external walls

survey would seek to identify the cause and scope of the defects and thereafter recommend specialist remedial repair or replacement works to address the defects.

In addition to borescopes, thermal imaging apparatus could also be useful and these normally deploy infrared cameras to detect 'cold bridges' through structures sometimes associated with water penetration and depletion of insulative materials within building elements, e.g., cavity wall insulation. Thermal equipment can also be useful in identifying parts of buildings that might be causing an excessive degree of heat loss to buildings and therein identify measures to improve energy efficiency and environmental sustainability. Figure 8.2 contains a photograph of one type of thermal imaging camera commonly used in building surveying.

Building surveys may also be linked with maintenance and preparing a planned preventative maintenance programme. Such programmes will predict what level of maintenance needs to be undertaken on a year-by-year basis to keep a building in good repair. Programmes of this nature seek to ensure that building issues do not evolve into building defects and cause ancillary damage to the structure and fabric of buildings. Planned preventative maintenance programmes and work undertaken in conjunction with them are sometimes undertaken by specialist facilities

Figure 8.2 Photograph of one type of thermal imaging camera commonly used in building surveying

management companies who are contracted by client bodies to take the burden of maintenance and facility management away from them.

Undertaking any type of building surveys calls for a forensic, methodical and systematic approach to the inspection process. This is especially the case during the process of defects diagnosis where the surveyor's skills, knowledge, expertise and experience will be called on to providing evidence-based findings, conclusions and recommendations. Desktop investigations prior to arriving at the property can sometimes be useful in identifying local problems in a particular area and this could include flooding and any underground mines. Information of this nature can then be referred to in the inspection reports and can also provide a reference point for closely assessing any problems with the building structures or fabric curtailing such potentially local conditions. In the case of an area synonymous for mining careful examination of brick walling could reveal diagonal cracking associated with differential movement and in more severe cases subsidence. Likewise, a history of local flooding could reveal signs of previous damp penetration at low levels in floors and walls. It is also useful wherever possible to request information such as plans and elevations of the property from the building owner in advance and any subsequent improvement or modifications that have been made to the property over the years. It is also worth checking with the building owner or occupier that there are no hazardous materials that the surveyor is likely to come into contact with during the survey process, e.g., asbestos. Other desktop information that may be useful could include previous building surveys of properties. In this way surveyors could review previous problems with buildings and ascertain whether these have been adequately and competently remedied in the past. If problems have not been addressed, defects may have led to further problems that surveyors will need to look out for on site.

Upon arriving at a property, every surveyor will have their own ways and means of carrying out building surveyors. It is wise for any surveyor prior to starting to write the report to have a good walk around the building both internally and externally. This is to assess the general overall condition of a building and to cross-check whether an external defect is causing internal problems and vice versa. It is also important on this initial walk around to assess any difficulties in carrying out the survey such as access problems associated with heights, depths and openings, confined areas or unsafe areas of the building. For health and safety reasons and to ensure that the employers insurance cover is not being compromised it is usual to carry out a very brief risk assessment and survey plan to confirm that by inspecting the property is not putting the surveyor in any imminent dangers. In carrying out health and safety risk assessments, surveyors should consider lone working, potential hazardous materials, any unsafe debris, unlit areas, uneven surfaces which could lead to trips and falls and any mechanical or electrical hazards, e.g., exposed electrical wiring. These can then be referred to in the inspection report. To ensure surveyors do not miss areas and elements of the building that they are inspective a methodical system could comprise inspecting the external areas first and from a top-down approach, working down to ground level. Sometimes high areas of building, for instance, roof areas are difficult to see from ground level and could require good quality high magnification binoculars such as the ones shown in Figure 8.3 to focus in on certain areas, e.g., roof ridge tiles. Other items of equipment may include a damp meter commonly referred to as a protimeter for checking moisture content in floors, walls and joinery (see figure 8.4).

Other equipment that surveyors could have in their kit bags could include overalls and face masks for accessing potentially dirty areas such as roof spaces and cellars, boots, crowbars and/or manhole keys for accessing manholes, knives for checking for timber decay and a camera for taking photos of buildings and recording defects and torches for accessing dark areas such as

Figure 8.3 Photograph of good quality high magnification binoculars

Figure 8.4 Photograph of damp meter commonly referred to as a protimeter for checking moisture content in floors, walls and joinery

lofts and under suspended floors. Sometimes a tape measure can be useful for recording measurement, e.g., cracking in brickwork. Other equipment could include a set of telescopic ladders for accessing high areas of the building, e.g., flat roof areas if it is safe to do so and these are illustrated in Figure 3.5. In some cases, ladders may not be adequate or safe for access to many areas. In such instances surveyors should agree with their clients for the hiring of access equipment or plants which could include tower scaffolding, or an access platform lift. Drones may be a more economical and safer alternative where conventional access systems are not financially viable or where there are specific known hazards. Sometimes dyes can be kept in surveyor kit bags for tracking drainage runs and ascertaining the source of roof leaks. Also, levelling and alignment equipment such as plum lines, spirit levels and lasers can be useful to check if the walls or roofs are deflecting and out of plumb.

Sometimes building surveys are undertaken in two stages, with the first stage comprising an initial inspection to get an overall high-level perspective on a particular defect, and to assess further investigatory works. The second stage would be more in depth and could include carrying out specialist investigatory works especially those hard-to-access areas or those unable to be viewed normally. For instance, if isolated subsidence is noted in one particular area of a building close to an adjoining drain, a camera drainage survey may be required to ascertain the condition of the drainage runs. Water leakage from collapsed or fractured drainage may in this way be revealed to be the underlying cause of the problem, and a drainage survey can accurately pinpoint the exact location of damage and therein the scope of the required remedial works. In some cases, the second stage may consist of lab reports curtailing a potential defect. Another example of a second stage process could be concrete testing following visible failure of concrete structures identified in the first stage inspection. Tests may be undertaken for the presence of high-alumina cement (HAC) which since its introduction in the 1960s has become a major latent defect, sometimes causing catastrophic structural failures especially in areas with high humidity levels such as roofs and walls of areas containing swimming pools. The modus operandi of this second-stage process is to gather enough information on the potential building defect to diagnose its cause and thereafter to be confident that remedial works can be competently designed and specified. In certain cases, in the second stages there may be a need to monitor remedial

Figure 8.5 Set of telescopic ladders for accessing high areas of the building

works and potential defects to see if stabilisation of problems has been achieved or if further problems still exist. This monitoring may be needed across a medium to long periods of time to ascertain whether problems have been solved or still require attention. An example would be in the case of movement cracking to external walling due to subsidence. Once remedial works to alleviate the source of settlement/subsidence have been completed it is important to verify whether any further movement is continuing. In such cases 'telltale plates which record whether existing walling cracks are stable or opening up can be installed across the length of cracking and these can be revisited regularly, and data collected as to the position of any further movement.

The exact way surveyors choose to conduct their surveys and record their inspection findings varies from one individual to another. Some surveyors choose to use a checklist of internal and external elements and write down their findings in preparation for typing their reports at a later stage. Others choose to dictate their observations in a Dictaphone as they proceed which an audio typist can incorporate into a written report. Technological advancements have now allowed for handheld devices to be used to record defects and the general condition of buildings. Using software on such devises allows data to be downloaded at a later stage and automatically populate a survey report with limited need for typing up and therein can offer practical and resource advantages in terms of time and cost.

8.5 Challenges and obstacles when undertaking building surveys

Every survey presents different pressures and challenges in the way they and planned and executed. Carrying out inspections in premises which are occupied and accessing all areas around live operations whilst trying to avoid disruption to end users can be particularly challenging and requires a careful degree of planning and the help of the owners or occupiers to negotiate around certain conditions and constraints. Conversely different types of challenges can present themselves when organising inspections of vacant potentially derelict buildings, where lone working and ensuring that a safe survey plan is established. Working in schools with requirements for safeguarding children may involve obtaining DBS (Disclosure and Barring Service) checks or alternatively surveyors being escorted around at all times. Schools, colleges and universities during term time can present problems for building inspections with special arrangements being required to access specialist, potentially hazardous areas such as laboratories and planning to access classrooms around timetabled teaching sessions. Hotels, retail and commercial premises can present similar challenges to be confident to be able to access all areas without unduly affecting and disrupting the end users. All types of property and categories of survey can present specific problems where intrusive and/or destructive investigations are required. This could involve opening up walls or floors to gain visibility or access for further investigations. Clearly, such works in their own right will create disruption and for this reason should be planned to take place out of normal working hours if at all possible. The most important thing to consider as part of the building surveying process, therefore, taking account of all the possible necessary arrangements, is to maintain excellent communications with clients throughout to manage their expectations, especially since abnormal arrangements and investigatory works could add significant additional costs to the survey work. In such cases clients may have to apply for additional funds to bolster their budgets prior to works proceeding.

Challenges when conducting building surveys could also emanate from establishing accurate measurements to assess the condition of building elements. Having apparatus that can be relied on to measure the moisture content of walls, floors and internal joinery can be an essential facet in assessing whether dampness is causing problems within buildings and providing the right conditions for mould growth, deterioration of building fabric and rot to thrive. Even with the right instruments calibrated to an accurate standard of measurement surveyors need to be careful when interpreting the results. One example could be that occasionally damp meters, manufactured to measure electrical resistance through fabric, cannot distinguish between soluble salts in building materials such as plaster and the actual moisture content of the actual materials that are tested. Accordingly, this once again could lead to a wrong diagnosis and expensive non-required remedial works being recommended to clients.

When carrying out building inspections it is important wherever possible for surveyors to define the severity of defects against predetermined metrics to assist them in making their assessment. This is to avoid cases where one surveyor regards a particular building problem as moderately severe, and another classifies the same problem as very severe. This is particularly critical if a team of surveyors are carrying out a programme of 'stock' building inspections for a client, e.g., housing association when consistency is required across many different building reports. By way of an example, in the case of movement cracking in brickwork walls, classifications could be agreed as to a scale or range of thicknesses and lengths of visible cracking, where 1 could denote hairline cracking below 2 mm and therein less severe and 5 could denote more than 15 mm which could be categorised as very severe. By utilising this methodology, building surveyors can build up a picture of defects on a particular property which might all be interrelated to a common cause or symptom. Using this method, it is also critical to record the configuration of cracking as this is normally a telltale sign of what is causing the cracking. Vertical cracking of external masonry walls predominantly at corners of buildings (but not always) is normally associated with thermal movement of the structure. Such cases normally are attributable to design deficiencies where movement joints have not been adequately designed, specified and built into the structure. Alternatively diagonal cracking could be associated with differential settlement of the walling possibly due to foundation failures. In severe cases subsidence that could be causing differential settlement across many different areas of a building and causing varying degrees of cracking across many locations. It is not always possible to monitor this problem by visual means and may involve instruments for measuring cracking or dampness. In the former case micrometres can normally give accurate dimensions (see 8.6) over time but sometimes telltales (see Figure 8.7) can be fitted across the cracks to measure movement.

Figure 8.6 Photograph of a micrometre

Figure 8.7 Photograph of a telltale

Another example where categorisation is important could be water penetration into buildings which could be causing extensive dampness to some areas of the building and limited moisture buildup in others. Being able to assign predetermined degrees of severity to each different area of dampness from minor to very severe is again a way of introducing consistency into the survey reports. In this case it may also be necessary to monitor the movement or dampness over a period of time to assess whether the defect is progressing or following remedial works whether the problem is now fully addressed and therein historical. To assist in this pursuit, damp metres can accurately record the moisture content of structures and photographic evidence and can record the degree of spread and exposure.

In all cases the aforementioned methods may only give surveyors insight as to whether the defects are still prevalent but not give enough information for diagnosis of the precise cause to allow remedial works to be specified. This dilemma may be common where there are potential defects in walls that are not visible. In such cases, opening up works may be required to inspect further into the structure. This is common on external cavity walls where cases of dampness or movement could be caused by defects within the cavity area. Dampness could for instance be emanating from bridging of the cavities caused by mortar deposits between the inner and outer leaves of brick and blockwork, whereas movement cracking could be resulting from corrosion of cavity ties or lintels. Normally this type of opening up works can be undertaken by a small builder who can sometimes reinstate the building structures and fabric back to their original form once the intrusive investigations have been properly carried out, recorded, costed, planned and managed. Examples of opening up for further intrusive investigation are also synonymous with the following cases as shown in Figure 8.8.

8.6 The importance of building investigations

When surveyors are undertaking building surveys as referred to earlier in the book it is essential for them to adopt logical and pathological methodologies. This can include the sequence of the inspections and the mode by which they record their findings. In undertaking diagnosis of building defects each potential defect should be considered carefully to forensically assess all the symptoms that have been presented and to examine all the possible or probable sources and causes of the problems. This process calls for systematic elimination of possible causes as usually the sources of the problems are not always initially obvious. It can involve the testing of

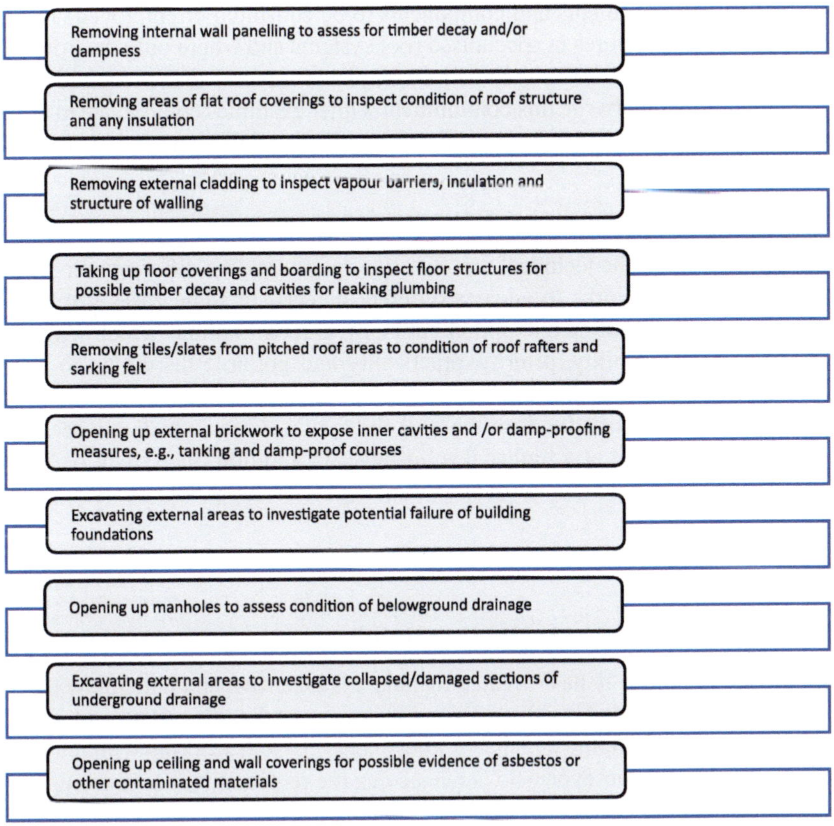

Figure 8.8 Examples of different types of opening up works for further intrusive investigations

different hypotheses relating to symptoms and possible causes. For instance, structural cracking could be resulting from expansion associated with thermal, moisture forces or conversely gravitational forces associated with foundation failures and/or ground subsidence. This process may lead surveyors to not just one possible cause but to a narrowing down of potential causes for further reflection. In this way the true cause of the defect can hopefully be identified and avoids the risk that a false diagnosis could be arrived at, leading to remedial works that are ineffective and wasteful. The successful diagnosis will ultimately allow a technical report to be prepared which will detail the extent of the problem, how the underlying causes have been arrived at and most importantly recommendations of all the necessary remedial works. Sometime the costs and timescales associated with the repair works can be outlined, especially if the surveyor is being appointed by the client to oversee the programme of remedial work. These technical reports will normally include details relating to insurances, liabilities, quality compliance; they will sometimes also include logistics for implementing the repairs in occupied premises where disruption to end users and health and safety are paramount. They could also offer recommendations or the names of specialist building contractors who would be appropriate and offer the necessary technical expertise to undertake all or some of the remedial works. In certain instances, there may be waivers in the technical report that call on more specialist reports from manufacturers or installers of specialist cements and components to be consulted or employed. This could be the case where there are failures in specialised roof systems and where only the original roofing manufacturer or installer is suitably qualified to complete the works to the required standard and provide a guarantee to underwrite this commitment. Other examples could relate to curtain walling systems or specialist areas of mechanical and electrical installations and plants.

8.7 Managing the remedial work process

As previously referred to, the technical report will normally outline different recommendations associated with remedial works to address different defects. In some cases, for each item of remedial works there will not be one option but many different options which will give clients choice in terms of affordability, priority, practicality and possible disruption to the building occupiers. Sometimes each option can relate to temporary repairs to monitor the problems or owing to availability of client's funds or the 'full and final' solution which will seek to cure the problem permanently, albeit at a higher level of cost. In addition, there may be intermediary options which provide a 'halfway house' in addressing the building defects. Take the following scenario as one example.

Asbestos contamination within a building

A technical building survey has revealed that there is extensive and potentially dangerous asbestos within an occupied building. Taking into account that the asbestos is widespread throughout the whole building and the building has many staff working within it the client has asked the building surveyor who is overseeing the remedial works to present a report on the options available to them. The following options have been outlined:

Record the full extent of the asbestos on a risk register and monitor for any damage

The asbestos is currently not in a state where it's an immediate threat to the health and safety of the building occupants. Some of the asbestos is the fibrous type and hidden behind wall lings and other areas contain the Amosite asbestos cement category which forms part of ceilings and wall panelling. All the visible asbestos is largely intact and not currently regarded as being in a condition which would cause release of fibres into the air and therein cause infection to building occupiers. Owing to the availability of funds for part or complete removal this scenario of recording the full extent of the asbestos on a risk register and monitoring for any damage would seem to present the best alternative. This position is predicated upon the alternative whereby removal works could be expensive, unaffordable and hugely disruptive to the building and business. Recording and monitoring the state of the asbestos could also be a suitable alternative where there are longer term plans to demolish and redevelop the building, whereupon the asbestos could be more affordably and safely removed in one stage once the building is vacated. With this alternative it is important to make clients aware that it requires continual inspection and recording of the condition of the asbestos on an asbestos register, as any interference or disruption to it could render it to be hazardous. Health and safety procedures will need to be put into place to record inspections and building maintenance operations will have to stringently adhere to strict protocol when being carried out in close proximity to affected areas of the building. In some cases, this may prove onerous for building owners in preventing them from carrying out interventions and works of alteration and improvement.

Undertake remedial works to remove all traces of asbestos

Where the asbestos is in a particularly dangerous state in areas and/or where clients want the full freedom to undertake improvement, alterations or refurbishment works unhindered by the presence of asbestos, this may present the best alternative. In this scenario it should be recognised that if the asbestos is extensive and requires intrusive/destructive opening up works to access it then it can be very expensive to remove. In addition, it is likely that the removal works will have to be carefully managed to be phased to particular areas of removal at one time whereby those areas to be accessed for the works can be safely isolated from the remainder of the building. This clearly will call for occupants to be decanted and moved as the works progress and can be hugely disruptive and disconcerting for them.

The previous example gives two far ranging and diametrically opposed options to addressing the problem depending on the constraints of the buildings, future plans and affordability. Notwithstanding this, there may be more intermediary measures between these two opposing solutions. There will be a range of works which could include a large degree of maintaining asbestos in its current state, inspecting, recording and managing as required, but a programme put in place to periodically remove those areas deemed to present a more imminent and important threat. Organisations like to adopt this hybrid approach to avoid too much immediate disruption and expense in favour of a managed rolling programme of removal as part of their respective long-term maintenance planning.

Similar scenarios for other building problems may exist for programmes associated with roof, window, door, curtain walling or replacement, where a more phased programmed approach may be more advantageous rather than be carried out as one project. Mechanical and electrical replacement could also be managed in this way to maintain levels of heating and servicing in buildings at all times.

8.8 Summary of chapter

Conducting building surveys calls for the forensic identification, collection and recording of evidence to support a data-led approach to defect diagnosis and a comprehensive knowledge of building typologies. In this pursuit it is vitally important for building surveyors to assess all the possible causes of defects and not simply jump to assumptions as relatively straightforward building problems can sometimes be misconstrued for more complicated and expensive building defects and vice versa. If a building defect diagnosis is incorrect it is almost certain that remedial works to address the problem will also be wrong, resulting in potentially unnecessary expensive repairs which will have little or no benefit in rectifying the problem.

There are many different types of surveys that can be carried out on buildings and each one will very much depend on the purpose of what is required from the survey. If clients are not clear which type of survey, they require that surveyors should outline their options, the respective approximate costs of each and make recommendations where necessary. Before commissioning building surveys, it is important to consider the purpose for which they are required, especially as they are designed as health checks for different types of buildings and should be tailored to the individual needs of clients. This is especially pertinent considering the cost of any type of survey and ensuring their remit covers the brief for which client requires in a particular instance.

In the UK most surveys are normally governed by the Royal Institution of Chartered Surveyors (RICS) as the professional body and many codes of conduct regulate for members as to how they should be conducted. Violation of such codes of conduct can result in members of the RICS facing disciplinary action from the professional body sometimes resulting in fines or in extreme cases their dismissal from the institution.

Mortgage valuation surveys are normally organised by the institution providing the mortgage for a particular property. In this sense it is important for prospective purchasers to note that these types of surveys are carried out for the lender's purposes only and not for the benefit of the customer. These surveys would confirm the value of the property, based on previous historical sales data in the area; the surveyor does not always have to inspect the property to undertake the valuation, and in a lot of cases the survey reports can be prepared 'desktop'. For this reason, mortgage valuation surveys can be prepared cheaply and quickly but as they don't consider condition are a riskier alternative to Homebuyers surveys. According to the RICS there are three types of Homebuyer surveys categorised as Level 1, 2 and 3. Level 1 is the most basic whereas Level 3 (formerly known as full structurer surveys) contain detailed information on all building elements. Level 2 is the most common form of Homebuyer and includes a reasonable amount of information on the condition of building and defects pertaining to a property.

A schedule of conditions includes details pertaining to the condition of a given property to be retained as a record for some time in the future and normally form part of an appendix to a lease. A well-constructed schedule of condition should benefit both landlords and tenants as terms of reference at the end of leases, and thereby protect their respective interests.

A schedule of dilapidations contains a description and record of defects and disrepair pertaining to a property normally towards the end of a lease. They can be categorised into interim, terminal and final schedules. Interim schedules are served within the lease period whereas terminal are served towards the end of the lease and final are after the lease has expired.

It is important for a contract to be entered into on the basis of this fee quotation prior to surveys being carried out and for surveyors to agree with their clients on the extent of opening up and making good access into and around the premises in order to conduct the inspections. Most building survey reports will exclude certain items which are considered as either too specialist for the type of survey being carried out and therein beyond the scope of the survey. This could include those areas that cannot be readily inspected, e.g., underground drainage or those elements which require further specialist advice, e.g., building services installations.

Building surveys may also be linked with maintenance and preparing a planned preventative maintenance programme. Such programmes will predict what level of maintenance needs to be undertaken on a year-by-year basis to keep a building in good repair. Programmes of this nature seek to ensure that building issues do not evolve into building defects and cause ancillary damage to the structure and fabric of buildings.

Undertaking any type of building surveys calls for a forensic, methodical and systematic approach to the inspection process. Furthermore, desktop investigations prior to arriving at the property can sometimes be useful in identifying local problems in a particular area and this could include flooding and any underground mines. Other desktop information that may be useful could include previous building surveys of properties. Sometimes building surveys are undertaken in two stages, with the first stage comprising an initial inspection to get an overall high-level perspective on a particular defect, and to assess further investigatory works. The second stage would be more in depth and could include carrying out specialist investigatory works especially those hard-to-access areas or those unable to be viewed normally.

On arriving at a property, every surveyor initially walks around to assess any difficulties in carrying out the survey such as access problems associated with heights, depths and openings, confined areas or unsafe areas of the building. For health and safety reasons and to ensure that the employers insurance cover is not being compromised it is usual to carry out a very brief risk assessment and survey plan to confirm that by inspecting the property is not putting the surveyor in any imminent danger. The exact way surveyors choose to conduct their surveys and record their inspection findings varies from one individual to another. Some surveyors choose to use a checklist of internal and external elements and write down their findings in preparation for typing their reports at a later stage.

There are many challenges and obstacles when undertaking building surveys. Carrying out inspections in premises which are occupied and accessing all areas around live operations whilst trying to avoid disruption to end users can be particularly challenging and requires a careful degree of planning and the help of the owners or occupiers to negotiate around certain conditions and constraints. All types of property and categories of survey can present specific problems where intrusive and/or destructive investigations are required. This could involve opening up walls or floors to gain visibility or access for further investigations. Clearly, such works in their own right will create disruption so should be conducted outside normal working hours if at all possible. When carrying out building inspections it is important wherever possible for surveyors to define the severity of defects against predetermined metrics to assist them in making their assessment. This is to avoid cases where one surveyor regards a particular building problem as moderately severe, and another classifies the same problem as very severe. Opening up works may be required to inspect further into the structure and fabric of buildings, and this can sometimes include taking up floor boarding, uncovering roof timbers, removing wall panelling and exposing ceiling and roof voids.

When surveyors are undertaking building surveys it is essential for them to adopt logical and pathological methodologies and this can include the sequence of the inspections and the mode by which they record their findings. In undertaking diagnosis of building defects each potential defect should be considered carefully to forensically assess all the symptoms that have been presented and to examine all the possible or probable sources and causes of the problems. This process may lead surveyors to not just one possible cause but to a narrowing down of potential causes for further reflection. Thereafter, the successful diagnosis will ultimately allow a technical report to be prepared which will detail the extent of the problem, how the underlying causes have been arrived at and most importantly recommendations of all the necessary remedial works. Sometimes options for different remedial works will be given. These can relate to temporary repairs to monitor the problems or owing to the availability of clients' funds or the 'full and final' solution which will seek to cure the problem permanently, albeit at a higher level of cost.

Reference

Higham, A., Challender, J. and Watts, G. (2021). *Introduction to built asset management*. Blackwell-Wiley.

Chapter 9

An innovative project to assist building surveying students

Designing a handbook as a practical teaching and learning guide for surveying buildings

9.1 Introduction and justification for the project

This chapter is based on the work of one of the book's authors Professor Jason Challender in developing a simple and easy-to-use handbook as a practical teaching and learning guide for surveying buildings. His justification for the handbook was to provide a practical tool to guide students through the different stages of surveying buildings. This was intended to provide a logical systematic pathology for conducting building surveys and therein ensure that all construction elements were covered and not missed. This justification was predicated on Professor Challender's reflection of his own education and to examine whether surveying students on leaving education are being fully prepared for employment and competent to undertake their future roles. His own personal experience, as a BSc building surveying graduate in 1989, was that on leaving university he was not equipped with the basic skills required to undertake a basic building survey, which was something that was expected of him in professional practice. The expertise and training that enabled him to undertake building surveys was gained whilst employed in industry.

This event has led him to ask the question as to whether there is a gap in provision as far as teaching and learning of students on surveying and construction-related courses are concerned, with particular emphasis on building surveys. Simple reflective practice from his own experiences, those of his past and present colleagues in industry and his recent interface with construction studies lecturers and students would appear to suggest that there is currently a large divide between classroom teaching and academic qualifications and the basic skills required in professional practice. He researched whether there is any data, case studies, reports or any other evidence that have been produced to support and bring meaning to his own opinions and findings on this matter. Several research documents presented themselves in this respect and will be discussed in this chapter and these will hopefully bring creditability to his argument and justify his views accordingly.

As a precursor to the teaching and learning aid that has been developed, this chapter will seek to explore why teaching and learning has not been adapted to meet the needs of the building surveying profession over the years. Furthermore, it will introduce the handbook as an innovative project that was developed and piloted onto an HND building surveying course at Stockport College. The chapter will explain how this project was devised, planned and developed and articulate how it is hoped that the handbook could seek to enhance teaching and learning for the benefit of the students and lecturers. Following its integration and implementation into the curriculum via an existing teaching and learning programme this chapter will discuss feedback which was received from students. It will then evaluate whether it has in fact made a difference and enhanced vocation-related training for surveying and construction students.

DOI: 10.1201/9781003358800-9

9.2 Reflective practice around a possible gap in teaching and learning provision for building surveying students

Professor Challender's observations from his experience in curriculum development and lectures to building surveying students reinforced his beliefs that not enough practical/site-based skills are being developed in students and too much emphasis is placed on construction theory in the classroom. These reflections also led him to the theory that a closer link to professional practice will motivate learners, who, although enthusiastic about future vocations, have found it hard to see the relevance of construction and surveying classroom teaching. To reinforce this theory there is research to show that when traditional teaching and learning methods are employed, students lack motivation and fail to progress beyond rote learning (Lavelle 1992, pp. 7–15). These findings indicated that students should be stimulated during the learning process in order to optimise their attention, performance and learning. For this reason, the project was designed to address this aspect specifically.

Following Driscoll's (three elements of reflection) model, the lead author's reflective practice also led to a reflective enquiry and triggered several questions that would need to be addressed. These included:

- Where is the evidence that justifies the idea that theory and practice need to come together in building surveying teaching?
- Why is there a gap in provision?
- What improvements could be made in teaching provision to bridge the gap?

9.3 Research, supporting evidence and justification

The next stage was to investigate whether there was any evidence to support and give credibility to Professor Challender's beliefs through research and answer some of the above reflective questions.

Initial research was quite alarming and revealed that:

- Eight percent of building surveying establishments in the Construction Skills Sector Skills Council area had faced hard-to-fill vacancies largely due to low numbers of applicants with the required skills (National Employers Skills Survey, 2003).
- Eighteen percent of building surveying practices in the Construction Skills Sector Skills Council area suffered from internal skill gaps, with staff being regarded as less than proficient (National Employers Skills Survey, 2003).
- Only 25% of students enrolling on Royal Institution of Chartered Surveyors RICS partnership courses in Building Surveying went on to become Chartered Surveyors (Mair, 2005).

Research has also identified that the gap in provision, in building surveying teaching, had existed for many years. Professor Mole's *Mind the Gap* report (Mole 1997) led to the Royal Institution of Chartered Surveyors Education Reforms of 2001 which sought to address the gap in practical based building surveying course content. This clearly did not go far enough as a study was carried out by Hoxley and Wilkinson in 2006 to review the RICS imposed changes. Their findings were that employers believed that 'graduates still lack sufficient knowledge' and lecturers seem to be 'teaching what they want to teach rather than what students need' (Hoxley and Wilkinson 2006).

Wilkinson and Hoxley (2005) concluded that concerns exist about the adequacy of provision of building surveying 'core content' within academic courses. They suggested that this could have been brought about through pressures related to the extent of practical surveying teaching that can be included within the curriculum. Wilkinson and Hoxley (2005) articulated that this had largely resulted from the increased pressure to maintain high staff/student ratios and increase modules sizes due to under-funding in the UK higher education sector. Their research paper also advised that there has been a genuine interest expressed by all employers to engage more with universities to address the gap in practical-based provision, but concerns were raised that universities were not receptive to approaches from industry in this regard.

The Royal Institution of Chartered Surveyors was made aware that 'students cite a lack of basic skills sets and are sometimes unable to make the adjustment from an accredited course to practice' but whilst welcoming diversity and not being prescriptive in the subject/modules of a course 'it expected a course to meet standards of quality, evident with the development of core surveying skills' (RICS 2010, p. 13).

As part of the research, Professor Challender liaised and consulted with various individuals who he had access to in his position as Project Director of Stockport College including professional practitioners and course leaders. There was a resounding belief from these individuals that there was still a long way to go, as far as improvements in teaching building surveying students are concerned, in order to meet the skills that are required by industry.

9.4 Planning and devising the project

As far as addressing the gap in building surveying provision as identified by the research, it was felt useful to develop an innovative and meaningful teaching and learning resource. This would then hopefully enhance the existing teaching programme for the benefit of building surveying students and fulfil the following SMART objectives that had devised for the project, namely:

1. Create a teaching and learning aid that will assist learners through the survey process and allow them to prepare a Home Buyers Building Survey Report.
2. Familiarise students with the Royal Institution of Chartered Surveyors Home Buyers Building Surveying Report format to better prepare them for professional practice.
3. Close current gap between education/training gained by students at College and University and those skills required by the industry.

A project plan proposal form (see Appendix 4) was prepared which mapped out the aims, objectives, resources and timescales and this was duly signed off by the course leader and head of programmes.

The project itself was essentially focused on the development of a simple and concise surveyor's handbook. As previously articulated, this was intended as a guide and template to assist students on construction and surveying-related courses in undertaking building surveys and preparing technical reports thereafter. The handbook incorporated concise elemental headings, following the standard format of the Royal Institution of Chartered Surveyors Homebuyers Building Survey Report introduced in 2001, to give structure and relevance to all student reports. Its main purpose was intended to steer students through a logical sequence as far as the various stages and the pathology/methodology of the survey process are concerned from start to finish. Furthermore, as students were generally not familiar with undertaking surveys of this kind, it was designed to prompt responses from the learners by way of a potential list of relevant statements related to description, condition, defect and remedial works which students can use as a checklist for their report.

The basis of this handbook, as a reference tool in carrying out the survey on site was designed to give all the necessary information that students require to enable them to write up their technical report at a later stage. It would also provide learners with a basic familiarity and understanding in undertaking surveys of this type to expand their experience in building surveying. This would then hopefully benefit them in their studies and better prepare and equip them for their chosen vocations in life.

Professor Challender utilised his extensive knowledge and experience of carrying out surveys of this type to create a multiple choice of potential defects for any particular element of the building and this has been incorporated into the handbook. This was designed to assist students when inspecting such elements as roofs, walls and floors and deciding which of the listed defects is most appropriate to represent the condition of that particular element.

9.5 Mapping against Lifelong Learning UK (LLUK) standards

In order to give authenticity to the project whilst providing further justification for its introduction, it was felt important to follow the standards set down by Lifelong Learning UK (LLUK) which is the body for professional development for those people working in further education.

It is especially important for LLUK standards to be followed in teaching whilst creating an effective stimulating environment to promote learning.

The new overarching professional standards for those who teach in the lifelong learning sector cover six domains namely:

- Professional Values and Practice
- Learning and Teaching
- Specialist Learning and Teaching
- Planning for Learning
- Assessment for Learning
- Access and Progression

These standards provide a consistent framework across the sector with the overall aim of increasing the standard and quality of teaching. They embrace important issues such as equality and diversity and maintaining an inclusive and motivating learning environment at all times for the benefit of students.

It was important to consider these standards when devising the project in order to meet as many of the professional values as possible. Accordingly, the project was devised to add value and enhancement to current building surveying teaching by utilising the lead author's own professional skills and creating an activity designed to motivate learners. Reflective practice, needs analysis, research, collaboration with others and creating and implementing an innovative tool for learning all demonstrated a commitment to the standards. Against some of the relevant standards set down by LLUK, Appendix 5 details the professional knowledge and practice that were met as part of this project and maps out where the added value created by this teaching and learning aid was.

9.6 Description of the project

The project sought to develop a simple concise handbook as a practical guide and template to assist students on construction related courses in undertaking building surveys and preparing survey reports. It will set down clear and concise elemental headings, following the standard format of the RICS Homebuyers Building Survey Report introduced in 2001, to give structure and relevance to all student reports. In addition, it was designed to attempt to steer students through a logical sequence as far as the various stages and the methodology of the survey process from start to finish. Furthermore, as some students are generally not familiar with undertaking surveys of this kind, it will prompt responses from the learners by way of a potential list of relevant statements related to description, condition, defect and remedial works which students can use as checklist for their report.

It was intended that the basis of this handbook, as a reference tool in carrying out the survey on site, would give all the necessary information that students require to enable them to write up the building survey report at a later stage. The handbook would also provide learners with a basic familiarity and understanding as far as undertaking surveys of this type are concerned. This will then hopefully benefit them in their studies and better prepare and equip them for their chosen vocations in life.

The handbook was essentially made up of three component parts. The first part as illustrated in Table 9.1 was designed as an introduction to the survey and to give context on general building information. It tasked students to complete property details including the address and information regarding the scope and extent of the survey being undertaken. For instance, it required details of whether the survey was simply an internal and external survey of the property or whether its scope included for inspection of building services and/or external areas and structures. The handbook also called for student surveyors to outline the restrictions of the inspection and those building elements such as drainage that would not be included in the survey report. Building details also were required to be populated into the required boxes including the type of house and its approximate age, alongside the construction of various elements including roof, walls, internal and external walls. It also called for a description of the internal accommodation including types and numbers of different rooms within the property. This section also included details on the locality of the property and circumstances related to the survey including weather conditions and access arrangements. At the end of this section the handbook requested a high-level assessment and summary of the building condition any prominent defects including movement and subsidence, timber decay, water penetration and dampness and the state of insulative properties of the building.

The second part of the handbook was designed to compile building condition information on the exterior of buildings. It was structured with six columns relating to element, description, condition, defects, remedial works and any additional remarks/information. By way of an example if the element was related to the roof structure and covering, the description might be given as a traditional dual pitched roof, covered with felt and interlocking concrete tiles. It may be deemed that the condition is poor owing to age of the roof or the absence of sufficient maintenance in the past. As such roof defects could include several damaged or slipped roof tiles, defective and missing mortar pointing to ridge tiles and excessive buildup of algae. All these particulars would be inputted by ticking the correct boxes and indicating in the remedial works additional remarks/information what the repairs would comprise. Repairs in this scenario may include partial reroofing, replacement of missing and defective ridge pointing and thoroughly cleaning down all roof areas to remove vegetation growth. The same type of assessment on other external elements could include rainwater goods, windows, doors and external walling.

Table 9.1 General property details

A: INTRODUCTION

Address of Property: ....
....
....

Name of Surveyor Undertaking Survey: ....
....
....

Scope of Survey:

Exterior of Property ☐
Interior of Property ☐
External Areas/Garages
Boundary Walls/Fences ☐
Building Services ☐

Restrictions of Survey:
Drains ☐
Under floor Areas ☐
Roof Void ☐
Other
(Name): . . . . ☐

B: THE PROPERTY AND LOCATION

B1 The Property

Type: Bungalow ☐
Terrace House ☐
Detached House ☐
Semi Detached House ☐
Flat ☐
Town House ☐
Other ☐
(Name) . . . .

Age: Less than 10 years ☐
Between 10-30 years ☐
Between 30-60 years ☐
Older than 60 years ☐
(enter approx age of property below)
. . . .

Construction:
Roof: Pitched ☐ Tiled ☐
Slated ☐
Hipped ☐
Other ☐
(Name) . . . .
Flat ☐ Bitumen Felt ☐
Asphalt ☐
Rubber ☐
Other ☐
(Name) . . . .

(Continued)

Table 9.1 (Continued)

Walls:	External Walls	Brickwork		☐
			Stone	☐
			Rendered	☐
			Cladded	☐
			Other	☐
			(Name	
Construction:				
Walls:	Internal Walls	Plaster/Stud		☐
			Blockwork	☐
			Brickwork	☐
			Other	☐
			(Name)	
Floors:	Ground Floors	Timber		☐
			Concrete	☐
			Other	☐
			(Name)	
	Upper Floors	Timber		☐
			Concrete	☐
			Other	☐
			(Name)	

Accommodation:

Lounge	☐
Kitchen	☐
Dining Room	☐
No. of Bedrooms	☐
No. of Bathrooms	☐
Utility Room	☐
Conservatory	☐

Garage and Grounds:

No Garage	☐
Single Garage	☐
Double Garage	☐
Front Garden	☐
Side Garden	☐
Rear Garden	☐

B: THE PROPERTY AND LOCATION

B2 The Location

Name of Village, Town, City: ....
....

Distance to Town, City: ....
....

Distance to and name of nearest Railway Station: ....
....

Distance to and name of nearest Motorway ....
....

(Continued)

Table 9.1 (Continued)

B3 Circumstances of the Inspection

Weather: Sunny ☐
Overcast ☐
Rain ☐
Snow ☐
Windy ☐

Access for Undertaking Inspection:
Inspection of Exterior linked to those parts
from ground level. ☐
Inspection of Floors limited to owing to
Floor finishes/coverings. ☐
Any other access issues
(Name) .... ☐

C: THE BUILDING

C1 Movement

Signs of Significant Structural Movement ☐
Specify location: ....

C2 Timber Defects

Signs of wet rot ☐
Signs of dry rot ☐
Signs of Beetle attack
Specify location: ....

C3 Dampness

Damp Proof Course Installed ☐
Signs of Rising Dampness ☐
Signs of Penetrating Dampness ☐
Signs of Condensation ☐
Any other forms of dampness ☐
give details: ....

C4 Insulation

Insulation present in roof space ☐
Cavity Wall Insulation ☐

Table 9.2 External elements of building survey

C5: The Exterior

Element	*Description*		*Condition*			*Defects*		*Remedial work*		*Remarks/additional information*
			Good	*Fair*	*Poor*			*Yes*	*No*	
Roof Structure and Covering	**Pitched Roof**					Loose Slates/Tiles		☐	☐	
	Traditional Slated	☐	☐	☐	☐	Missing Slates/Tiles	☐	☐	☐	
	Clay Tiled	☐	☐	☐	☐	Deflection/Sagging	☐	☐	☐	
	Concrete Tiled	☐	☐	☐	☐	Moss Growth	☐	☐	☐	
	Other	☐	☐	☐	☐	Defective Pointing	☐	☐	☐	
	(Name)					Split/Holes in covering	☐	☐	☐	
	Flat Roof					Blisters/ripples	☐	☐	☐	
	Bitumen Felt	☐	☐	☐	☐	Pounding of Water				
	Asphalt	☐	☐	☐	☐	Blocked Outlets	☐	☐	☐	
	Lead	☐	☐	☐	☐	Others	☐	☐	☐	
	Aluminium	☐	☐	☐	☐	(Name)				
	Other	☐	☐	☐	☐					
	(Name)									
Chimneys	Farefaced Brickwork	☐	☐	☐	☐	Flashing Loose	☐	☐	☐	
	Stone	☐	☐	☐	☐	Flashings Missing	☐	☐	☐	
	Rendered	☐	☐	☐	☐	Defective Pointing	☐	☐	☐	
	Other	☐	☐	☐	☐	Damaged Brick/Stone	☐	☐	☐	
	(Name)									
						Missing Cowl/Pot	☐	☐	☐	
						Defective	☐	☐	☐	
						Other	☐	☐	☐	
						(Name)				

(*Continued*)

Table 9.2 (Continued)

C5: The Exterior

Element	*Description*	*Condition*				*Defects*	*Remedial work*			*Remarks/additional information*
			Good	*Fair*	*Poor*			*Yes*	*No*	
Rainwater Fittings (Gutters & Downpipes)	Plastic	☐	☐	☐	☐	Misalignment of Sections	☐	☐	☐	
	Cast Iron	☐	☐	☐	☐					
	Timber	☐	☐	☐	☐	Missing Sections	☐	☐	☐	
	Aluminium	☐	☐	☐	☐	Leaking Joints	☐	☐	☐	
	Steel	☐	☐	☐	☐	Blockage	☐	☐	☐	
	Other	☐	☐	☐	☐	Corrosion	☐	☐	☐	
	(Name)					Fungal Decay				
						Others	☐	☐	☐	
						(Name)				
Main External Walls	Cavity Construction	☐	☐	☐	☐	Defective Painting	☐	☐	☐	
	Solid Construction	☐	☐	☐	☐	Spalling of Finish	☐	☐	☐	
	Brack Faced	☐	☐	☐	☐	Crazing of Render	☐	☐	☐	
	Render Finish	☐	☐	☐	☐	Structural Cracking	☐	☐	☐	
	Cladded Finish	☐	☐	☐	☐	Moss Growth	☐	☐	☐	
	Other	☐	☐	☐	☐	Bulging/Movement	☐	☐	☐	
	(Name)					No Signs of DPC	☐	☐	☐	
						Other	☐	☐	☐	
						(Name)				
External Joinery	Timber	☐	☐	☐	☐	Timber Decay	☐	☐	☐	
Fascias and Soffits	UPVC	☐	☐	☐	☐	Missing Sections	☐	☐	☐	
	Metal	☐	☐	☐	☐	Loose Sections	☐	☐	☐	
	Other	☐	☐	☐	☐	Other	☐	☐	☐	
	(Name)					(Name)				

(Continued)

Table 9.2 (Continued)

C5: The Exterior

Element	*Description*	*Condition*				*Defects*	*Remedial Work*			*Remarks/additional information*
			Good	*Fair*	*Poor*			*Yes*	*No*	
Windows	Timber	☐	☐	☐	☐	Condensation	☐	☐	☐	
	UPVC	☐	☐	☐	☐	Timber Decay	☐	☐	☐	
	Metal	☐	☐	☐	☐	Missing Furniture	☐	☐	☐	
	Other	☐	☐	☐	☐	Warped/Damaged	☐	☐	☐	
	(Name)					Frames/Casements	☐	☐	☐	
						Corrosion	☐	☐	☐	
						Others	☐	☐	☐	
						(Name)				
Doors	Timber	☐	☐	☐	☐	Defective/Missing Furniture	☐	☐	☐	
	UPVC	☐	☐	☐	☐	Timber Decay	☐	☐	☐	
	Metal	☐	☐	☐	☐	Damage/Warping to Frame/ Door	☐	☐	☐	
	Other	☐	☐	☐	☐	Corrosion	☐	☐	☐	
	(Name)					Loose Fitting	☐	☐	☐	
						Other	☐	☐	☐	
						(Name)				
External Decoration	All Parts of Exterior	☐	☐	☐	☐	Minor Flaking/Blistering of Paintwork	☐	☐	☐	
						Major Flaking/Blistering of Paint work	☐	☐	☐	
						Paintwork discoloured	☐	☐	☐	
						Other	☐	☐	☐	
						(Name)				

(Continued)

Table 9.3 Internal elements of building survey

C6: The Interior

Element	*Description*	*Condition*				*Defects*	*Remedial work*			*Remarks/additional information*
			Good	*Fair*	*Poor*			*Yes*	*No*	
Ceilings	Plasterboard/Skin	☐	☐	☐	☐	Crazed/damaged Plaster finish	☐	☐	☐	
	Lath/Plaster	☐	☐	☐	☐					
	Timber	☐	☐	☐	☐	Deflections to ceiling.	☐	☐	☐	
	Other	☐	☐	☐	☐					
	(Name)					Evidence of dampness/water Penetration.	☐	☐	☐	
						Other (Name)	☐	☐	☐	
Floors Ground Floor	Solid concrete	☐	☐	☐	☐	Uneven surface	☐	☐	☐	
	Solid Stone	☐	☐	☐	☐	Rising dampness	☐	☐	☐	
	Suspended timber	☐	☐	☐	☐	Structural cracking	☐	☐	☐	
	Not Accessible	☐	☐	☐	☐	Deflection/bulging	☐	☐	☐	
	Other	☐	☐	☐	☐	Timber decay	☐	☐	☐	
	(Name)									
						Loose/springing floor	☐	☐	☐	
						Other	☐	☐	☐	
						(Name)				

(*Continued*)

Table 9.3 (Continued)

C6: The Interior

Element	*Description*	*Condition*				*Defects*	*Remedial work*			*Remarks/additional information*
			Good	*Fair*	*Poor*			*Yes*	*No*	
Floors First Floor	Suspended timber	☐	☐	☐	☐	Uneven surface	☐	☐	☐	
	Suspended concrete	☐	☐	☐	☐	Structural Cracking	☐	☐	☐	
	Not accessible	☐	☐	☐	☐	Deflection/bulging	☐	☐	☐	
	Other	☐	☐	☐	☐	Timber decay	☐	☐	☐	
	(Name)					Loose/Springing floor.	☐	☐	☐	
						Other	☐	☐	☐	
						(Name)				
Internal Walls and Partitions	Plasterboard/Stud	☐	☐	☐	☐	Rising Dampness	☐	☐	☐	
	Solid brickwork	☐	☐	☐	☐	Penetrating dampness	☐	☐	☐	
	Solid blockwork	☐	☐	☐	☐	Structural Cracking Movement	☐	☐	☐	
	Other	☐	☐	☐	☐	Missing plaster	☐	☐	☐	
	(Name)									
						Damaged plaster	☐	☐	☐	
						Other	☐	☐	☐	
						(Name)				
Internal Joinery Internal Doors	Hollow Core Timber	☐	☐	☐	☐	Hinges Loose	☐	☐	☐	
	Solid Core Timber	☐	☐	☐	☐	Defective/missing furniture	☐	☐	☐	
	Glazed	☐	☐	☐	☐	Damaged/warped sections.	☐	☐	☐	
	Other	☐	☐	☐	☐	Other	☐	☐	☐	
	(Name)					(Name)				

(Continued)

Table 9.3 (Continued)

C6: The Interior

Element	*Description*		*Condition*			*Defects*		*Remedial work*		*Remarks/additional information*
			Good	*Fair*	*Poor*			*Yes*	*No*	
Skirting/Architrave	Timber moulded	☐	☐	☐	☐	Sections missing	☐	☐	☐	
	Plain Timber	☐	☐	☐	☐	Loose sections	☐	☐	☐	
	Other	☐	☐	☐	☐	Split/damaged sections	☐	☐	☐	
	(Name)									
						Wet rot/dry rot affected	☐	☐	☐	
						Evidence of warping	☐	☐	☐	
						Other	☐	☐	☐	
						(Name)				
Internal Decorations	All internal areas	☐	☐	☐	☐	Minor flaking/blistering of paintwork	☐	☐	☐	
						Major flaking/blistering of paintwork	☐	☐	☐	
						Paintwork discoloured	☐	☐	☐	
						Other	☐	☐	☐	
						(Name)				

The third part of the handbook was designed to compile building condition information on the interior of buildings. It was structured in the same format as the second part with six columns relating to element, description, condition, defects, remedial works and any additional remarks/information. By way of an example if the element was related to internal ceilings, the description could be inputted as plastered and painted. It may be deemed that the condition is poor owing to cracked and damaged plasterboard, water damage caused through leakage from plumbing above ceilings and generally poor decoration to ceiling areas. Again, all these particulars would be inputted by ticking the correct boxes and indicating in the remedial works additional remarks/information what the repairs would comprise. Repairs in this scenario may include taking down and replacing areas of plasterboard to ceiling, prior to plaster surface skimming and redecoration. Other remedial work may include attending to leaking pipework causing previous damage to ceiling and again redecoration.

For the purpose of this teaching and learning aid the survey was restricted to the exterior and the internal areas of buildings. It could and in normal practice would be extended to include other elements such as external areas and building services and adapted accordingly. In the latter case this could include electricity installations including wiring, gas and oil supplies and ancillary installations, water supply and installations, drainage, heating and common services.

9.7 Implementing the project

Professor Challender agreed with the course leader of the building surveying module (as part of the building surveying course) at Stockport College that the project could be piloted with students using the handbook as a template to assist them in undertaking building surveys of their own homes. This would then form part of their course assignment. In this regard he prepared some guidance notes in Table 9.4 and conducted a teaching session with the students to fully

Table 9.4 Guidance notes to assist students using the Building Surveying Handbook as a teaching and learning tool

Guidance notes for students using the handbook
1. This handbook is designed to assist students in preparing a domestic building survey following the standard format of 'The Homebuyers Building Survey Report' set down by The Royal Institution of Chartered Surveyors.
2. The handbook will take students through the various stages of the survey process and familiarise them with the scope and extent of information that should be included within their report.
3. Completion of the forms within the handbook is designed to act as a checklist on site and not intended to substitute the preparation of a full written report which students will be required to complete as part of their assignment task.
4. Students are advised to undertake a brief inspection of all internal and external areas before completion of any details on the forms to give them a feel for the building and its condition and to familiarise them with the shape, accommodation and layout of the property.
5. Only Sections A to C of the RICS Homebuyers Report have been included as part of this handbook which covers the property details, and report on the condition of the external and internal elements of the building. The remaining sections D to G which cover such items as mechanical and electrical services, external areas and legal information have been excluded at this stage to keep the report within a reasonable and realistic scope for the students.

explain the methodology behind the handbook and how to make best use of it (see Appendix 6 for session plan).

The teaching session consisted of an introduction to explain the usefulness of the Student Handbook for undertaking Building Surveys and that its purpose had been devised as a teaching and learning aid. The objectives of the session were explained, and these were predicated on students at the end being able to:

1. Name each section of the RICS Homebuyers Survey
2. Explain how the handbook will assist them in compiling building information
3. Complete the handbook information requirements as part of their student assignment

Each section of the handbook and how it relates to an RICS Homebuyers Building Survey in its content was thoroughly explained during the teaching session. Students were tasked to peruse and evaluate the handbook against realia survey documentation to compare and contrast the difference in survey format. The advantages of the handbook and the reasons it has been devised to assist them in carrying out their own survey reports were then explained to the students. They were then asked to suggest their own advantages and benefits that the handbook can create for them on a personal level. Following this, it was explained to students how to complete the various sections of the handbook and the level of information was requested from them under each section/heading. Thereafter, students were tasked to read and analyse each section of the handbook to become familiar with it, in order to complete the contents at a later date, as part of their module assignments. In addition, they were requested to complete a questionnaire for feedback on it for the next session.

9.8 Feedback and evaluation

At the end of the pilot study, once students had completed their building surveys, a questionnaire developed by the lead author was completed by participating students and lecturers to obtain feedback as to the usefulness of the survey handbook. The questions on the questionnaires had been carefully worded to enable an evaluation to be carried out against a list of predetermined criteria that had been devised as measurable outcomes. These were in turn related to the three main objectives in Section 9.4 and would seek to assess the usefulness of the handbook. In addition, a session was conducted with students to review their completed feedback on the questionnaires and to discuss how useful they found the handbook and the session plan is contained in Appendix 4. A similar exercise was undertaken using the questionnaire with college course leaders and professional practitioners in the industry to obtain feedback from a different perspective. The questionnaire is included in Table 9.5 and its purpose was to assist in the evaluation of the project as a teaching and learning tool.

Positive feedback received from all participants included the following:

1. The handbook was considered clear and unambiguous.
2. The checklist and multiple-choice format of the handbook had allowed building information to be easily compiled on site.
3. The teaching and learning guide promoted a logical and methodical approach to carrying out building inspections.
4. As a guide it ensured that each stage of the survey process was undertaken, and no building elements were inadvertently left out.

Table 9.5 Questionnaire to gain feedback from students and lecturers to test the usefulness of the handbook as a teaching and learning guide

Feedback Questionnaire to assess the extent to which the Building Surveying Student Handbook is a worthwhile Teaching and Learning Resource

Name

. . . .

Status (e.g. Student, Teacher, Professional Practitioner)

. . . .

Organisation

. . . .

Please answer the following questions:

1. Do you feel that the survey handbook was clear and unambiguous as far completing the building survey information?

   . . . .

2. Was the checklist and multiple choice format user friendly when you were carrying out the survey and easy to compile the building information?

   . . . .

3. Do you consider that the guide helped you progress each stage of the survey process and thereby ensured that elements were surveyed in a logical and methodical manner?

   . . . .

4. Do you believe that you are now more prepared and able to progress the full written survey report using the information that you have compiled within this teaching and learning resource?

   . . . .

5. Would you have found it more difficult to prepare the full written report if you had not completed this initial task or had the benefit of the handbook guiding you through the survey process, element by element?

   . . . .

6. Would you recommend this handbook as a teaching and learning resource for other students carrying out similar building survey reports?

   . . . .

7. Can you suggest any ways to improve or further develop any aspects of the guide? (e.g. contents, format, scope, presentation)

   . . . .

   . . . .

   . . . .

8. Please outline any other general thoughts or comments as far as the handbook is concerned below:

   . . . .

   . . . .

   . . . .

   . . . .

   . . . .

   . . . .

   . . . .

Signature . . . .

In addition to the advantages from an educational perspective, the students reported that they felt that they were now more prepared and able to progress the fully written survey report using the information that they had compiled within the handbook. Without the guide steering them through the survey process element by element, they confirmed that they would have found it more difficult in preparing the full written report for their assignment.

The professional practitioner felt the guide was probably more conducive to teaching and learning than as a tool for industry in its existing simplistic form. This individual did, however, feel that the handbook could potentially save a lot of surveyors' valuable time on site in collating information more swiftly rather than relying on conventional note taking if developed further.

Some development points were, however, forthcoming from the feedback and they included:

1. Certain questions on the handbook had only one box to tick and could have benefitted from two boxes to record yes or no.
2. Certain questions could benefit from being reworded for clarity.
3. If the handbook was to be developed for professional practice it would need to be reformatted to include more technical details which would normally be required as part of an RICS Homebuyers Report. This would almost certainly include building services and external areas which were excluded from the Building Surveyors handbook.

In addition, with hindsight it was felt that a more quantitative format could have been created for the feedback questionnaires by introducing degrees of responses. For instance, rather than a response of yes or no answers, ask the participants to indicate whether they strongly agree, agree, disagree or strongly disagree to a particular statement. This would have developed and facilitated a more detailed pattern and percentage of responses and this will be something I will seek to introduce in future questionnaires of this kind. Another example of constructive criticism could be that there should have been a traffic light system to deal with the severity of some defects. This could have been predicated on the following:

- Red – indicating defects that are serious and require to be urgently replaced, repaired or investigated, e.g., roof leak
- Amber – indicating defects that are not as serious as those in the Red category but nevertheless need to be replaced, repaired or investigated. Normally these are considered to be non-urgent, but the property must be maintained in the normal way. One example in this category could be missing roof insulation
- Green – Indicating that no repair is currently needed but the property must be maintained in the normal way.

A further category could include NI representing those areas that were not inspected such as drainage installations or underground services.

Other potential inclusions in the scope of the handbook could have included a section on risk from things discovered during the inspection. These would normally revolve around the condition of certain items or building elements which if they were to fail or deteriorate could cause further risks to buildings, people or external areas.

As far as the desired objectives for the project were concerned it is believed the first two were successfully achieved to a satisfactory level as the handbook clearly assisted learners through each stage of the process and familiarised them with the format of the RICS Home Buyers Report. As far as the third objective is concerned, it was felt that the handbook went some of the way to bridging the current gap between education/training gained by students at college and those skills required by industry. However, this was established as being only the tip of the iceberg as it was deemed that much more development of practical skills and learning aids need to be introduced into construction-related curriculum generally to better train and prepare students for employment in the future. It was also established that there also needs to be more consultation between the Royal Institution of Chartered Surveyors, as the professional body for building surveying, and academic institutions, including colleges and universities, to advise on further teaching and learning improvements and be aligned on accredited course content.

9.9 Summary

This chapter has provided a descriptive account of how a simple and easy-to-use handbook was created and developed as a practical teaching and learning guide for surveying buildings. This was intended to provide a logical systematic pathology for conducting building surveys and therein ensure that all construction elements were covered and not missed. In this way it was designed to bridge a gap between academia and professional practice. It was designed specifically for teaching and learning of students on surveying and construction related courses with particular emphasis on building surveys. Justification from observations of the book's co-author Professor Challender and from his experience in curriculum development reinforced his beliefs that not enough practical/site-based skills are being developed in students and too much emphasis placed on construction theory in the classroom. These reflections also led him to the theory that a closer link to professional practice will motivate learners.

Investigations whether there was any evidence to support and give credibility to Professor Challender's beliefs and research data revealed low numbers of building surveying applicants with the required skills resulting in employers having internal skill gaps in their companies. In addition, surveying graduates were regarded as being less than proficient and, possibly most worrying, only 25% of students enrolling in Royal Institution of Chartered Surveyors RICS partnership courses in building surveying went on to become Chartered Surveyors. Accordingly, despite Professor Mole's *Mind the Gap* report (Mole 1997) and the subsequent Royal Institution of Chartered Surveyors Education Reforms of 2001 employers believed that graduates still lack sufficient knowledge.

As a means of addressing the aforementioned gap in building surveying provision as identified by the research, it was felt useful to develop an innovative and meaningful teaching and learning resource via a Building Surveying Handbook. This was designed to assist learners through the survey process and allow them to prepare a Home Buyers Building Survey Report. Furthermore, it would familiarise students with the Royal Institution of Chartered Surveyors Home Buyers Building Surveying Report format to better prepare them for professional practice and therein close the gap between education/training gained by students at college and university and those skills required by industry. The handbook was designed to have elemental headings, following the standard format of the Royal Institution of Chartered Surveyors Homebuyers Building Survey Report introduced in 2001, to give structure and relevance to all student reports. Its main purpose was intended to steer students through a logical sequence as far as the various stages and the pathology/methodology of the survey process are concerned from start to finish. The basis of this handbook, as a reference tool in carrying out the survey on site, was designed to give all the necessary information that students require to enable them to write up their technical report at a later stage. It would also provide learners with a basic familiarity and understanding in undertaking surveys of this type to expand their experience in building surveying. This would then hopefully benefit them in their studies and better prepare and equip them for their chosen vocations in life.

The handbook was essentially made up of three component parts. The first part was designed as an introduction to the survey and to give context on general building information. The second and third parts of the handbook were designed to compile building condition information on the exterior and interior of buildings. It was structured with six columns relating to element, description, condition, defects, remedial works and any additional remarks/ information. The handbook became part of a building surveying module and became part of a pilot study for college students using the handbook as a template to assist them in undertaking building surveys of their own homes. This would then form part of their course assignment. Some guidance notes for students were created to assist them in this assignment task and a teaching session was delivered to fully explain the methodology behind the handbook and how to make best use of it

At the end of the pilot study, once students had completed their building surveys, a questionnaire developed by the author was completed by participating students and lecturers to obtain feedback as to the usefulness of the survey handbook. From an educational perspective, the students reported that they felt that they were now more prepared and able to progress the full written survey report using the information that they have compiled within the handbook. Without the guide steering them through the survey process element by element, they confirmed that they would have found it more difficult to prepare the full written report for their assignment. It was felt that the handbook could potentially save a lot of surveyors' valuable time on site in collating information more swiftly rather than relying on conventional note-taking if developed further and was deemed an important teaching and learning tool for students.

9.10 Conclusions and recommendations

In general terms the handbook was felt to be a useful, simple and innovative teaching and learning aid for worthy inclusion within the curriculum. This is an opinion that was shared by the course leader at Stockport College who planned to incorporate it as part of the course programme. It was felt that it could offer students a useful tool as relative beginners to the task of carrying out complex building surveys and hopefully narrows the gap between traditional classroom teaching and those skills required by professional practice. From experience building surveys of this kind can be bewildering when undertaking them for the first time and it is believed that the handbook will hopefully make the process more comprehensive for students in the future.

It has future development opportunities for professional practice as much for education in that it presents a more effective and efficient means of collating building information on site. It has the potential to become integrated by way of handheld computers or smartphones for instance to electronically input details on site rather than rely on hard copies. This technology could be networked and downloaded on to a software system that could effectively prepare a report automatically. This would make the whole process so much more efficient and greatly reduce the resources and time currently expended by surveyors and administration staff in preparing the completed report. This is particularly relevant in the current economic climate, in construction and surveying especially, where companies are seeking alternative ways to reduce costs and become more competitive accordingly.

In terms of developing the handbook further, feedback from the pilot study and constructive criticism from participants was sought. Several recommendations emerged from this feedback for future development including:

1. Reformat to include more technical details which would normally be required as part of an RICS Homebuyers Report particularly including building services and external areas which were excluded from the Building Surveyors handbook.
2. Introduce degrees of responses to certain questions.
3. Include a possible traffic light system to deal with the severity of some defects.

Furthermore, a potential inclusion to the scope of the handbook could add value to it with a section on risk from things discovered during the inspection. These would normally revolve around the condition of certain items or building elements which if they were to fail or deteriorate could cause further risks to buildings, people or external areas.

References

Construction Skills. (2003). National employer skills survey. Available at: www.constructionskills.net/research/researchacticity/nationalandregionalreports (Accessed 28 April 2010).

Hoxley, M. and Wilkinson, S. (2006). RICS education reform and building surveying: The employers' view. In: Sivyer, E. (ed.) *COBRA 2006, Proceedings of the Annual Research Conference of the Royal Institution of Chartered Surveyors*, University College London, 7–8 September. RICS.

Lavelle, D. (1992). Promoting student learning and motivation via a systematic introduction to the use of IT. In: Sloan, B., and Schofield D. (eds) *Proceedings of the Conference: Computer Based Training in Property and Construction*. University of Salford.

Mair, W. (2005). *Built environment group, APC competency review, building surveying faculty*. RICS.

Mole, T. (1997). *Mind the gap-report on the rationale and expectation of the building surveyors division in the education and training of chartered surveyors*. RICS.

RICS. (2010). *Keep learning – a framework for building surveyors*. RICS.

Wilkinson, S. and Hoxley, M. (2005). The impact of the 2001 RICS education reforms on building surveying. In: Sidwell, A.C. (ed.) *The Queensland University of Technology Research Week International Conference*, Brisbane, 4–5 July. QUT.

Chapter 10

The relationship between maintenance management and environmental sustainability

10.1 Introduction to chapter

This chapter will start by articulating a context for the need to rethink maintenance management strategies around environmental sustainability, to assist in the commitments made by many countries around the world to net-zero carbon targets. In this regard it will outline the contributions made to promoting 'green agendas' by the lead author in his many different conference presentations. Environmental sustainability and its relationship to future estates and maintenance management strategies will then be discussed. Some of the building accreditation schemes in the UK will be referenced including Passivhaus and the British Research Environmental Assessment Methods (BREEAM). In addition, some of the issues and areas that can reduce environmental impact will be examined around building designs, reduction of waste, ecology and biodiversity.

The importance of energy efficiency will then be covered alongside the importance of generating accurate consumption data for utilities through building management systems and inductions and training for maintenance teams in use of such systems. The potential energy saving initiatives will then be discussed including examples of LED lighting, different methods of construction and renewable technologies. The chapter will include a short case study of a research facility, namely Energy House 2.0, which the lead author procured for diagnostics testing on buildings within a controlled environment. It will examine how this type of facility will benefit buildings in the future in energy conservation and therein generate cost and carbon savings.

Education and training for staff including building occupant will be covered to ensure that buildings are used correctly and to improve the way environmental strategies are implemented. In addition, this area will address cultural barriers that might exist within organisations associated with old and established ways of working, fears and anxieties of staff and how to overcome these issues. Life cycle and longevity of buildings will be analysed and the notion of 'future proofing' buildings. This area will touch on the importance of quality control in design and construction processes and the need to procure buildings which are flexible and versatile enough to be converted for alternative uses in the future. Following on from this, embodied carbon will be examined from a building perspective. It will reference a software tool which the lead author was involved within namely 'CarboniCa' and explain how this can assist building clients and decarbonise local communities. Furthermore, life cycle costings and the need to rethink financial decision-making will be debated and the importance of considering OPEX costs alongside CAPEX costs in decision-making.

Environmental impact assessments will be covered, and the chapter will describe what they comprise, how they can be used and their importance in considering the impact of redevelopment. The different steps in how they can be prepared will be described. Following on from this,

DOI: 10.1201/9781003358800-10

the various issues and technologies around sustainable building design will then be introduced. This will cover available renewable technologies such as photovoltaics, active solar panels, green roof, wind turbines, ground and air source heat pumps, rainwater harvesting and grey-water recycling. The chapter will then look at building envelope considerations for sustainable construction including passive solar design, window positioning and design to reduce heating and cooling, thermal massing, daylighting and indoor environments. Following on from this the important area of waste management in maintenance and estates strategies will be introduced. This will explain alternative approaches including avoidance and reduction, reuse and recycling, energy recovery, disposal or treatment.

Retention of buildings for creative reuse will be analysed as well as the benefits that retaining existing buildings and repurposing them for alternative accommodation can bring. Some of the integral aspects of intelligent buildings will be described and the advantages that these could bring for more effective space management and utilisation. Reference will be made to a case study of the Spine Building in Liverpool and where technologies deployed within its design can have the greatest impact on the comfort and well-being of its occupants.

Finally, the ways and means by which living buildings and outdoor environments and public realm can have on reversing carbon emissions, alongside the well-being benefits for individuals will be identified and discussed.

10.2 Context for the need to rethink maintenance management strategies

The agenda for change to reduce carbon emissions and attain global targets on such initiatives as net-zero carbon cannot be disentangled for future maintenance management strategies. The reason for this is that environmental sustainability and energy efficiency will have to be at the heart of any future maintenance management strategies if they are to be successful. Organisations have been facing for many years inflationary rises in utility, construction and maintenance costs, affecting their profitability. This predicament could be compounded by any future government interventions to levy companies that do not meet minimum environmental thresholds. Notwithstanding this premise, some current practices in maintenance management, certainly in the UK, are still currently out of step with national aspirations and targets related to wastage reduction and use of renewable 'green' energy sources. Hopefully some of the suggestions, ideas and recommendations included in this chapter will present the impetus and catalyst for rethinking current practices and a move to more environmentally sustainable ways of working.

10.3 Brief background to the contributions of Professor Challender towards the agenda for change in environmental sustainability

The author has been a life-long advocate of environmental sustainability. This is demonstrated in his roles in both academia and industry. As a visiting professor, he has for many years educated students on the need for change, the current technologies available to industry and the alternative estates and maintenance strategies that can be deployed. He also has participated as a key speaker at many UK and global conferences to present and discuss his various research papers related to environmental sustainability. Two international examples included World Renewable Energy conferences held in Ajman, United Arab Emirates in November 2022 and Kuala Lumpur, Malaysia, in July 2023.

For the Ajman conference he developed a collaboration toolkit designed to assist organisations specifically tailored around estates, construction and maintenance strategies to improve their current practices. This was predicated on managing organisational relationships with their supply chains and external partners to assist them in developing and meeting new benchmarks in environmental sustainability. The paper, entitled 'Achieving sustainability and environmental enhancements through a collaboration toolkit to deliver new sustainability strategies for The Emirate of Ajman, UAE', was very well received at the conference. In addition, he received an award for his academic contributions, which was presented to him by His Excellency, the prince of Ajman and this is included in Figure 10.1.

Figure 10.1 Sustainability award being made to the lead author by the prince of Ajman in November 2022

Figure 10.2 Presentation by Professor Challender at the World Renewable Environmental Conference (WREC) in Kuala Lumpur in July 2023

For the Kuala Lumpur conference in 2023, namely the World Renewable Environmental Congress (WREC), he was invited to speak about his paper 'Case study analysis of improving environmental ethics using a collaboration toolkit' (Challender 2023). This included an analysis of some of the work he had undertaken at the University of Salford, as Director of Estates. Alongside his environmental sustainability he was instrumental in developing a new strategy for the university, which was branded '*Striving for a More Sustainable Salford*' and is included in Appendix 7. Amongst other initiatives this involved appointing an external energy partner to assist the Estates Department with meeting many different environmental targets. This case study will be included in the next chapter. He also spoke at the same conference about the importance of environmental ethics for organisations and them meeting their corporate social responsibility commitments and charters. The presentation in Kuala Lumpur was welcomed by many renowned delegates and academics across the world and a photograph of the event is included in Figure 10.2.

10.4 Environmental sustainability and its relationship to future estates and maintenance management strategies

The need to create more energy efficiency and improve environmental sustainability to reduce emissions is clearly part of a wider global agenda to reduce carbon generation. There has never been a better time when the built environment can play a pivotal and instrumental part in this pursuit, given that it accounts for approximately 38% of annual global carbon emissions. Certainly, in the UK there has been for many years a noticeable increase in participation in accreditation schemes linked to energy efficiency and sustainability and these have included Passivhaus. This scheme was designed to deliver net-zero solutions for developments whilst providing a high degree of comfort for building occupants using relatively very little energy for cooling and heating. This is crucial in meeting carbon targets set by organisations. This represents a huge challenge for the industry in changing a long history of antiquated working practices and old outdated traditional forms of construction coupled with utilisation of high carbon-emitting materials. Alongside Passivhaus, the British Research Environmental Assessment Methods (BREEAM) is another assessment tool used to accredit how environmentally sustainable buildings are. For many years it has become the standard deployed by a lot of local authorities and other public sector organisations for new buildings and refurbishments. The accreditation involves the appointment of a BREEAM assessor at the early design stages of buildings. Their role is to assist their clients in gaining BREEAM credits in ten different categories according to the performance of buildings. These categories are related to environmental principles and applicable to both commercial and domestic properties. Points are allocated to buildings against each category, and these are totalled to produce a single overall score. The score mark then dictates the level of BREEAM accreditations which vary between Pass, Good, Very Good, Excellent and Outstanding. The scheme has been widely regarded as a recognition kitemark or badge of honour for employing environmental principles in both the way they have been constructed and their future environmental impact.

Some countries and organisations around the world have committed to net-zero carbon pledges in a concerted effort to address the growing problem of global warming, and therein reverse the current trend. Considering the increasing impetus to uphold their reputational and ethical credentials and standings from an environmental perspective and meet targets for reduced carbon emissions, organisations are looking towards new strategies. Such strategies should address a whole array of issues around environmental impact, not purely for new building but considering alterations, maintenance and refurbishment works on existing buildings. According to Thomas (2013) there are many ways and measures to achieve significant carbon reductions, in the built environment, and these are illustrated in Figure 10.3. Wherever possible construction professionals including surveyors, architects and engineers should consider including facilities managers at the design stages for buildings and having conversations with those individuals tasked in maintaining buildings. This could help them to understand the implications for maintenance on some aspects of design or specification.

The lead author was involved in developing a major change in environmental strategy using some of the initiatives referred to in Figure 10.3, during his employment as Director of Estates with the University of Salford. This was to raise the credentials of the university and meet its target for net-zero carbon by 2038.

10.5 The importance of energy efficiency in future maintenance management strategies

10.5.1 Reducing carbon generation and meeting net-zero carbon emission targets

When considering maintenance management, it is imperative to consider energy efficiency as part of any policy, strategy or maintenance regime. As referred to at the start of the chapter, the built environment accounts for over 38% of global carbon output. Accordingly, one cannot consider maintenance management of any large estate without looking for a radical step change.

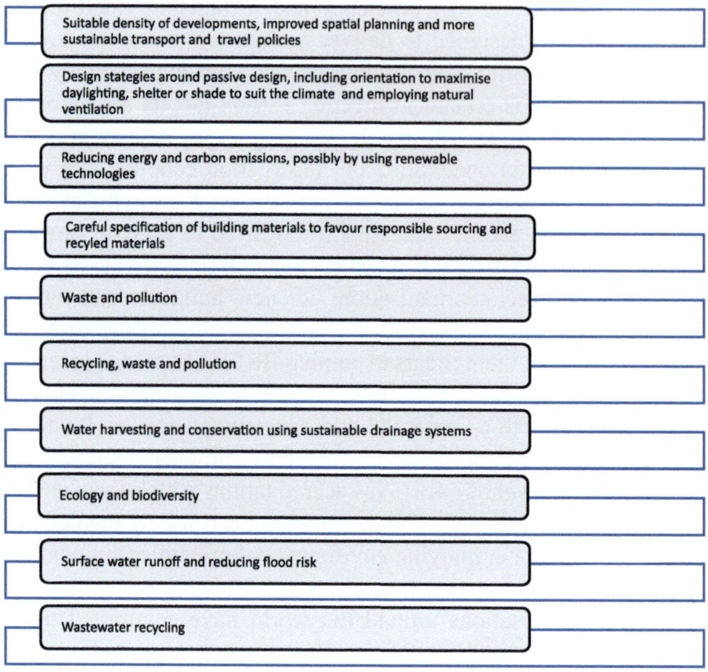

Figure 10.3 Some of the issues and areas that can reduce environmental impact

This is especially pertinent in the case of the UK, given that the government has committed to a new zero-carbon pledge by 2050. Meeting such a pledge will have its challenges given that in the case of housing stock in the UK, Green Building Council has estimated that approximately 80% of homes has already been built. For this reason, energy measures and technological improvements should not simply apply to new buildings but to a programme of upgrading and retrofitting of existing buildings across the country.

10.5.2 The importance of reliable energy data

Energy management represents a key objective in the pursuit of environmental sustainability for any organisation and should therefore influence the way buildings are designed, constructed and used throughout their occupancy. This objective has been accentuated by the continual rise in energy costs have driven innovation to create alternative building and engineering technologies, with an increased focus on renewable energy supplies. Notwithstanding this drive for innovation and improvement, organisations need the tools to be able to accurately calculate, monitor and predict their energy consumption as a measure of performance. This sometimes relies on investing in building management systems, commonly referred to as BMS, that can give accurate data on how much energy buildings have used or are projected to use over a period for cost comparison and future budgetary planning. The data should highlight whether certain buildings are using proportionally more energy than other buildings over the same periods of time. Variances in energy consumption could be explained by the construction of certain buildings whereby older buildings tend to suffer more from heat conservation and may incorporate old and inefficient building service installations.

10.5.3 Inductions and training on new energy management systems for maintenance teams

In some cases, however, this data may pinpoint problems that would otherwise not have been uncovered, sometimes on newly commissioned buildings where facilities managers have not been fully trained or inducted in the correct operation of building services. This is particularly the case where newly installed building services are sophisticated and complex which can present many challenges from a maintenance management perspective. In such cases, facilities and maintenance managers may be running systems in an inefficient manner, potentially leaving certain plants operating when not required, owing to a lack of knowledge and awareness of system controls. The lead author has come across this scenario on many occasions when cooling plant was being operated in the coldest of months and heating plant left on during heatwaves.

10.5.4 Potential energy-saving initiatives

If buildings are constantly performing badly from an energy efficiency perspective, measures should be considered to invest in measures, small and large, in a concerted attempt to improve energy management. Figure 10.4 illustrates some of the measures that could be adopted as part of estates and maintenance management.

In considering some of the energy efficiency measures included in Figure 10.4, it is important for organisations to progress research and development opportunities wherever possible. This could include new methods of construction and forms of renewable energy systems to

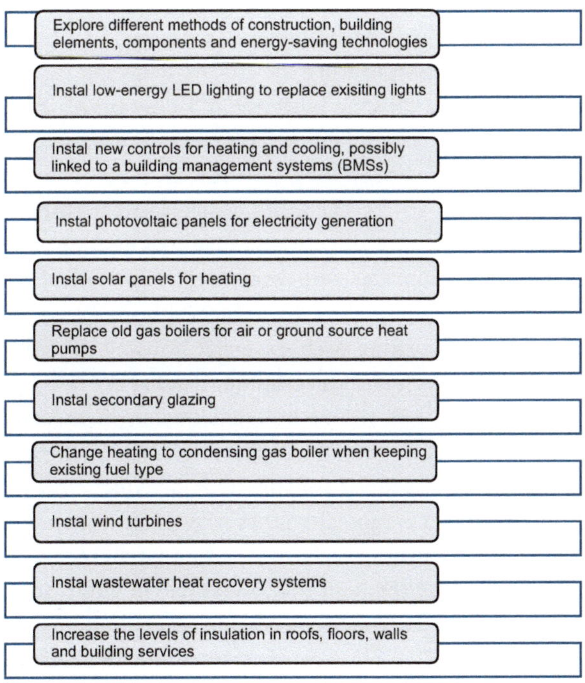

Figure 10.4 Energy efficiency measures

investigate the effects that any potential new innovations and technologies can have for reducing energy consumption of buildings. One such research and development project which the lead author was responsible for procuring related to the Energy House 2.0 project at the University of Salford and will be included next as a short case study.

10.5.5 Case study: Energy House 2.0 project

Project Value: £16m
Client: University of Salford (Estates Department)
Architects and design engineers: Aecom
Contractor: Bowmer and Kirkland
Contract period: October 2021 – March 2022

The project known as Energy House 2.0 was completed in March 2022 by the Estates Department of the University of Salford, under the directorship of the lead author. It took a further ten months to fit out and officially opened in January 2023. It was mostly funded directly by the university, but external funding was received by the Higher Education Funding Council (now Office for Students) and the European Regional Development Fund (ERDF).

The facility represents the largest research centre of its type in the UK in providing controlled environments for testing different types of buildings, structures and technologies. In this regard the building was designed to be able to undertake environmental testing with temperature variations between −20 degrees C to +40 degrees C, with a required tolerance with +/− 0.5 degrees C. Furthermore, it was designed to simulate other controlled weather conditions associated with wind, snow, solar radiation and rain over long periods of time. The building took over two years to design owing to its technologically sophisticated design especially around building services to create the correct levels of heating and cooling required for testing. It essentially comprises two very large super insulated environmental chambers, each able to accommodate two detached houses.

The university has been working with external housing partners involved in the scheme, to deliver the research and development outputs. Two houses in 2023 had already been constructed by the house builders in one of the chambers and their designs have incorporated many new innovative construction technologies. These houses have been fitted with special sensors that provide valuable information via a central database on how certain elements are performing under the extreme conditions. The walls of both houses have been constructed with super insulated envelopes to reduce heat loss through the structure. In addition, there were other energy-saving building services deployed within each building including environmental heat pumps and building management systems. Over a period of time, the testing of these houses within the chambers will reveal how these technological advancements can assist in energy conservation for homes. It is then expected that the home builders can use the data, to generate new improved homes in their future housing programmes. In this way it is hoped that the facility can spearhead a revolution in the design of new buildings to greatly improve energy efficiency for the future and meet the UK's net-zero carbon commitments.

This important project, since its completion, has won many awards including Building Project of the Year in 2022, for the North West Region of the UK. It also has been useful from a teaching and learning advocacy perspective and the lead author has himself escorted students around the facility to educate them in technologies and research. Please see Figure 10.5 showing a visit by the lead author and university students to the facility in 2023.

Figure 10.5 The lead author and university students visiting Energy House 2.0

10.6 Education and training for staff

Applying environmental policies especially around energy conservation does not purely rely on some of the physical measures referred to in Figure 10.4. It is important that building users are influenced to preserve energy in their workplaces. There are however several barriers to be considered in this pursuit. These barriers relate to lack of awareness of the benefits of energy conservation, lack of information and distrust in new information and adoption of new technologies. To address these barriers, estates and maintenance management strategies and policies should seek to educate building occupants in the correct use of buildings. In simple terms this may involve stressing the importance of switching light switches off when rooms are empty. It could also relate to asking staff to turn off computers and other electrical equipment in offices not being utilised at certain times. Furthermore, in the case of estates maintenance staff any energy conservation strategies should properly equip them with the training and knowledge that they require to run their estates efficiently. This is crucial in allowing estates staff to manage their maintenance programmes more effectively through continual professional development. This may entail human resource investment in both short and long courses for estates staff to equip them with new competencies and qualifications in environmental sustainability. In this way they may be more averse in exploring and implementing some of the improvement measures, linked to estates and maintenance management, in Figure 10.6.

Sometimes cultural barriers associated with old traditional ways of working may be difficult to break through in this regard, with some staff feeling uncomfortable and threatened by new

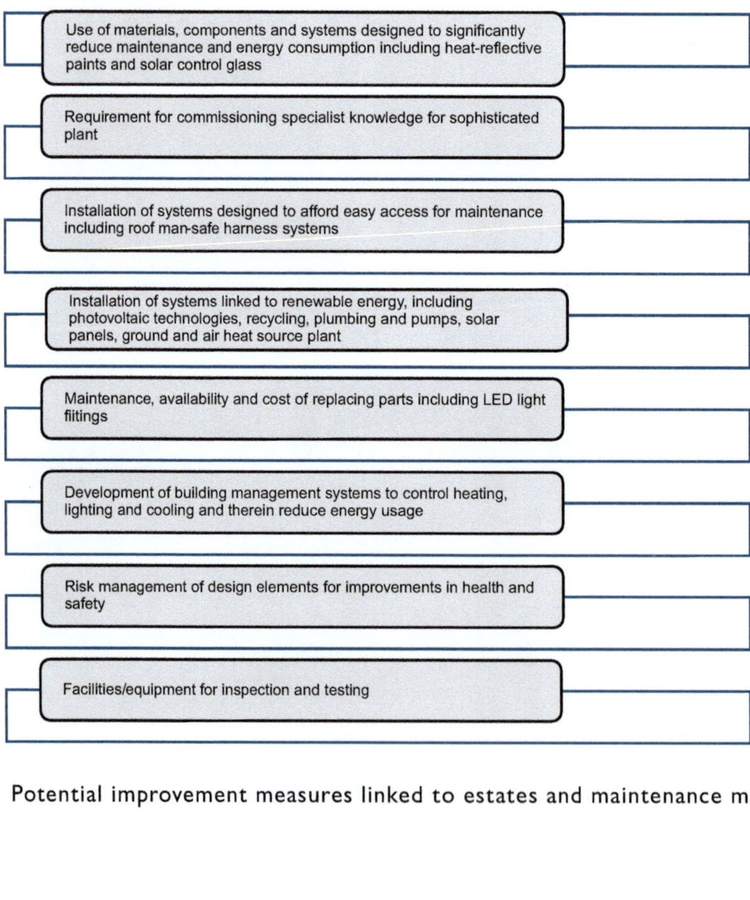

Figure 10.6 Potential improvement measures linked to estates and maintenance management

initiatives and technologies. Continuous professional development of estates staff, to educate them on the benefits and equip them with the technical knowledge to manage such systems, is therefore imperative.

10.7 Life cycle and longevity of buildings

The life cycle of buildings relates to the period from their inception at the start of the conceptual design process to their demolition at the end of life. In this regard it is regarded as a 'cradle to grave' approach which considers all stages in the life of buildings. The lifespans of buildings can vary tremendously and will depend on many different factors. These include the extent and quality of repairs and maintenance that they will receive which can affect their condition and influence their deterioration and longevity. The quality of their design and construction will also play a major part in how they perform and the duration of their occupation in use prior to demolition. Poor-quality buildings will tend to present many more difficulties and costs associated with maintenance management. They may also be expensive to operate from an energy management perspective if they have not been designed with any environmental sustainability considerations. Moreover, those buildings which have been constructed with high-quality design, materials and integral elements, components and systems should be able to withstand the test of time to a greater degree. However good building design and construction of buildings is only one factor which can affect their life spans. The degree to which buildings are flexible and versatile enough to deal with changing requirements made of them throughout their lifecycle is very important. An example may be a commercial building, which owing to market conditions in a particular area is no longer viable in its current use. There may therefore be pressures to change its use to residential to present a more attractive yield for their building owners. One example of this is the resurgence of city centre living in the UK in recent years and where residential markets have dictated higher demand and returns. This has predicated the creative reuse and conversion of many former office buildings into residential flats to take advantage of the market shift away from commercial accommodation. Considering the issues raised, estates and maintenance management teams should be mindful when proposing recommendations for new developments, refurbishments and alteration works to ensure that future flexibility, adaptability and versatility is strongly considered.

Over the course of the lifecycles of building there are a lot of natural resources consumed in the design and construction period but especially during the occupation periods. In addition, buildings during their life cycle generate significant amounts of waste and carbon emissions. For these reasons the construction industry has made concerted efforts in recent years to curb the consumption of raw materials, reduce their 'embodied energy' in the construction process and the energy used during their occupation periods. The term embodied energy according to Guidetti and Ferrara (2023) relates to total past expenditure of non-recoverable energy that buildings have consumed in their construction. The calculation of embodied energy considers other factors other than just construction, and these include the manufacturing processes associated with building materials, elements and components coupled with transportation costs. In summary embodied energy encapsulates the quantity of energy required with all the activities associated with the production process, for example energy to quarry, transport and manufacture building materials plus energy used in construction.

It is important to understand how much embodied energy buildings have consumed during their development stages, as reflection for future improvements, and to consider more energy efficient modes and methods of construction. In this pursuit some organisations have developed their own software which can measure the embodied energy of a particular building at the design stages. These advances in information technologies can allow designers to amend their designs and specifications according to calculations of embodied energy to compare and contrast which forms of construction yield the lowest levels. One such system was developed by the Morgan Sindall Group in the UK and was aptly named ‘CarboniCa’. They described it as an in-house digital carbon reduction tool that will enable them to assist their clients and decarbonise local communities. This tool is particularly useful in that it can measure carbon emissions across the whole life of buildings, to ensure that any potential carbon outputs can be correctly managed and reduced during the life cycle of buildings. This system was demonstrated to the lead author as a client of Morgan Sindall using live projects which he was overseeing at the University of Salford. It was evident during this time that small design changes can lead to significant savings in carbon in some cases. These efficiencies have led to more naturally sourced materials being selected as opposed to concrete, glass and steel which incur large amounts of energy in their manufacturing processes. The CarboniCa software also looks at recycled materials in design specifications as these may assist the overall embodied energy assessments. The software can also differentiate between certain building systems associated with high maintenance costs and increased energy consumption. This enables reflection on the utilisation of different building methodologies and technologies in a concerted attempt to avoid those that generate higher carbon emissions.

10.8 Life cycle costings and the need to rethink financial decision-making

In estates maintenance management, decision-making should be predicated on not just considering capital costs, commonly known as CAPEX costs, associated with new build construction, refurbishments, improvement, retrofit and alteration works. There should also be close consideration of operation and maintenance costs, commonly known as OPEX costs, as these can amount to most of the costs incurred, across the lifetime of building assets. Some OPEX costs are included in the iceberg illustrations in Figures 10.7 and 10.8 and these will include all ruining costs such as maintenance, energy, water, remodelling and eventually demolition. Life cycle costings (LCC) in this way involves looking at all capital costs associated with construction and commissioning buildings but also all operational costs that are projected to be incurred through the life spans of buildings. Notwithstanding this premise, sometimes the financial structures and budget setting within organisations are too closely geared to considering only CAPEX costs associated with carrying out development or estates activities. In this way estates projects are approved or rejected on business cases predicated on their initial development costs or timescales for implementation. In such scenarios, there is little, or no consideration given to the overall cost of these propositions in terms of life cycle cost analysis or environmental sustainability. This is a major deficiency and one which needs more focus and attention, especially at a time when organisations are facing increasing energy costs and looking for more sustainable estates strategies.

10.9 Environmental impact assessments

In estates and maintenance management there will be instances where the environmental effects on decision-making need to be understood and considered. This may relate to a programme of redevelopment where a master plan is being created or simply when considering plans for a new build or refurbishment project. In such cases it may be useful and possibly a planning authority requirement that an Environmental Impact Assessment be prepared.

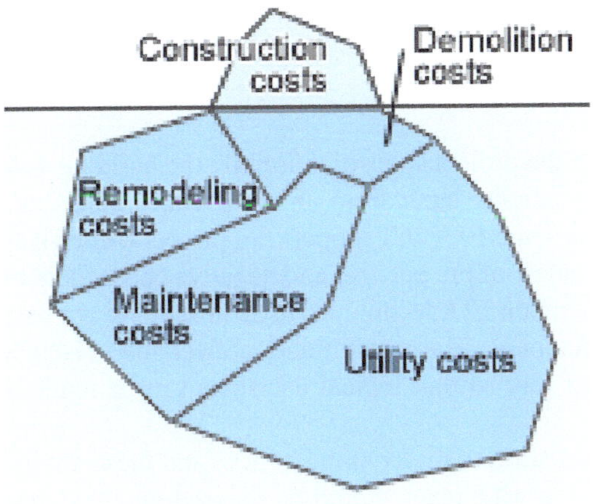

Figure 10.7 Range of operation and maintenance costs

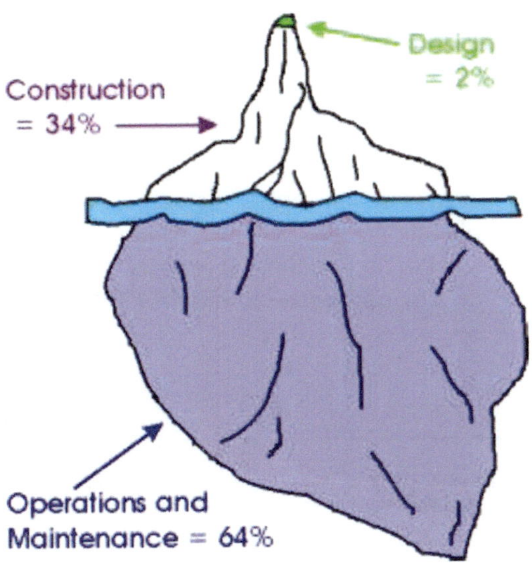

Figure 10.8 Comparison between operation and maintenance costs to design and construction costs

An Environmental Impact Assessment (EIA) is a tool used to assess the significant effects of a project or development proposal on the environment. EIAs make sure that project decision-makers think about the likely effects on the environment at the earliest possible time and aim to avoid, reduce or offset those effects. The main goal of EIA is to conserve the environment and bring out the best combination of economic and environmental costs and benefits. An Environmental Impact Assessment is important for the following reasons:

- To disclose significant environmental effects of proposed projects to decision-makers and the public

- To identify ways to avoid or reduce environmental damage
- To prevent adverse environmental impacts by requiring implementation of feasible alternatives or mitigation measures

It is the responsibility of the project sponsor, normally the building owner, to make the EIA but they normally require specialist input from an experienced consultant. Accordingly, it is the consultant's role to review the EIA and support the project owner if necessary. The EIA shall comprise a systematic review of the positive and negative consequences of the proposed project on the environment. The main EIA techniques used in scoping are baseline studies, checklists, matrices and network diagrams. These techniques collect and present knowledge and information in a straightforward way so that logical decisions can be made about which impacts are most significant.

There are normally six steps to undertaking an EIA and these are included in Figure 10.9.

The characteristics of an EIA team should be focused on achieving forecast environmental impacts early in the project planning and design process. They should tailor projects to the local community and be aware of any negative impacts and identify ways to mitigate these. In addition, they should present the predictions and options to decision-makers at all stages throughout projects. The characteristics of a good EIA will normally include the scope of items included in Figure 10.10.

In terms of scoping EIAs checklists should be used to avoid any stages not being considered. Preparing these checklists will involve drawing up a list of activities and then looking at the possible areas within the project area that will be affected by the activity. The scope of an Environmental Impact Assessment usually includes a survey and review of prior studies done by others to describe the physical, chemical and biological conditions of the study area, followed by an evaluation of their individual impacts. Following on from this, it is then important to characterise the environmental aspects within that area that will be impacted (primary impact).

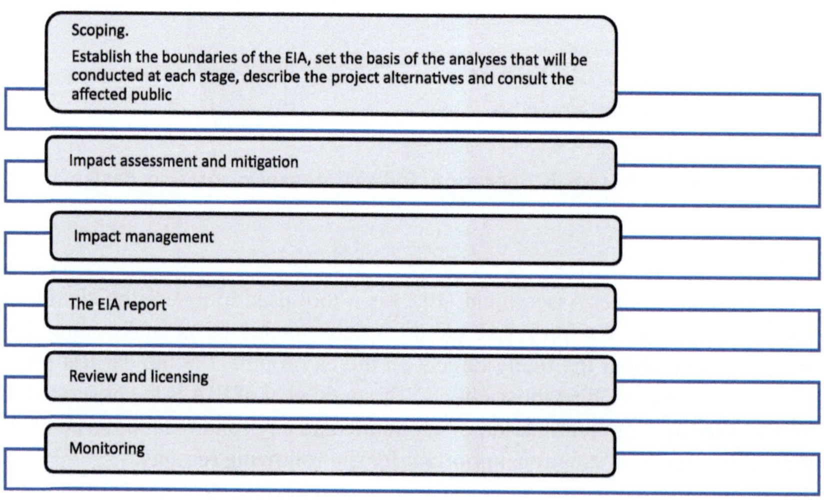

Figure 10.9 The six steps to undertaking an EIA

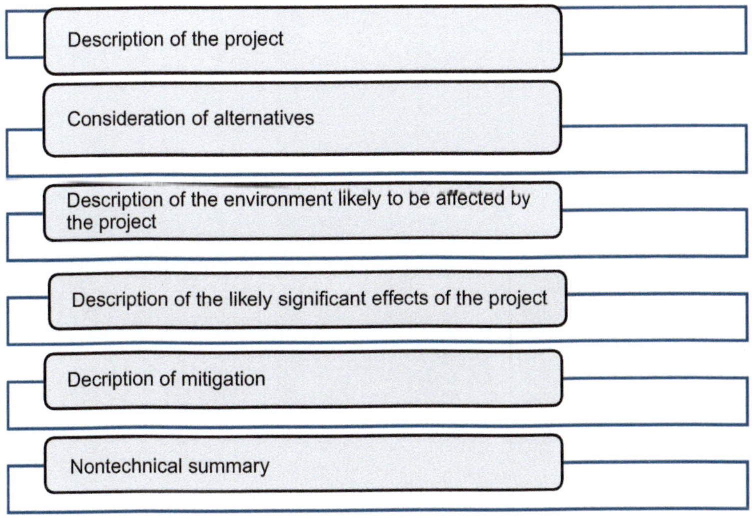

Figure 10.10 Scope of items included in a good EIA

10.10 Sustainable building design

When organisations are considering their programmes for capital investment and redevelopment, they should be mindful of the many different ways and means that they can improve their estates through good environmentally sustainable design. The objectives for sustainable building design include energy and resource efficiency, pollution prevention, harmonisation with the environment and to provide an integrated and systematic approach to environmental management. Resource efficiency tends to focus on the ways in which materials are used can optimised throughout the life cycle of buildings and can be categorised into three main areas of manufacture, operation and disposal.

The manufacturing category looks at waste reduction, pollution prevention, use of recycling materials and embodied energy reduction whereas the operation category looks at energy efficiency, water treatment and conservation, nontoxic uses, renewable energy resources and attaining a greater longevity for buildings. Finally, the disposal category is concerned with end-of-life sustainability for buildings at the demolition stage. This includes looking at materials from demolition becoming biodegradable, reusable and recyclable waste.

The inclusion of many renewable energy and sustainability technologies should be considered wherever possible. Some of these are listed next.

(1) Photovoltaics

Photovoltaics using solar cells convert radiation energy from the sun to generate electrical power. Photovoltaic panels can be mounted on roofs and sometimes walls of buildings or installed as freestanding structures on the ground. In the latter case 'solar farms' have been developed in many countries around the world to provide a valuable power source for their national grids or local communities. An illustration is included in Figure 10.11.

Photovoltaic cell technology works by converting solar energy into a flow of electrons. This electrical energy can be then used to charge batteries or power equipment. They are

Figure 10.11 Photovoltaics

manufactured from very thin semiconductor materials, protected and sandwiched by layers of glass and plastics to form panels. They have become more efficient in recent years and used extensively across the globe which has made them become a cost-effective solution for both commercial and domestic applications.

(2) Active solar panels

These can sometimes be confused with photovoltaics, which are frequently referred to as solar panels also. However, active solar panels are very different in the respect that their technology is not as sophisticated as photovoltaics, and they convert thermal energy from solar radiation into heat rather than electricity. Their installations use solar energy to heat a fluid and then transfer the solar heat from the liquid into the internal spaces or alternatively a storage system for later use. In this regard they do not seek to compete with photovoltaics but instead can present a complementary solution where both are installed on buildings. Like photovoltaics they represent a cost-effective solution especially in domestic uses where government grants have in the past been available to cover some of the supply and installation costs. An illustration is included in Figure 10.12.

(3) Green roofs

Green roofs are becoming more common around the world in recent years owing to the many sustainable advantages related to biodiversity and ecology. They are normally constructed from sedum, but sometimes moss or sempervivum, and laid on new or existing roofs in a mat or carpet formation (see Figure 10.13).

As well as the benefits for wildlife through their ecological and biodiversity credentials, they also offer an opportunity to save energy through insulation. In addition, they can also give many more benefits as listed in Figure 10.14.

Despite the advantages of green roofs there can be sometime be practical considerations and problems to overcome. They can add considerable additional loading to new and

Figure 10.12 Active solar panel installation

Figure 10.13 Green roofs

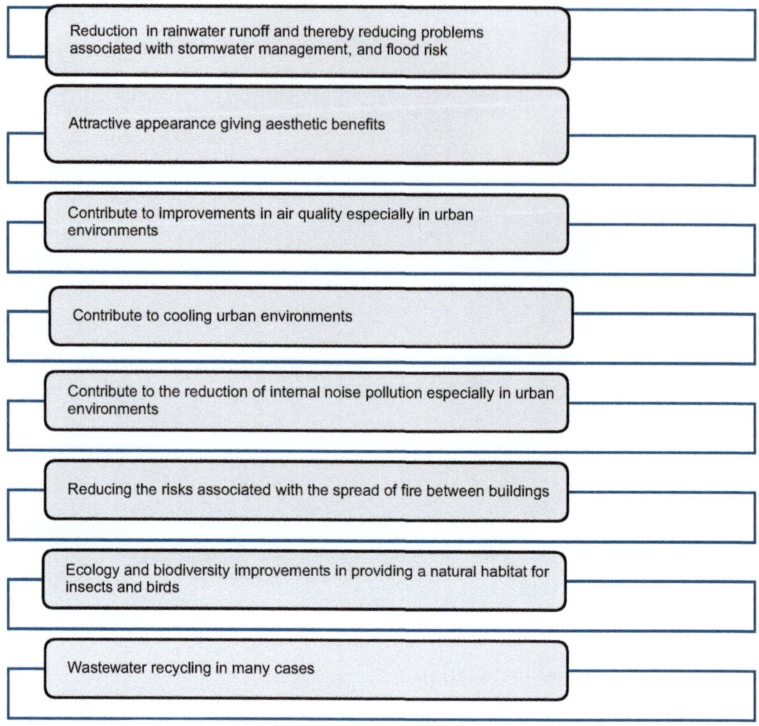

Figure 10.14 Benefits of green roofs

existing roofs, owing to their reliance on a large buildup of soil and drainage substrates in their construction. Furthermore, this can add considerable cost to new buildings and when fitted retrospectively on existing roofs which may not be able to withstand the additional loading in their present forms. Structural engineers' advice will therefore be required to assess the feasibility in such cases and where additional structural support may be required.

(4) Wind turbines

Wind turbines are pieces of apparatus which can take on many different forms, small and large and can be installed as free mounted structures or attached to buildings. They work by converting the kinetic energy of the wind into electrical energy. This is facilitated as the wind energy turns propeller blades around rotors which are connected via shafts to generators which create the electricity. They can be installed as large installations on land or offshore and governments especially in the UK have been promoting alternative clean energy produced to meet their environmental sustainability commitments. There has been controversy in some instances where objections have been received from communities on the grounds of their appearance, especially in areas of natural beauty, and the potential threat to wildlife. However, the UK government is pressing ahead with many development programmes related to wind energy and using the planning system to support this proposition in most cases. An illustration is included in Figure 10.15.

(5) Ground and air source heat pumps

Ground source heat pumps operate by extracting heat from the grounds of a building and transferring this to heat hot water cylinders, radiators or underfloor heating. They work by

Figure 10.15 Wind turbines

circulating fluid which flows through pipes buried underground in coiled loops, which can be configured horizontally in trenches or vertically within boreholes. Heat is transferred from the ground and absorbed into the fluid which then traverses through into a heat pump via a heat exchanger. This process increases the fluid's temperature and then the additional heat is transferred to water. In the UK grants have been available in the past to promote the installation of this clean technology and to phase out traditional gas boilers to meet its 2050 net-zero carbon targets. In some cases, this has made it a more affordable solution given that the technology can save considerable amounts of energy.

Air source heat pumps work in a similar way to their ground source contemporaries but transfer heat to water from outside air rather than the ground. They are unlikely to save as much energy as ground source heat pumps, but installation costs are normally significantly less. An illustration is included in Figure 10.16.

(6) Rainwater harvesting

Rainwater harvesting, according to the environmental industry, represents a very efficient way to reduce water consumption. In its simplest form, rainwater harvesting incorporates relatively basic technologies which makes it a cost-effective solution in most cases. It works by collecting and storing rainwater from roofs and other surfaces of buildings that would otherwise be fed into drainage systems as runoff. Rainwater collected is stored in tanks which can be located underground or overground. Water is not suitable for drinking, but it can be pumped back into buildings for a multitude of purposes without being treated. Uses include flushing toilets, washing clothes, cleaning the exterior of buildings, washing cars or watering external areas. Rainwater harvesting installation can be scaled up or down to meet different property requirements.

Figure 10.16 Ground and air source heat pumps

(7) Greywater recycling

Greywater is terminology relating to nonindustrial wastewater from non-toilet plumbing systems which include showers, baths, washing machines, sinks (but not normally kitchen sinks) and handwashing basins. These sources differentiate it from rainwater, and it is sometimes referred to as sullage. Greywater typically has an inferior water quality to rainwater as it can sometimes contain bacteria, detergents and other pollutants. For this reason, in most cases it will require treatment to make it suitable for reuse without the health risks associated with being untreated. Whilst such treatment can be conducted locally, for small applications it is normally easier to recycle, through sewerage treatment plants. Greywater can be used for toilet flushing but its application is more conducive to commercial buildings where dedicated maintenance teams can monitor water quality.

10.11 Building envelope considerations for sustainable construction

The terminology 'building envelope' refers to external roofs and walling/cladding systems of buildings. The design building envelopes represents a major factor in determining the amount of energy buildings will use in their operation. Building envelopes must balance requirements for ventilation and daylight while providing thermal and moisture protection appropriate to prevailing climates. In addition, they should consider climate and activities inside buildings. Some of the ways building envelopes can improve energy efficiency and comfort levels within buildings are explained next.

(1) Passive Solar Design

Passive solar design can reduce energy consumption by using the sun's energy for the heating and cooling of living spaces. In this approach, buildings themselves or some element of them take advantage of natural energy characteristics in materials and air created by exposure to the sun. Passive systems are simple, have few moving parts, require minimal maintenance and no mechanical systems.

When sunlight strikes a building, it can reflect, transmit or absorb solar radiation. Heat from the sun causes air movement which if harnessed correctly can provide heating and cooling effects in buildings. This will rely on appropriate design of building elements and internal spaces including the choice of materials. One example would be the 'stack effect' of design atrium in buildings where warm heated by solar energy rises, within a plenum, to create a suction effect lower down providing natural ventilation. This effect has been practiced in many hot countries around the world and dates back to ancient times.

(2) Window positioning and design to reduce heating and cooling

Other ways in which passive solar design can be achieved include positioning and configuring buildings with appropriate solar orientation. Examples of this include elongating the east–west axis of buildings. Interior spaces requiring the most light and heating should face the sun whereas less used spaces should be away from the sun. In addition, designing roof overhangs or incorporating solar shade by brise-soleil installation will shade those elevations exposed to the sun to reduce the effects of heat gain. Using appropriate natural ventilation through correct and appropriate choice and location of opening windows can also be useful. The author was the project director of a college building in Manchester namely the Harpurhey Sixth Form Centre, which went in to win Sustainability Building of the year in 2006. One of the many features incorporated into the building design included mechanically operated windows. These were automatically opened during hot periods to allow a through flow of cross ventilation and closed during the evenings in colder times of the year with shutters to preserve or 'lock in' heat energy overnight.

Another way that windows can save energy is through the use of solar reflective glass, which is particularly effective in hotter climates. This type of glass can reflect a large percentage of the sun's radiation and prevent heat buildup in internal spaces.

(3) Thermal massing

The Harpurhey Sixth Form Centre, referred to previously, was also designed to incorporate 'thermal massing' within its envelope. This entailed constructing a very heavy concrete envelope that would absorb heat during the summer, to prevent overheating of internal spaces. The heated envelope would then give back thermal heat from it during the winter to reduce heating. Thermal massing in this way can be compared to using heavy envelopes as mass storage heaters in this regard.

(4) Daylighting

Daylighting is the controlled admission of natural light into spaces through windows to reduce or eliminate electric lighting. By providing a direct link to the dynamic and perpetually evolving patterns of outdoor illumination, daylighting helps create a visually stimulating and productive environment for building occupants. The cooling required to counter waste heat generated by lights can amount to up to 30% of total energy use. Daylighting reduces the need for electrical light sources, cutting down on electricity use by as much as 5% of total energy, representing significant savings on operating cost and carbon generation.

(5) Indoor environments

Buildings should be designed to meet basic human requirements for healthy and comfortable indoor environments whilst at the same time ensuring low energy consumption. Building envelopes can be designed to counter air and noise pollution which is particularly important for urban developments. Acoustic interventions to envelope design through insulation or double skin construction can prove useful in certain applications.

10.12 Waste management in maintenance and estates management

Estates and maintenance management teams in organisations should strive for policies to reduce their waste production as part of their environmental sustainability strategies and programmes. Any sustainable waste management programmes should therefore seek to maintain materials in use for the longest time, in their present or recycled forms before being incinerated or disposed of in landfill. Notwithstanding this premise considerations regarding waste should be considered before the manufacturing of products and the ease by which it can be recycled or disposed of easily and responsibly through biodegrading. For this reason, the life cycle of products used in the built environment should be an area for reflection by maintenance teams, to enable them to lower any negative and detrimental social, environmental and financial impacts associated with modern day consumption. This may involve a focus on both end-of-life waste and attempting to design waste out of production cycles at design stages. Sustainable waste management practices should be reliant on systems and policies which consider avoidance, reduction, reuse, recycling, energy recovery and, lastly, disposal or treatment. These considerations are commonly referred to as the waste management hierarchy and each one will be discussed next:

(1) Avoidance and reduction

The priority for organisations and individuals is to avoid or reduce the creation of waste wherever possible as the main priority. To achieve this, it may involve either reducing consumption or alternatively increasing efficiency. In practice this may entail organisations and individuals selecting products and materials which need the least amount of resources to produce them. In addition, wherever possible, they should avoid the use of disposal single use products. The materials that make up such products are extracted, processed through manufacturing and then transported to become waste in a very short period of time. In an estates management context this could involve elimination of single-use plastics, commonly used in most organisations e.g. plastic coffee cups and cutlery in canteen areas.

(2) Reuse and recycling

If avoidance or reduction of waste is problematic, then the next priority should be to only purchase products or materials where their waste can be reused and education and training of methods to responsibly reuse waste products should be conducted for staff. Reusing waste is a preferred option rather than recycling because it can be carried out without processing any new materials which can incur energy, resources and cost. If reuse is not feasible then recycling may be the only option. Recycling, like reuse, maintains materials in a cycle without the requirement to use new raw materials and resources for manufacturing. One example for maintenance and facilities staff may be to consider ways and means to recycle food waste from their canteens, for compost or for collection by specialist companies for the creation of energy through incineration.

(3) Energy recovery

Energy recovery involves the creation of fuel, electricity or heat from waste materials. It can be created through processes such as incineration, digestion, gasification and landfill gas. In this way it also reduces the volume of waste that needs to be transported as land waste. Furthermore, the energy created by these recovery processes offsets what would otherwise be required by the mining of new fossil fuels.

(4) Disposal or treatment

Disposal or treatment represents the least attractive option in the waste management hierarchy, as it generally refers to incineration without energy recovery or landfill. Accordingly sustainable waste management strategies should look to defer this stage for long periods of time wherever possible.

When maintenance and facilities teams are looking to make improvements to waste management there are some practical ways and means that they can adopt. Some of these may be relatively straight forward and easy to deliver and therefore represent 'low-hanging fruit' for organisational change. For instance, scrapping single-use items for reusable ones can be a quick win in increasing their green credential by reducing their waste. Other proactive initiatives could be to provide recycling bins across organisations for employees to separate paper, plastic bottles, aluminium cans and food waste. This will make it easier for estates and maintenance departments to process waste for recycling thereafter. Separation of waste in this way should reduce the risk that a certain category of waste will contaminate another whereby all the waste becomes land fill. Switching to digital ways of working in lieu of traditional paper-based practices can save organisations a considerable amount of cost and waste associated with paper and printing.

In practice this could include replacing paper-based practices such as minutes of meetings and invoices with a digital alternative with electronic records.

10.13 Retention of buildings for creative reuse

In the past it has been commonplace for organisations when buildings fall into disrepair, and their use for a particular purpose is no longer required, to simply demolish them and replace with a new build. This can clearly involve generating considerable waste through demolition, some of which will be transferred to landfills, coupled with the embodied carbon of all the new materials, products and components associated with a new building. This may be unavoidable in certain circumstances if a particular building must be relocated owing to the sale of land for instance. However, careful thought and reflection by estates teams should consider creatively reusing existing buildings, for different purposes wherever possible. The feasibility of this alternative proposition will be affected by the flexibility, adaptability and suitability of existing buildings for reuse. For most commercial buildings which are constructed from steel or concrete frames structures and have few load bearing compartment walls, this may present a financially viable and carbon friendly alternative to demolition. The author has procured new facilities in this way over many years. In some cases where existing buildings are in poor condition the only element of them that can be salvaged may be their foundations and superstructure. Notwithstanding this premise, retaining the foundations and structural frames of building can save up to a third on construction costs. Even greater savings can be generated through the reduction of carbon associated with these types of innovative estates repurposing. Sometimes a combination of retention of buildings coupled with new build extensions may be required. This could be appropriate where existing buildings are considered too small or the layout not conducive to fulfil functional operational requirements for a given new facility.

10.14 Intelligent buildings

When redevelopment of an estate is being planned, organisations at early stages in the master planning and design processes should consider wherever possible the introduction of intelligent buildings.

An intelligent building is one that incorporates innovative technologies and processes which creates new facilities that can be deemed more efficient, comfortable and productive for their clients. Some of the integral aspects of them are illustrated in Figure 10.17.

Characteristics and features of intelligent buildings are that they are adaptive, responsible and flexible to meet changing user requirements in terms of building usage and environmental challenges. This means they can save energy and lower wastage, leading to lower carbon emissions by managing occupied space. They can also make efficient use of staff through centralised control and integration of information from different sources. In addition, they can manage the changing status of accommodation demands that such changes create over time, through reprogramming. Furthermore, they can assist in better communications between groups and individuals over time. Examples of such groups could be end user occupants and maintenance teams where faults and defects are being reported through a computer-assisted facilities management (CAFM) system. Other advantages are included in Figure 10.18.

Intelligent budlings can be differentiated from traditional buildings in that the former can personalise such aspects as lighting, temperature and ventilation based on the occupancy of spaces and in this way utilise space more effectively. This can clearly have consequential benefits

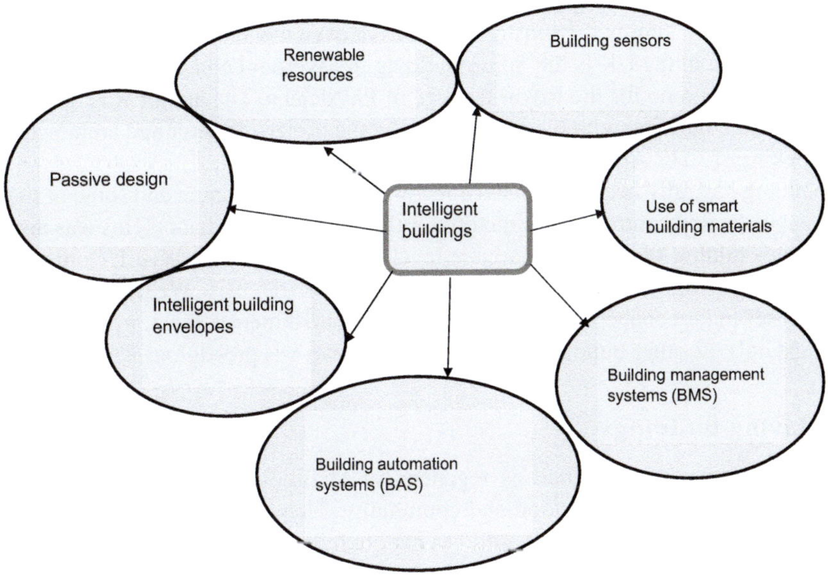

Figure 10.17 Some integral aspects of intelligent buildings

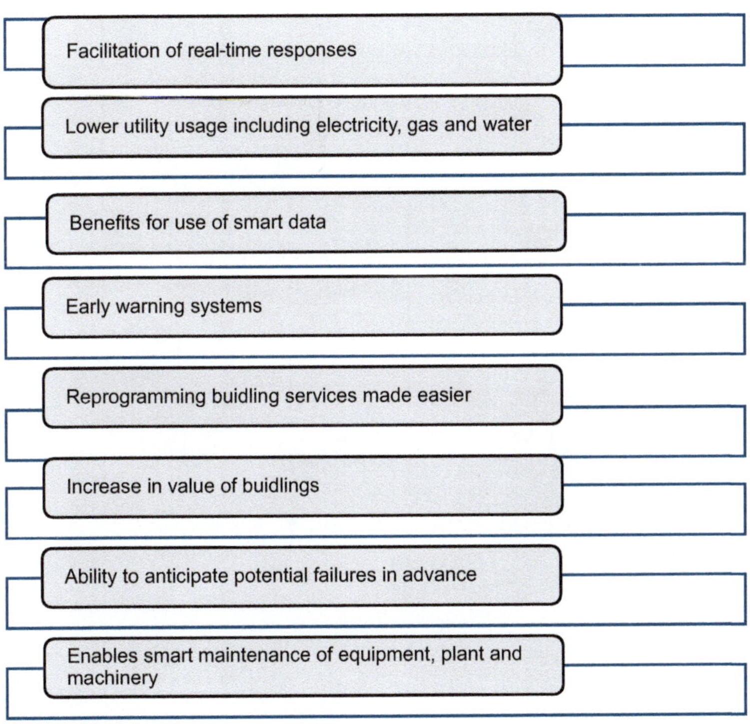

Figure 10.18 Other advantages of intelligent buildings

for energy usage and benefitting organisational environmental strategies. In contrast traditional buildings may rely on simply thermostatic controls or even manual operation. An example of an intelligent building in the UK is the Spine Building in Liverpool completed in 2022, and which has become a new home for the Royal College of Physicians. The author was privileged to be invited to visit the building with his estates team by the project director and architect, Professor Robert Hopkins of AHR Architects in November 2022.

The building had only been completed a few weeks before this time and some of the features included sophisticated controls on air quality for the benefit of occupants. This was instrumental in earning the building a converted platinum level accreditation in the WELL Core certification, issued by the International WELL Building Institute (IWBI). This WELL Building Standard investigates seven key concepts which are light, air, nourishment, fitness, mind and comfort. It sits amongst only 12 other buildings in the UK to achieve this prestigious accreditation.

10.15 Living buildings

Living buildings can be described as regenerative facilities which provide occupants with connections to nature, air, light, food and community. They are commonly regarded as being self-sufficient and not reliant on any resources remote from their sites. Figure 10.19 provides an example of one in Singapore.

Figure 10.19 Living building in Singapore

They have been proven to be able to create positive impacts on both natural and human systems that have interaction between them. There are seven categories of performance, otherwise known as petal flowers, for living budlings and these are:

- Happiness
- Health
- Materials
- Place
- Energy
- Water
- Equity and beauty

In consideration of the numerous benefits of living buildings, estates departments should consider them when deemed appropriate for a particular site. There are clearly barriers to their feasibility in some cases based on financial affordability or the suitability of a particular site. However, the long-term benefits for environmental sustainability and health and well-being of building occupants may outweigh the associated increased capital costs of construction.

10.16 Outdoor environments and public realm

Landscaping and the public realm has been long considered to have human millesimal qualities and can therefore provide health and well-being benefits for people. In addition, landscaping has other benefits associated with environmental sustainability in improving biodiversity, ecology and providing a natural habitat for winds and wildlife. What is less understood is that carefully planned public realm can contribute to reductions in energy consumption for building as careful landscape planning can reduce cooling and/or heating costs by up to a third. Trees, grass and shrubs through evaporative cooling can reduce air temperatures near to buildings. Trees in particular can provide shade, reduce the surface temperature of buildings and prevent direct heat gain through windows. Deciduous trees specifically can provide shade in summer and admit light in winter when the leaves fall, whereas evergreen trees provide year-round sun and wind protection. Windbreaks created by trees can reduce wind within a distance of three times their height. Maintenance teams, taking account of these, should look wherever possible to use natural landscaping to their advantage for reducing energy consumption and providing quality recreational areas for use by building occupants and members of the public.

10.17 Summary of chapter

The agenda for change to reduce carbon emissions and attain global targets on such initiatives as net-zero carbon cannot be disentangled from future maintenance management strategies. Accordingly, environmental sustainability encapsulating energy management should be at the heart of any future estates strategies if they are to be successful. This is especially important for organisations which have been facing for many years of inflationary rises in utility, construction and maintenance costs, affecting their profitability. This chapter has responded to this challenge and presented many ideas and recommendations hopefully to create the impetus and catalyst for rethinking current practices and a move to more environmentally sustainable ways of working.

The need to create more energy efficiency and improve environmental sustainability to reduce emissions is clearly part of a wider global agenda to reduce carbon generation. Certainly, in the UK there has been for many years a noticeable increase in participation in accreditation schemes linked to energy efficiency and sustainability and these have included Passivhaus and BREEAM. Such schemes have sought many ways and measures to achieve significant carbon reductions, mostly through good building design. Measures have included passive design, use of renewable technologies and responsible specification of building materials to reduce waste and increase recycling.

When considering maintenance management, it is imperative to consider energy efficiency as part of any policy, strategy or maintenance regime. This chapter has suggested that a major step change is required, especially in the case of the UK, given that the government have committed to a new zero carbon pledge by 2050. Energy management should therefore represent a key objective in the pursuit of environmental sustainability for any organisation. It should influence the way buildings are designed, constructed and used throughout their occupancy. Potential environmental initiatives could include new methods of construction, control systems, energy-efficient LED lighting, and technologies for harnessing wind and solar power. However, organisations need the tools to be able to accurately calculate, monitor and predict their energy consumption as a measure of performance, and building management systems are one way of enabling this. Notwithstanding this premise, on all new systems introduced it is important for inductions and training on new energy management systems for estates and maintenance teams.

A case study of Energy House 2.0 was presented in this chapter which the lead author was heavily involved with. The facility represents the largest research centre of its type in the UK in providing controlled environments for testing different types of building, structures, and technologies. It is hoped that this award-winning project can spearhead the way forward for more energy-efficient technologies to be introduced in construction and maintenance of existing estates.

Education and training for staff is important especially for estates and maintenance teams and building users who should be influenced to preserve energy in their workplaces. However, sometimes cultural barriers associated with old traditional ways of working may be difficult to break through in this regard, with some staff feeling uncomfortable and threatened by new initiatives and technologies. To respond to this challenge a programme of continual professional development may be required within organisations.

The life cycle of buildings relates to the period from their inception at the start of the conceptual design process to their demolition at the end of life. Life cycles are affected by the extent and quality of repairs and maintenance that estates and buildings will receive which can affect their condition and ultimately influence their deterioration and longevity. The quality of their design and construction will also play a major part in how they perform and the duration of their occupation in use prior to demolition. The degree to which buildings are flexible and versatile enough to deal with changing requirements made of them throughout their life cycle is very important as any creative reuse could improve their longevity. Estates team should therefore be mindful to plan for future proofing buildings at early design stages.

The construction industry has made concerted efforts in recent years to curb the consumption of raw materials, reduce their 'embodied energy' in the construction process and the energy used during their occupation periods. This chapter has an example of a system, aptly named

CarboniCa, which is a digital carbon reduction tool and can measure carbon emissions across the whole life of buildings. In this way designers can choose those materials and forms of construction which yield the lowest embodied energy levels. There should also be close consideration of operation and maintenance costs, commonly known as OPEX costs, alongside capital costs, as these can amount to most of the costs incurred, across the lifetime of building assets. These will include all ruining costs such as maintenance, energy, water, remodelling and eventually demolition.

In estates and maintenance management there will be instances where the environmental effects on decision-making need to be understood and considered and an Environmental Impact Assessment (EIA) is a tool used to assess the significant effects of a project or development proposal on the environment. The different stages, scope and characteristics of such assessments have been described in this chapter which will allow estates policies to consider the overall effects and implications of developments and maintenance strategies at early design stages.

Organisations should be constantly looking to improve their estates through good environmentally sustainable design *to* include energy and resource efficiency *and* pollution prevention. *Accordingly*, the inclusion of many renewable energy and sustainability technologies should be considered wherever possible, including photovoltaics, green roofs, wind turbines, rainwater harvesting, ground and air source heat pumps and active solar panels. In addition, the design of building envelopes represents a major factor in determining the amount of energy buildings will use in their operation. They should consider such factors as passive solar design, window positioning, thermal massing, daylighting and internal environments.

Estates and maintenance management teams in organisation should strive for policies to reduce their waste production as part of their environmental sustainability strategies and programmes. These policies should consider avoidance, reduction, reuse, recycling, energy recovery, treatment or disposal.

When redevelopment of an estate is being planned, organisations at early stages in the masterplanning and design processes should consider wherever possible the introduction of intelligent buildings which can incorporate some of the environmentally sustainable technologies referred to in this chapter. These can include passive design, building automation/management systems, renewable resources and technologies, smart building materials, and intelligent envelopes. They can save energy and lower wastage, leading to lower carbon emissions by managing occupied space. Other benefits include facilitating real-time responses, providing early warning systems, the ability to anticipate potential failures in advance and enabling smart maintenance of equipment, plant and machinery.

Living buildings can be described as regenerative facilities which provide occupants with connections to nature, air, light, food and community. They have been proven to be able to create positive impacts on both natural and human systems. Benefits include improved health and well-being of all occupants, reduction of carbon emissions, no pollution and generation of no waste.

Finally, landscaping has other benefits associated with environmental sustainability in improving biodiversity, ecology and providing a natural habitat for winds and wildlife. Furthermore, it can also provide health and well-being benefits for people and reduce energy consumption if designed to avoid buildings overheating by providing cooling and shade.

References

Challender, J. (2023). Case study analysis of improving environmental ethics using collaboration toolkit. In: *World Renewable Energy Congress*, Kuala Lumpur, 16–20 July.

Guidetti, E. and Ferrara, M. (2023). Embodied energy in existing buildings as a tool for sustainable intervention on urban heritage. *Sustainable Cities and Society*, 88.

Thomas, M. (2013). *Managing building pathology and maintenance construction managers' library*. Leonardo da Vinci.

Chapter 11

Case studies related to embedding environmental sustainability into estates practices

11.1 Introduction

This chapter will follow on from Chapter 10, in providing two case studies in which the lead author was responsible for devising and implementing new environmental sustainability strategies.

The first case study will relate to a college in the North West of England, where the lead author was instrumental in developing a new environmental policy. This was related to his role as Project Director in delivering a £200m master plan for the college. To address some issues that became apparent in an environmental sustainability audit, the new policy included pledges linked to corporate social responsibility and measures required to improve the organisation's green credential. This covered many different aspects including reduction in carbon, a new transport policy and wastage.

The second case study will detail a new environmental sustainability policy and strategy which was developed by the lead author and his sustainability team at a North West university in the UK where he was Director of Estates. The chapter will give a detailed account of some of the initiatives and measures that the university instigated as part of this new strategy and the successes and benefits that this generated.

11.2 Case study one: the development of a new environmental sustainability strategy for a college in the North West of England

11.2.1 Introduction to case study

The lead author, Professor Challender, was employed as Project Director of the college between 2006 and 2011. During this period, he was responsible for the master planning of the college which involved four major phases of redevelopment of the existing campus. The value of the redevelopment works was estimated to be in excess of £200m and was to include the demolition of a number of campus buildings and replacement with new facilities. It also involved the refurbishment of existing building and the creative reuse of a listed hospital building it had recently acquired from a North West hospital trust. As part of the master planning strategy, the college wanted to reflect on its existing environmental sustainability credentials. The initial stage involved undertaking benchmarking by independent consultants to audit and assess how it was currently performing against agreed environmental and sustainability agendas and existing legislation. The findings were very disappointing, and a report concluded that the college was failing in both its existing environmental policies and not fully delivering on its corporate social responsibilities. Clearly improvement was required and accordingly the lead author was

DOI: 10.1201/9781003358800-11

therefore tasked with collaborating with colleagues from different departments to develop a new environmental policy and strategy for the college. This involved extensive consultation with staff and external stakeholder, and the appointment of external consultants to understand how the college could improve its environmental credentials. This case study will give a summary of some of the background to the work that was carried out and some of the environmental pledges and commitments that were generated from this.

11.2.2 Context and background on the college's environmental sustainability aspirations

Following the college's new build construction and engineering workshop completed under the leadership of the lead author in 2011, it had been successful in gaining an excellent BREEAM accreditation. BREEAM represents the world's leading sustainability assessment method for master planning projects, infrastructure and buildings. It recognises and reflects the value in higher performing assets across the built environment life cycle, from new construction to in-use and refurbishment. BREEAM does this through third-party certification of the assessment of an asset's environmental, social and economic sustainability performance, using standards developed by the British Research Establishment (BRE). This means BREEAM rated developments are more sustainable environments that enhance the well-being of the people who live and work in them, help protect natural resources and make for more attractive property investments.

The college did seek, alongside external design consultants, to promote sustainable construction methods, mindful that the construction industry contributes to almost half of the world's energy consumption and that there is the possibility of running out of fossil fuels in the next 40 years (Roaf 2001).

The college's proposed master plan redevelopment plans for further new energy-efficient buildings on the campus were cancelled following the Learning Skills Council withdrawal of funding in 2011. However, this did not preclude the college from looking at simple, cost-effective solutions to increasing building energy efficiency. These included improving heating controls, increasing thermal insulation, installing energy-efficient light fittings/bulbs, fitting door closers, managing classroom utilisation more effectively, fitting automatic controls and displaying energy meters so that students and staff can measure and monitor improvements. In addition, a waste-management strategy involving different types of waste for recycling could be introduced quickly and at relatively little cost. This would then hopefully comply with EU Directive 2006/11/EC of the European Parliament which states:

> In order to achieve a high level of environmental protection, Member States should, in addition to taking responsible action to ensure disposal and recovery of waste, take measures to restrict the production of waste particularly by promoting clean technologies and products which can be recycled and reused.

11.2.3 Implications and impacts for the college and the need for change

The college considered that unless it was to adopt a sustainable waste management policy the changing agenda could present negative outcomes for the college on ethical, public relations, political and social grounds. Furthermore, on an economical basis the financial operating costs for the college could increase as time passes through the introduction of government initiatives such as the Landfill Tax Escalator. This taxation point is reinforced by the fact that 'further adoption of the 'polluter pays principle' could see future differentiated taxes introduced by government seeking to remove or reduce emissions of carbon dioxide into the atmosphere.'

The dilemma for this organisation was not to assess the affordability and feasibility of sustainability as a choice of whether they chose to participate or not. The real question for the college was whether they could afford not to change the way they were currently operating and place sustainability high on their agenda.

11.2.4 The college's corporate social responsibility

Given the importance of the sustainability agenda the college looked for solutions that presented environmental and social advantages and that would benefit the college in doing so. Some would argue that this corporate social responsibility does not exist in the real world. Banerjee (2007), in support of this view, presents a cynical representation of this ideology for most organisations and explained: 'This ideology is a convenient, yet dangerous myth and this 'win-win' rhetoric is nothing but a mask, made up of glossy brochures, beguiling speeches and 'media-savvy' partnership initiatives.' One could, however, argue that the college was unlike most organisations as its primary function was not to make a profit but to serve the community that we work within through education and training, and this alone presented an important reason for their active participation in the sustainability agenda. On ethical grounds they considered, as a major public sector organisation, this was an environmental policy that could pave the way for cultural change and set a good example for other institutions in the sector.

11.2.5 The need to embed sustainability

The college had for many years now concentrated more on economic factors (especially in light of the real reductions in funding over the years) and educational excellence in meeting government targets on student achievement and retention rather than the bigger global issues of the environment. Michael Gove, former UK Education Secretary, conceded this point and argued that 'we need to embed sustainability into the heart of our academic establishments' (Guardian 2011). This view was underlined at the time by the former Secretary of State for the Environment, Caroline Spelman: 'Genuine sustainability will never again be an add on or afterthought. It will be at the heart of everything we do, from making policy to running government buildings or purchasing goods and services' (DEFRA 2011). The college accepted that a long-term cultural change at the highest level was required within the organisation rather than considering the initiatives as individual short-term projects. This view is supported by the work of Hjorth and Bagheri (2006) who opined that 'sustainability should not be perceived as a 'project' that has an end point, but as an ongoing process that needs to be regarded as part and parcel of everyday work.'

11.2.6 Potential measures to address the sustainability agenda

Table 11.1 presents some potential recommendations that were designed to allow the college to address the sustainability areas. It lists a combination of *dematerialisation* and *substitution* measures. Dematerialisation would include reduction in energy usage, waste, paper etc. and substitution could be achieved through alternative energy sources. In some cases, innovative solutions could fulfil both objectives at the same time. For instance, the college could have reduced the extent of wood waste through recycling (dematerialisation) and utilised the waste

Table 11.1 Main environmental activities of the college in 2011 and the possible ways and means by which improvement measures introduced could mitigate the existing risks

Activity of College	*Creative Recommendations to Address Legislation and Mitigate Impacts of Environmental Risks*
Environmental Management	The college could: Promote sound environmental management by communicating procedures, guidelines and regulations via induction, staff training programmes and tutorials. Encourage and facilitate feedback and suggestions on ensuring good practice via a network of operational groups and team meetings across the college. Develop a management system based on a sound understanding of current performance and the adoption of realistic targets to improve environmental performance.
Waste Reduction	The college could: Implement sustainable resource management practices based on 'reduce, reuse, recycle' principles. Reuse and recycle materials wherever practicable. Reduce the amount of waste being disposed of in landfill sites, by segregating different types of waste. Develop a programme of target setting to reduce waste. Prevent pollution. Comply with the legislation and meet the requirements of the waste regulators. Reduce litter.
Water Consumption	The college could: Make efficient and responsible use of water, including making opportunities for water reuse and rainwater harvesting.
Carbon Management	The college could: Implement a carbon management strategy to ensure the efficient use of energy. Ensure the uptake of low-carbon technology in buildings and equipment. Improve the Energy Performance Certificate assessment grade of buildings from the base year, October 2008.
Transport	The college could: Implement sustainable transport practices across all activities, as indicated in College Travel Plan. Encourage the use of public transport, walking, cycling and vehicle sharing, where appropriate. Possible discounts to incentivise train and bus travel. Charge students and staff for car parking. Reduce the need to travel especially by car e.g. working from home. Extension of bus routes and increasing bus frequency. Encourage multipurpose/linked trips. Engage with public transport providers to maximise use of existing transport infrastructure. Design the campus to encourage cycling to work e.g. more cycle storage, dedicated changing areas and showers.

(*Continued*)

Table 11.1 (Continued)

Activity of College	*Creative Recommendations to Address Legislation and Mitigate Impacts of Environmental Risks*
Ethical and Sustainable Procurement	The college could: Promote the use of sustainable resources whenever practicable. Promote lifecycle thinking in the procurement of goods and services. Promote the development of central purchasing strategies to ensure efficient and sustainable purchasing practices. Embed sustainability into selection criteria, processes, procedures, business cases, budget allowance and early market engagement when purchasing goods and services. Manage relationships with suppliers to raise awareness and allow suppliers to respond to additional or different measures in some cases.
The College Estate	The college could: Ensure that, wherever practicable, all new building works meet the sector standards for sustainability. Promote a responsibility culture to improve the internal environment of all college rooms. Promote a responsibility culture to improve the external environment of the college. Promote the efficient use of all buildings and facilities to meet sector standards for space utilisation. Manage the college estate to enhance biodiversity wherever possible.
Education for Sustainable Development	The college could: Develop links with local and national groups which encourage education for sustainable development. Endeavour to embed education for sustainable development into all curriculum areas. Encourage students to participate actively in environmental projects, locally and globally.

to fuel biomass boilers for heating as an alternative zero carbon energy solution (substitution). Table 11.1 includes the main environmental activities of the college in 2011 and the possible ways and means by which improvement measures introduced could mitigate the existing risks.

11.2.7 Economic, political and social factors influencing the evaluation of the measures

Detailed evaluation of the financial viability and environmental benefits of each measure was required to be undertaken by the college. This approach to business decisions around sustainability is advocated by Ding (2006) who explained that 'Environmental issues and financial considerations should go hand in hand as parts of the evaluation framework of a project.'

Some of the recommendations in Table 11.1 were intended to save money especially when considering life cycle costs. Ayres (2008, p. 289) supported this point and states that 'it may be possible, in principle to cut costs and pollution at the same time in some sectors . . . although it is rarely accomplished in practice.' Sant and Carhart (1981) also endorsed this view and suggest evidence through 'least cost' studies that very significant opportunities for savings through sustainability measures have been neglected at the national level. The converse argument is made by Appleton (2006) who claimed:

> An increasing number of corporations and governments have recognized that practices like green buildings, alternative ways of storm water management and certain change to land use practices offer them chances to make additional profits at little downside cost or risk.

The college was required to identify projects which should take priority over others and develop overall timescales for instigation of strategies. Factors included complexity, cost resources, funding incentives and the political appetite for change in certain areas. One argument could have been made for priority to be given to those that stand a relatively good chance of yielding a good return on investment. Graham (2003) presented a converse argument in that 'sustainability should not be influenced to such an extent by cost alone but by all the environmental and social benefits that generate from it.'

11.2.8 *Sustainable procurement process*

Using the sustainable procurement processes for the purchase of goods and services the college attempted to implement measures and realise benefits from those initiatives very quickly. The Office for Government and Commerce (OGC) announced in 2015 new commitments for 'greening' public sector procurement to understand and reduce the impacts of its supply chain. It is hardly surprising with an annual public sector spend in excess of £236 billion that the government regards procurement as a powerful lever that can reduce emissions and drive innovation. Government also supports the National Procurement Programme which is delivering training on sustainable procurement to public procurers including the FE and HE sectors as well as interest-free loans for setup costs. In addition, there were already procedure, processes and documentation from previous models of good practice as templates that could be utilised for this purpose, which would have precluded the need for a bespoke solution. Financial and resourcing demands would therefore have presented few barriers to changing the procurement strategy of the college in the short term, given the aforementioned incentives for change.

11.2.9 *Sustainable transport policy*

There are other examples of sustainability initiatives such as changes to transport and the introduction of a green travel plan which could take much longer to initiate. Any changes would have had to involve a considerable consultation period and extensive negotiations with partners such as the local authority (as part of their Agenda 21 Sustainable Transport 1 initiative) and public transport providers in the borough. In addition, the need to undertake physical alterations and build new additional facilities on the campus was a consideration. Such measures could have included creating additional changing rooms, bike storage and showers for cyclists could take some time in planning and construction. Whilst this is was a valid argument the transport agenda was a major focus for the college. After all, according to the DTI, domestic transport accounted for 22% of UK greenhouse gas emissions. In addition, the introduction of the government's Local Transport White Paper in 2010 detailing £560 million enabled local bodies to deliver sustainable transport solutions. This should have provided an incentive for the college to start the process and planning earlier in their environmental sustainability programme, to ensure implementation within a reasonable timescale.

11.2.10 The need for effective consultation and communication

In planning long-term changes to the college's sustainability strategy their success was dependent on undertaking extensive consultation and dialogue with their students and staff as well as third-party stakeholders such as their local authority. Innovative ideas were generated to assist the college in developing environmental strategies and sought to discard the former 'identify-and-repair' approach and instead adopt an 'anticipate-and-prevent' approach (Al-Sharrah *et al.* 2010).

At each stage the college shared and communicated their sustainability strategy as it developed and unfolded. This was aimed at generating maximum support and to be as transparent as possible on the issues and their sustainability objectives. This improved public relations for the college and was intended to enable transparency which could open up avenues of possibilities that would be difficult to achieve without it.

11.2.11 The need for commitment, cultural and organisational change

The college decided that success of their sustainability strategy was predicated on securing full commitment from the college's senior management team and board of governors. For this reason, all college managers were given responsibility and duties to support the implementation and monitoring of the policy. It was recommended that a new role should be created as head of environmental services, who would be required to report to the principal, senior management and the governing board and provide regular progress reports against sustainability programmes and implement an annual review of all environmental policy.

Finally, in terms of cultural change, educational institutions, such as the college, did not simply regard themselves as academic institutions that enabled students to obtain qualifications and training. Instead, they proclaimed that they offered and created an excellent opportunity to encourage people to adopt more sustainable lifestyles. In addition, the college believed that the increasing regulation and sustainability agenda affecting organisations like theirs should not be regarded as a negative influence to adversely affect their business operations. Moreover, they professed that they should be encouraged to push the boundaries in the creation of more environment-friendly communities (Architecture Week 2011).

11.2.12 Summary of case study

This case study hopefully provides an insight into an organisational sustainability agenda, which in this instance was a North West college in the UK. This example has articulated the analysis and evaluation of the implications and impacts from ever increasing regulation alongside the benefits that could be realised from participation in the green agenda. Furthermore, the sustainability principles and initiatives which were proposed by the college could have been regarded as embedding a culture of responsibility and a need for change within their organisation and everyday activities that they carry out. The college's corporate social responsibility on ethical grounds alone was designed to promote its reputation as a public sector body. Furthermore, it was intended that the proposed innovative measures would provide the tools and motivation for implementing the necessary sustainability strategies and policies towards promoting a cleaner sustainable environment.

11.3 Case study two: the development of a new environmental sustainability strategy, policy and implementation programme for a North West University in the UK

11.3.1 Introduction to case study

This case study relates to achieving sustainability and environmental enhancements through a collaboration and partnering 'toolkit.' The measures undertaken and successful outcomes at the university will be explained and how the development of a carbon management plan assisted the organisation in achieving its environmental target.

The lead author whilst employed as the Director of Estates at the university between 2017 and 2024 and worked alongside his sustainability team around improving environmental performance of the organisation.

11.3.2 Use of a collaboration toolkit

A 'collaboration toolkit' was formulated through the author's research to assist client bodies to adopt a more collaborative model of procurement that integrates the supply chain into the environmental agenda at early stages of projects. Plans were in train for the toolkit to be piloted on a major project as part of the university's master plan and capital programme.

The focus for the toolkit was finding environmental solutions through collaboration with industry and delivering a new sustainability strategy for the university. In this regard the following measures and initiatives were developed and planned to be deployed as part of the case study from many different environmental perspectives including:

- Pursuing a holistic approach across the supply chain to meet the demands of the carbon neutral agenda. All bodies (client end users, design consultants, contractors, subcontractors and other external stakeholders) are being encouraged to collaborate early in design stages on alternative construction methods, systems or solutions to generate more sustainable solutions.
- Reducing embodied carbon in buildings, with academics and industry working collaboratively to formulate pragmatic solutions through workshops.
- Introducing an energy partner to work with the university to assess the ways and means of them reducing its gas, electricity and water consumption.
- Reviewing which renewable technologies are going to give a significant and practical reduction in carbon and are cost effective, respective to site logistics, decanting and appraisal.
- Creating a new sustainable transport policy linked to reducing car travel.

11.3.3 Energy, water and carbon management plan

An energy, water and carbon management plan was developed by the university and this is in Appendix 8. It gives context and background to the plan and contains an energy and water review. It also incorporated baseline data and key targets for future energy and water usage alongside an action plan and risk assessment. Furthermore, it provided month-upon-month results from energy audits conducted with data on water and energy usage. In addition to this document, an environmental sustainability strategic plan was prepared by the university which sets out further targets linked to wider sustainability aims and objectives, including procurement, transport and carbon generation (included in Appendix 9). The targets linked with these measures, as part of the case study in 2020, three years after implementing the plans, were as follows:

- Carbon emissions on target for reduction of 81% by 2030 (from a baseline of 2011)
- Reductions of energy and water consumption by an average of 5% over last three years
- Environmental accreditations achieved in last two years

  1. Green Flag
  2. Eco Campus Platinum Status

- Reduction in waste of 20% over last three years through recycling
- Increase of EV charge points by 30%
- Flexible (Sustainable Procurement) Framework Level 4 achieved
- Robust energy, water and carbon management plan developed consisting of:

  1. Energy audits
  2. Training
  3. Automated systems
  4. Measurement
  5. Switch it off campaign
  6. Water leak surveys
  7. Low flow sanitary fixtures
  8. Rainwater and grey water harvesting
  9. Electric vehicles
  10. LED lighting
  11. Green IT
  12. Cycle to work initiatives
  13. BREEAM Excellent for new builds and refurbishments

- Communication and engagement on sustainability measures and outcomes

  1. Sustainability linked into the curriculum
  2. Active member of the Greater Manchester Green Impact Committee
  3. Active member of Go Green Salford
  4. Green Campus Group at the university

The previous successes highlight the measures and consequential beneficial outcomes for the university and how the collaboration toolkit played an integral part in delivery. This will no doubt be of interest to those organisations that are planning for long-term transformation change and raising their sustainability credentials and accreditations such as the EcoCampus Scheme.

11.3.4 Practical use of the collaboration toolkit to target and deliver environmental sustainability objectives

The collaboration toolkit provided the practical guidance and the platform for systematic partnering at all design and construction stages with the supply chain. This was intended to provide the conduit for improving outcomes on projects as part of strategies designed around environmental sustainability. Notwithstanding this premise the following considerations needed to be integrated into environmental sustainability strategies and policies.

i) **The need for an effective environmental management system**

As 'part and parcel' of the sustainable procurement of construction and engineering projects an environmental management system was put in place as a way for the organisation to demonstrate that they have addressed and minimised the environmental impacts of development. In this way it could oversee and manage legal compliance of development whilst striving for continuously improvement on environmental sustainability performance. Furthermore Figure 11.1 shows the environmental measures which should be incorporated into the environmental management system as part of the project management process.

In addition to the measures in Figure 11.1, having energy, water and carbon management plans in place greatly improved the success of environmental sustainability projects. The collaboration toolkit could be tailored to ensure that these plans area developed by the projects team, assisted by the supply chain, and then become embedded into construction and engineering programmes.

ii) **Waste reduction**

On all projects through the collaboration toolkit there should be integrated measures to ensure that waste from construction and engineering activities is minimalised. Every effort should be made to recycle waste materials and thus reduce the overall amount of arisings disposed off-site. This will serve to reduce the embedded carbon of the build process thus lowering the carbon footprint. Furthermore, measures to improve the green credentials and status of projects will have many advantages for those organisations involved in them, and these revolve around corporate social responsibility, community relations and reputation.

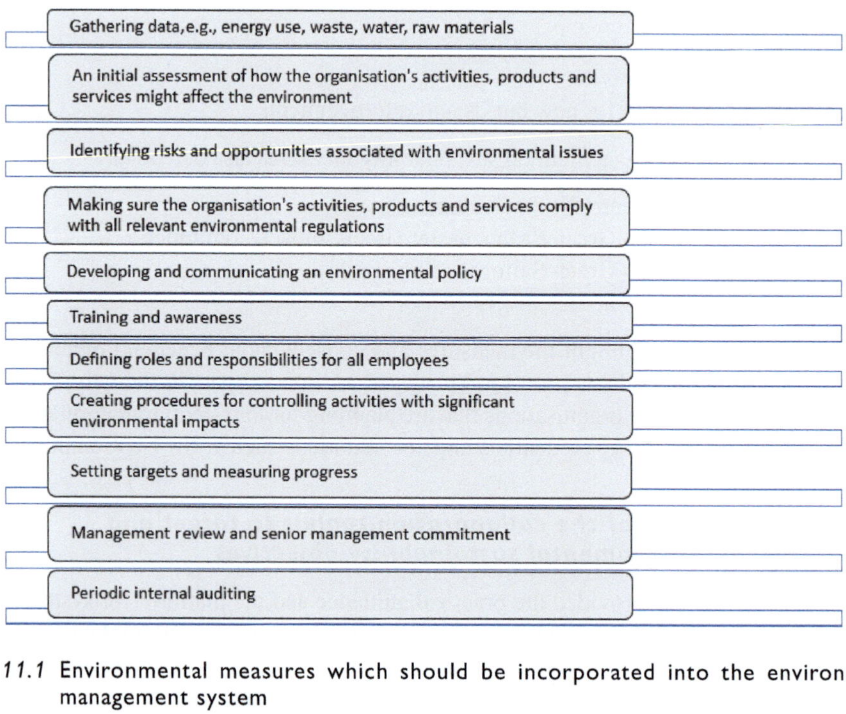

Figure 11.1 Environmental measures which should be incorporated into the environmental management system

iii) **Sustainable and ethical procurement**

In a construction and engineering context sustainable procurement philosophies should be deployed as part of the collaboration toolkit when appointing consultants, main contractors, subcontractors, suppliers and other specialists. In the UK there are various purchasing consortiums, who ensure that "checks and balances" are conducted on those organisations that are included as part of their framework. Such checks will ascertain the ethical and sustainability credentials of organisations and ensure that only those firms which meet a required minimum standard are included on the framework. In this way, employers can have the comfort factor that companies who are commissioned on projects have already been through a rigorous vetting process to assess their environmental sustainability credentials. Furthermore, organisations should ensure that all tender evaluations required when appointing the supply chain include sustainability weighting as a key quality factor alongside other criteria such as price. Developing and applying ethical and sustainable procurement policies in this way will ensure social and environmental impacts of those employed on construction and engineering projects are appropriately considered.

iv) **Sustainable construction and engineering designs and technologies**

The collaboration toolkit was able to assist project teams in making collaborative decisions with their supply chains on the use of environmentally sustainable building technologies in the construction process. This entailed careful and protracted early discussions on the use of renewable technologies. Reducing embedded carbon in construction and engineering projects is a topical issue in the UK and cross discipline dialogue with the supply chain has encouraged more sustainable materials and construction processes to be implemented within projects. Accordingly, sustainability as the core principle of the any project and should embed a high standard of adherence to best practice philosophies.

v) **Community engagement**

Communication and engagement, encouraged and facilitated through the collaboration toolkit at all construction and engineering stages, are regarded as good practice on most construction and engineering projects. Processes around consultation with the local community into environmental issues and corporate social responsibility to maintain or enhance their well-being, environment and facilities can prove extremely fruitful in most cases. In addition, becoming accredited with schemes such as the Considerate Contractors Scheme in the UK can pay dividends and enhance the reputation of the project team.

Once again, the collaboration toolkit played an important part on bringing parties together, formally or informally to present innovative ideas and solutions in this regard.

vi) **Education around environmental sustainability**

Engineering and construction projects should 'promote sustainability and social responsibility' through teaching, learning and research. The collaboration toolkit encouraged industry partnerships and therein can encourage teaching and learning. As part of construction and engineering projects using the toolkit, strategies should be focused on the environmental sustainability of all operations and teaching, learning and research around this agenda through education programmes and continuous professional development should be encouraged. In this way construction and engineering projects could be regarded as 'Living Laboratories' especially when partnering with specialist energy partners.

11.3.5 Successful outcomes that the collaboration toolkit has had on delivering environmental sustainability

Although projects where the collaboration toolkit was being piloted were at relatively early stage in terms of delivery, they were already proving to deliver many beneficial outcomes from an environmental sustainability perspective. For the client body these include those benefits shown in Figure 11.2.

11.3.6 Conclusions and recommendation from the case study

The collaboration toolkit has been presented and as previously outlined forms part of action research as part of an ongoing case study where it is being piloted on UK construction and environmental sustainability projects. It has so far, since it was introduced in 2018, achieved very successful outcomes for the university in this case study by facilitating and encouraging many different environmental sustainability initiatives and policies to be brought forward. In this way it has assisted this organisation to keep on target with their environmental sustainability

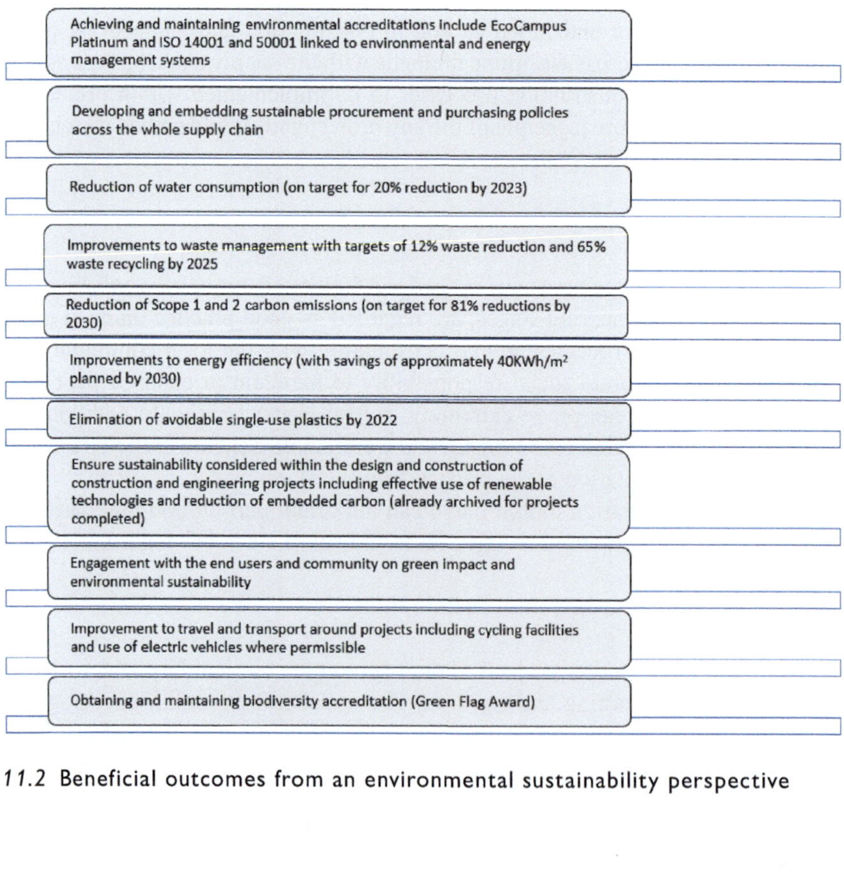

Figure 11.2 Beneficial outcomes from an environmental sustainability perspective

objectives. Furthermore, it has created the collaboration and partnering ethos that is so important for innovation and delivery strategies linked to environmentally sustainable construction and engineering projects and processes that could otherwise have been problematic.

The benefits of utilising the toolkit have been clearly articulated in this paper. Accordingly, it is hoped that the toolkit can provide the right platform for construction and engineering projects and therein achieve similar environmentally beneficial outcomes for many other countries other than the UK. For this reason, it is recommended that the principles and mechanics of the toolkit be deployed when embarking on environmental projects as part of global aspirations for a more sustainable future.

11.4 Summary of both case studies

Two case studies have been presented relating to a college and university in the North West of England, where the lead author was responsible for devising and implementing new environmental sustainability strategies. These have provided an important piece of action research for projects linked to environmental ethics and in the second example using a 'collaboration toolkit'. They articulated the benefits of developing successful sustainability outcomes with environmental ethics at the heart of decision-making. In this regard there has been many policies and initiatives developed which have assisted the two case study organisations in meeting their environmental sustainability aims and objectives and to be able to constantly monitor and track their progress against predetermined goals and targets. Accordingly, it has created the partnering and collaboration ethos which is considered vitally important for delivery strategies and innovation, linked to engineering processes and projects around environmentally sustainable construction. Without this it could otherwise have been problematic and challenging in attaining the successful environmental sustainability outcomes. It is hoped that the toolkit can provide the right platform for construction and engineering projects in globally and therein achieve similar environmentally beneficial outcomes for many other organisations.

References

Al-Sharrah, G., Elkamel, A. and Almanssoor, A. (2010). Sustainability indicators for decision-making and optimisation in the process industry: The case of the petrochemical industry. *Chemical Engineering Science*, 65(2), pp. 1452–1461.

Appleton, A.F. (2006). Sustainability: A practitioner's reflection. *Technology in Society*, 28(1–2), pp. 3–18.

Architecture Week. (2011). Architectural global warming. Available at: www.architectureweek.com/2011/0218/environment_1-1.html (accessed 5 April 2011).

Ayres, R.U. (2008). Sustainability economics: Where do we stand? *Ecological Economics*, 67(2), pp. 281–310.

Banerjee, S.B. (2007). Corporate social responsibility, the good, the bad and the ugly. *Journal of International Management*, 15(1), pp. 118–119.

DEFRA. (2011). Embedding sustainability in government. Available at: www.sd.defra.gov.uk/2011/02/embedding-sustainability-in-government (Accessed 28 April 2011).

Ding, K.C. (2006). Sustainable construction: The role of environmental assessment tools. *Journal of Environmental Management*, 86(6), pp. 451–464.

Graham, P. (2003). *Building ecology. First principles for a sustainable built environment*. Blackwell Science Ltd.

Guardian. (2011). The Guardian home page. Available at: www.guardian.co.uk/politics/2011 (accessed 3 May 2011).

Harvard Institute for International Development. (1992). Report of the United Nations Conference on Environment and Development. UNCED Report A/CONF.151/5/Rev.1, 13 June.

Hjorth, P. and Bagheri, A. (2006). Navigating towards sustainable development: A system dynamics approach. *Futures*, 38(1), pp. 74–92.
HM Government. (2011). *Carbon plan*. DECC Publications.
Roaf, S. (2001). *A design guide*. Butterworth-Heinemann.
Sant, R.W. and Carhart, S.C. (1981). *8 great energy myths: The least cost energy strategy*. Carnegie-Mellon University Press.

Chapter 12

The role of information technology and digital twins in asset and maintenance management

12.1 Introduction to the chapter

The chapter will start by assessing the importance of asset and maintenance information as part of estates management strategies. This will cover the diverse nature of different sources of data and documentation which can be captured and utilised as part of digital estates management systems. It will then discuss the growing challenges that are faced by the built environment associated with increasing complex estates, fragmentation of services, low underachieving productivity levels and the introduction of new building technologies. The role of digitalisation will then be covered and how this can respond to these challenges and therein assist asset and maintenance management.

The chapter will then provide a background and context to the use of digital twins and their potential to solve some of the current estates challenges. In this regard it will assess the ability for digital twins to monitor and control real-time asset outputs and how this technology can boost the any number of adaptable services. This will lead into a discussion of the benefits of digital twin technologies for estates from the perspectives of efficiency and streamlined information workflows in asset operational management, health and well-being, security, building automation and predictive maintenance.

Finally, the chapter will make recommendations on strategies which can be introduced to derive value from digital twins in managing complex estates.

12.2 The importance of asset and maintenance information

Having information, which is accessible, comprehensive and accurate is regarded as essential for the strategic and operational management of property estates. Any information systems require to be constantly updated to ensure that data is current and reflects the status quo of estates at a given point in time. This calls for users to continually replenish data on systems that can provide facilities and estates management teams to have convenient 24-hour access to information. Any information system will need to be user friendly and flexible enough to cater for future changes. Furthermore, they should constantly seek to allow processes, systems and procedure to be improved and adapted as needs develop. This can call on significant financial and human resources in developing, implementing and maintaining such systems. The use of mobile devices in collecting data can be particularly useful especially when conducting condition survey reports referred to in earlier chapters of the book. This allows surveyors to download data remotely whilst on site, and therein save time and resources later having to input information separately. Such devices can be integrated with software systems which are designed to

DOI: 10.1201/9781003358800-12

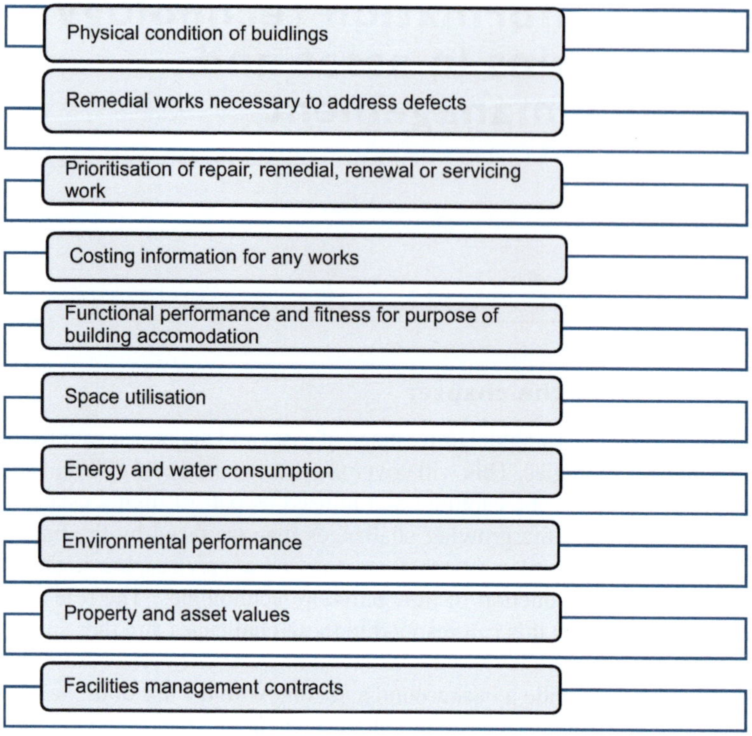

Figure 12.1 Different aspects of information gathering

automatically transfer data into reports. Information downloaded can include a variety of different aspects including those in Figure 12.1.

In addition to these areas, other information can be used as to form estates management systems which seek to assist with asset planning. Examples of such information could include floor plans, certificates, guarantees, planned preventative maintenance programmes, building occupancy data, type of building construction and health and safety risk assessments. Building assets can sometimes be assigned QR codes which detail their maintenance histories including past repairs and servicing. This is particularly useful for electrical and mechanical plants as a 'logbook' to record previous issues, historical data and any ongoing faults. Other information that is compiled as an estates management system could be generated from computer-aided facilities management (CAFM) systems. These systems could be utilised as help desk facilities where building occupiers are encouraged to report faults or work request through a central estates administration team.

12.3 Digitalisation and data management to tackle challenges in the built environment

The built environment sector is now very much a data management business, where concepts like building information modelling (BIM) are now essential to driving information-led productivity in estates projects and programmes. Customers understand the value of purchasing not only a physical asset but also the associated data and information sets. Accordingly, the use of data, emerging technologies, materials and instruments has enormous prospects. Despite this, managing complex estates presents one of the most significant developmental challenges. It results in an increased demand for the built environment to rethink and change the adverse effects such as real-time monitoring, big data management, climate change, resource scarcity and highly complex technologies. The built environment is the critical player in retrofitting, rehabilitation and renovation of existing property assets.

Globally, the built environment is one of the most fragmented sectors, which hampers sustainable development, particularly in urban settings. It contributes to a myriad of industrial challenges in the UK covering low and underachieving productivity levels as compared to other developed economies and industries such as manufacturing, aerospace and automotive industries. The industry is prevalent with outdated technologies incapable of meeting industry needs, leading to client dissatisfaction, inefficiencies in process and operations, safety and carbon footprint.

The digitalisation of the built environment is an opportunity to tackle industry's challenges by taking advantage of the general availability of best practices from other sectors and of advanced techniques and tools, digital workflows and technological expertise to shift to a higher level of efficiency and to become a digital industry. Despite the vast potential of the industry, increased efficiency and productivity can only result from digitisation, new construction techniques and innovations.

12.4 The digital revolution

The digital revolution is a generic term to describe the phenomenon of usage of information and automation technologies to allow collaboration between products and their process and supply chain. Over the past decade, digital technologies have transformed vast industries, setting in motion the digital revolution. Digital transformation in the built environment is the practice of how digital technologies can allow the sector to develop what could not be created before, to construct new forms of building, to transform estates and respond to growing asset management and sustainability challenges. The digital revolution in this regard promotes growth and development through its efficiency and versatility towards productivity. For this reason, most companies, from different sectors ranging from service to manufacturing, have transformed over the last couple of decades and have strived to put estates process innovations at the centre of their operations.

Dramatic shifts in climate, technology, culture, global production and value chains are forcing governments, companies and professionals to reconsider not only how they interact with each other but also how they produce, distribute and consume the environment and its built assets. The advent of broad deployment of emerging technologies, such as artificial intelligence, sensor technology, cloud computing, open data, the Internet of Things and blockchain, have reinvented the way we can handle development and adapt to these challenges. Accordingly, built environments need to use innovative technologies to allow them to develop, sustain growth and exploit high-tech products to serve their people better (ESPON 2020).

Building information management (BIM) and developing information management in the building industry is a significant move forward particularly for the UK. Notwithstanding this premise, it is essential to exploit the full potential of emerging digital technology such as the Internet of Things, digital twins, artificial intelligence, and advanced data analytics throughout the built environment.

12.5 Digital twins for streamlining the operation of building assets

12.5.1 Description of a digital twin

Digital twins represent open, trusted and reliable digital representations of elements and dynamics of a physical system or built assets such as a building, urban network or asset portfolio and can be referred to as a cyber-physical system. Traditionally they contain feedback loops that allow big data transportation and synchronisation between the virtual replication and physical environments with near-real-time Internet of Things (IoT) sensors. In this way they are embedded to multiple data points, to relay structured and big unstructured data flows from physical to virtual. They also offer the ability to incorporate analytics, including machine learning and simulation algorithms, to provide an understanding to the elements and dynamics. In this way they offer real-time corrective triggers back to the physical asset or responses, current insights and predict various potential future scenarios, to promote better decision-making on a built asset or an ecosystem of connected digital twins.

12.5.2 The digital twin concept for estates assets

A digital twin is a cycle flow between physical and virtual realities (mirroring or twinning), where data flows from the physical to the virtual environment, and information and processes respond from virtual to physical state (Jones *et al.* 2020).

Digital twins harness data to prepare future digital operational life of built assets. At the handover stage, digital twins provide a single source of truth, through a complete and reliable set of information about buildings and estates. This creates a uniformed smart built asset database, saving money and time having not to recreate these applications. During the operational stages of projects, smart assets gather data from their surroundings using Internet of Things networks and sensors, so they can adapt to their environment and change their operating performance accordingly. This enables them to evolve over the lifespan of the asset using machine learning and operational data AI to recognise trends in usage, recognise where building or network activity does not meet expectations. This not only increases the efficiency, sustainability and security of buildings and infrastructure but also transforms the management of facilities. Operations decision-making of this kind led by smart live and static information management can extend the life cycle of built assets and improve their performance. The use of digital twins also allows asset owners and operators to have greater leverage over their buildings and service networks, with the capacity to enforce and test against specific targets. An example of this could be carbon footprint monitoring and energy controls in buildings and across estates portfolios.

12.5.3 Building automation

Digital twins have the power for building automation operations. They can integrate complex physical systems and leverage twin and sensor data for an asset to be intelligently operated with transparency and providing valuable insight into the owners and operators. They bring together previously unconnected physical systems into real-time Internet of Things integrated systems. These include smart heating, ventilation and air-conditioning (HVAC) systems, digitised pneumatic thermostats, smart plugs, intelligent elevator controls, security sensors, environmental responsive kinetic facades and smog-eating facades. This result enables a unified building database with optimised processes. Building ventilation systems became more critical following the COVID-19 pandemic. Smart ventilation systems and sensors can be integrated into the physical building and linked to the digital twin to provide simulation, monitoring and control of airflow to improve air quality and control the level of virus infections. The application of digital twins for building automation enhances aspects such as human comfortability, air quality and overall asset productivity.

12.5.4 Predictive maintenance

Digital twins can continuously monitor assets and activities, and identify abnormal behaviours, allowing human operators to react rapidly and reduce downtime to manage predictive maintenance. Predictive maintenance incorporates machine learning and sensors to enable the digital twin to inform the facility management team that, based on actual user data, a component needs servicing or is near to the end of its life. Sensors provide new insight into the digital twin performance and predictive maintenance of the asset and its integration with smart systems. For this reason, they can link with users through visualisation web dashboards and provide new insights into improving built asset utilisation and performance. This gives users greater control over their own spaces and environment, thus improving the user experience and reducing maintenance costs.

12.5.5 Remote assistance technology

Digital twins can operate autonomously just by being able to evaluate a scenario, formulate tailored solutions and bring them into a corrective action through the power of remote assistance technology. Remote assistance technology enables estates departments to remotely monitor and control of built assets, plants and machinery. This technology in digital twins can provide decision support and alert to maintenance managers to any potential malfunction, incident or anomaly to an asset or an ecosystem. In estates and facilities management, remote monitoring can provide real-time monitoring to troubleshoot distant sensors systems and perform remote maintenance, enabling operators to fix the issue at hand instantly. Furthermore, remote assistance technology can accelerate the adoption of digital twins, especially in the aftermath of COVID-19. They can achieve this firstly by minimising the need to have people working from offices, by allowing remote monitoring and controls of assets via an Internet of Things platform.

12.5.6 Building information modelling (BIM)

Digital twins can be seen as a natural evolution of building information modelling (BIM) which is already beginning to reflect the built environment digitally (BSI Group 2018). BIM is a descriptive term for the advanced, collaborative and information-centred process used to drive the built environment design, construction and operations. As a revolutionary technology, it optimises the construction industry's conventional way of operating, by harnessing early collaboration to increasing efficiency beginning from the design stage to the construction stage with the greatest advantages in the operational phase of the asset. BIM is a value-generating collaborative process during a built asset life cycle, focused on the creation, compilation and exchanging of shared 3D models and the integration of intelligent and structured data.

A BIM model forms a core subset and enabler of a digital twin. The BIM model constitutes persistent data collaboratively gathered and updated information at key phases of a project. The model is a finely tuned tool for more accurate phases of a building's life in design, collaboration, visualisation, costing and construction sequencing. The primary function of BIM is to design and construct a building and to provide a digital record of a completed project after completion (Roper 2020).

BIM should be regarded as a core enabler to the progress of digital twins for the built environment. The primary role of it is to design and construct an asset collaboratively and to have a digital record forming the Asset Information Model (AIM) upon completion of a finished project. This is useful in the operational phase of the built asset and continuous improvement of the model provides background for the digital twin. The digital twin harnesses the Asset Information Model and links it to the physical built asset enabled through the Internet of Things (IoT) and geographical information system (GIS) to provide a digital context. Connected devices and sensors are attached to the physical asset to enable real-time operational data collection. Such devices provide feedback to the AIM to integrate the operational and digital twin data to provide real-time analytics of the physical asset. This forms a basis for common interoperable knowledge base on the cloud, useful in operational performance of the built asset through aspects such as predictive maintenance and what-if analysis test cases.

Digital twins are designed to monitor and improve the operational performance of a physical asset and to allow predictive maintenance (Khajavi *et al.* 2019). BIM is also not designed to operate with real-time data but mostly used to design, build and maintain tasks and interoperability does not inherently require real-time capabilities. It is, however, a strategic facilitator for product life cycle management (PLM) for both buildings and infrastructure assets. PLM was a central focus of the introduction of the digital twin concept and is the process of managing a product's entire lifecycle from conception, through design and production, to service and disposal. It connects people, data, processes and systems and provides an information backbone for organisations and their extended enterprise. A fusion of BIM into the digital twin can promote product life cycle management of built assets by enabling product function analysis, performance prediction, product monitoring, ergonomics analysis, layout planning, prediction of the status of equipment, process analysis, operation optimisation and forecast of maintenance.

Building information modelling (BIM) is a rich dataset for built assets and its integration with geographical information systems (GIS) helps bring more efficient project delivery. Integrating the two systems can add more advantages to datasets including a combination of large-scale and small-scale built environments. They can achieve this by looking at BIM models on a broader sense of 3D geographic location, providing a more realistic representation of urban environments including buildings, trees, property, road networks and agents.

12.5.7 Internet of Things (IoT)

Digital twins explicitly work with data fed by digital devices or sensor systems to monitor and analyse a physical asset's structural and environmental parameters to perform highly precise digital twin simulation and real-time data analysis (Khajavi *et al.* 2019). Sensors are added to the physical object or one or more of its connected IoT components. The sensors gather data on the object in real-time and send it to a cloud-hosted system for processing of the data. This allows the digital twin to digitally replicate the tangible object, offering real-time analytics that are automated through machine learning. In this way it detects trend behaviours and probabilities in built assets for potential project performance improvements or well-informed decision-making. In addition, it offers insights into life cycle operational functionality, whilst predicting any potential problems on how the asset will change or behave in the future (Dentons 2020).

Linking digital twins with the Internet of Things (IoT) provides the data required to understand how the physical counterpart conducts and interacts in its operational environment (Madni *et al.* 2019). The integration of IoT and digital twins enhances preventive maintenance and artificial intelligence-based physical system and operational process optimisation. IoT transmits performance, maintenance and safety data and other data from the physical twin to the digital twin, serving as a link between the physical and virtual worlds. IoT facilitates predictive modelling by transmitting multi-sensor information to the digital twin (Madni *et al.* 2019). The integration of digital twin and IoT makes it possible for an organisation to gain insight into how consumers use the products. These insights will allow customers to optimise the schedule of maintenance and use of resources, proactively anticipate possible product failures and reduce system downtimes.

12.5.8 Big data

Big data refers to vast volumes of heterogeneous, multisource data defined by the 5 Vs, namely: high volume, variety, velocity, veracity and value. Big data analytics offer a modern way of understanding the physical realm. A vast amount of data is generated during the operation of a digital twin and to retrieve valuable raw data and information, advanced data analytics and fusion technologies are required. By integrating big data with developments in artificial intelligence (AI) and machine learning (ML) prospects in digital twins can be enhanced.

12.5.9 Cloud computing and data analytics

Cloud computing and data analytics integrated into the digital twin offer insights into building assets and their performance, enabling users to take timely and well-informed action, promoting feedback mechanisms that lead to product and process improvements. As the sensors continue to proliferate, edge computing is another significant technology gaining traction as an advancement to cloud computing. It involves processing data on a network's periphery. At a time when more and more of our data is being processed in the cloud, the prospect of continuous feedback loops seems daunting, but edge computing can mitigate these problems by storing data at or near the source. The next generation of autonomous, driverless vehicles are a prime candidate to both gains from edge computing and 5G networking, which together would promote instant reactivity in an ever-changing built environment.

Through linking objects to a cloud-based network that processes the collected data, the performance of entities in the physical world can be used to evaluate virtual model performance. Digital twins create efficiencies by optimising processes and workflows by interacting with virtual objects and systems to enhance asset management proactively. In this way they can provide decision support to users (Madni *et al.* 2019). Furthermore, enabled by remote assistance technology, they can alert maintenance managers of any potential malfunction, incident or anomaly in their estates services. Such a concept is aimed at addressing outdated methods for troubleshooting, enabling operators to instantly fix the issue at hand. In this way the digital twin can operate autonomously by being able to evaluate a scenario, formulate tailored solutions and bring them into action (Iberdrola 2020).

12.5.10 Smart buildings

Smart buildings are an excellent example of applications that benefit from digital twin machine learning capabilities. The use of machine learning in a digital twin includes the supervised learning of operator or user preferences and involves setting up priorities in a simulation-based, controlled test bed. This can be achieved through supervised learning of objects and patterns using clustering techniques in virtual and real-world environments. It can also entail improving environmental systems and learning in unpredictable situations and operational environments which are only partly measurable (Madni *et al.* 2019). The University of Salford's Energy House 2.0, referred to in Chapter 10 of the book, is one example of research and development into smart buildings and one which should be promoted within the built environment.

12.6 Summary of chapter

Having information, which is accessible, comprehensive and accurate, is regarded as essential for the strategic and operational management of property estates. However, any information system will need to be user friendly and flexible enough to cater for future changes. Such information can be used to form estates management systems which seek to assist with asset planning. Examples of such information could include floor plans, certificates, guarantees, planned preventative maintenance programmes, building occupancy data, types of building construction and health and safety risk assessments.

The built environment sector is now very much a data management business, where concepts like building information modelling (BIM) are now essential to driving information-led productivity in estates projects and programmes. The digitalisation of the built environment is an opportunity to tackle industry's challenges by taking advantage of the general availability of best practices from other sectors and of advanced techniques and tools, digital workflows and technological expertise to shift to a higher level of efficiency and to become a digital industry.

Digital transformation in the built environment is the practice of how digital technologies can allow the sector to develop what could not be created before, to construct new forms of building, to transform estates and respond to growing asset management and sustainability challenges. Accordingly, built environments need to use innovative technologies to allow them to develop, sustain growth and exploit high-tech products to improve their estates.

Digital twins can be utilised for streamlining the operation of building assets. They can harness data to prepare future digital operational life of built assets. In this way they can create uniformed smart built asset databases, saving money and time by not having to recreate these applications. The use of digital twins also allows asset owners and operators to have greater leverage over their buildings and service networks, with the capacity to enforce and test against specific targets.

Digital twins have the power for building automation operations and assisting with prediction of future maintenance requirements. Furthermore, they can assist with remote assistance technology enables estates departments to remotely monitor and control of built assets, plants and machinery. Cloud computing and data analytics integrated into the digital twin offer insights into building assets and their performance, enabling estates teams to take timely and well-informed action, promoting feedback mechanisms that lead to product and process improvements.

Finally, smart buildings are an excellent example of applications that benefit from digital twin machine learning capabilities. The University of Salford's Energy House 2.0, referred to in Chapter 10 of the book, is one example of research and development into smart buildings and one which should be promoted within the built environment.

References

BSI Group. (2018). Digital twins for the built environment. *Sustainability*, 17(2307), p. 56.

Dentons. (2020). Digital twins and the Internet of Things (Internet of things): Utilising data. Available at: www.dentons.com/en/insights/alerts/2020/january/24/digital-twins-and-the-internet-of-things-utilising-data (accessed 3 July 2020).

ESPON. (2020). *Digital innovation in urban environments*. ESPON.

Iberdrola. (2020). Iberdrola. Available at: www.iberdrola.com/innovation/digital-twin (accessed 17 July 2020).

Jones, D., Snider, C., Nassehi, A., Yon, J. and Hicks, B. (2020). Characterising the digital twin: A systematic literature review. *CIRP Journal of Manufacturing Science and Technology*, 29, p. 36.

Khajavi, S.H., Motlagh, N.H., Jaribion, A., Werner, L.C. and Holmström, J. (2019, October 7). Digital twin: Vision, benefits, boundaries, and creation for buildings. *IEEE Access*, pp. 147407–147419.

Madni, A.M., Madni, C.C. and Lucero, S.D. (2019, January 22). Leveraging digital twin technology in model-based systems engineering. *MDPI*, pp. 1–13.

Roper, J. (2020). Unissu-defining, the digital twin for buildings. Available at: www.unissu.com/proptech-resources/Defining-the-Digital-Twin-for-Buildings (accessed 3 July 2020).

Chapter 13

Implications in practice for maintenance management in the built environment

13.1 Introduction

This chapter is largely focused on considering the implication of the proceeding chapters of the book, underpinning the main themes, and looking at potential recommendations for future practice. In this regard it will reiterate the traditional poor and outdated maintenance management practices which are still prominent in estates management. The consequences and implications of these for organisations and the built environment will be discussed. These will include reputation damage and image considerations for organisations alongside the influence on their long-term business dealings and performance. In consideration of these problems and challenges, there is a need to 'turn the tide' and overcome poor and outdated traditional working practices and behaviours in the pursuit of more successful business outcomes. Throughout the book the authors have established and discussed the relationship between maintenance management and business performance within organisations. Utilising this relationship the book has sought to adopt an innovative and alternative approach to maintenance management and therein infill the current gaps in knowledge. This chapter will recommend various initiatives and improvement measures to promote best practice which will include the following areas:

1. The need to rethink the concept of maintenance management
2. Fully integrating asset management strategies into operational planning
3. Raising environmental sustainability credentials for organisations as part of asset and maintenance management
4. Improving health and safety in maintenance management
5. Raising awareness of how successes, failures and key risks within organisations can be influenced by maintenance management practices
6. Promoting collaborative working and partnering principles in maintenance management
7. Using the surveyors practical guide to form the basis of a practical toolkit for guidance and teaching of maintenance management in the built environment

Finally, education, training and continued professional development (CPD) around asset and maintenance management will be covered and an articulation of how construction professionals can learn to move away from traditional outdated and inefficient practices.

13.2 Implications for organisations and the reputation of the built environment from poor maintenance management

The book has articulated that maintenance management practices can take a negative toll on the performance and ultimately outcomes of estates projects and programmes. This could lead

DOI: 10.1201/9781003358800-13

to assertions that estates and asset management in the built environment are inefficient, unsafe, wasteful, compromised on quality and cannot deliver to time and budgetary constraints. Other implications and consequences could include reputational damage and image considerations for organisations. The book has given examples and presented case studies where poor asset and estates management practices and behaviours have come to light. This can be extremely damaging especially when where health and safety risks are exposed in the public domain. In this regard, some high-profile press coverage in the past has led to widescale public condemnation and a poor image of the built environment and a perception that they are not sufficiently managed or regulated.

13.3 Initiatives and measures to improve maintenance management

The earlier chapters of the book have identified various ways that organisations can improve maintenance management within their respective businesses. The benefits were articulated from perspectives of both property and business performance. The authors have reflected upon these and in an attempt to promote change have made recommendations for the future. These have been captured next and are designed as a model of innovative good practice in maintenance management:

13.3.1 The need to rethink the concept of maintenance management

Maintenance management is not a concept that is new. Moreover, its roots have gone back thousands of years to the times of our ancient forefathers in maintaining religious temples, places of burial and landmarks e.g. Egyptian pyramids. In recent years, however, the world has been witnessed to many different challenges in maintaining buildings and facilities which are deemed safe, compliant and conducive to user needs. Sometimes such challenges have resulted in failures where buildings have been rendered potentially unsafe leading to building closures and expensive remedial works. This controversy has damaged the reputation of the built environment and all those organisations affected. Recent examples in the UK have included building risks associated with reinforced autoclaved aerated concrete (RAAC) and use of aluminium composite cladding (ACM) that has been deemed unsafe and potentially hazardous cladding for buildings, following the 2017 Grenfell Tower Disaster in London.

To avoid such cases in the future there is a growing need to elevate the importance of asset and maintenance management within organisations on par with their core business activities. In the past some organisations have perceived maintenance of property and estates as desirable rather than essential. In addition, for many years business organisations have focused mainly on strategies linked to making increased profits and creating value for shareholders. Accordingly, asset and maintenance management has had to take second place amongst other corporate strategies linked with operations and business performance. This can present a very costly and risky strategy for companies especially where cases of noncompliance have emerged.

Another dilemma for maintenance management is that very little attention has been focused on tailoring decision-making and everyday practices around corporate strategies and processes of business planning. Estates strategies have instead been concentrated on only the property benefits rather than the business benefits for employers, employees and customers and in areas such as gaining operational efficiencies. In some cases, silo mentalities have emerged around estates issues and maintenance of assets and a clear disconnect from operational planning and core business activities. Clearly there is a call for change needed to respond to this predicament. Raising awareness of asset management has been a main theme of this book which will hopefully serve to press the need for reforms and strategic rethinking. A change in perception of asset and maintenance management is therefore crucial. Buildings and estates should then be regarded as business enablers rather than expensive assets which burden companies with repairing liabilities.

Notwithstanding the aforementioned dilemmas, in recent years companies are realising that they need to focus on nonfinancial strategies alongside their economic goals. This involves maintaining a minimum level of quality standards for their buildings and estates. Businesses have become aware that providing good, well-maintained and legally compliant working environments for employees can improve their reputations and in turn increase their success in the marketplace.

13.3.2 Fully integrating asset management strategies into operational planning

The area of maintenance management has in the past has not received the prominence it has deserved as part of corporate strategies. Moreover, it has been regarded as a secondary consideration and one which has been detached from core business activities. To respond to this dilemma, the book has presented a strong case for fully integrating maintenance management into operational planning. Accordingly, decision-making around business propositions and proposals should consider how the assets and estates of organisations can help them perform more successfully, efficiently and reduce disruption to business activities. Examples may include creating additional new space to cater for their expansion plans and therein increase productivity, or adaption of existing accommodation to improve its functional performance.

13.3.3 Raising environmental sustainability credentials for organisations as part of asset and maintenance management

The book has considered why environmental considerations are so relevant for businesses today and how can these by integrated within maintenance management strategies. For this reason, organisations as part of their maintenance programmes, asset management and business planning should embrace all aspects of environmental sustainability. This may involve considering alternative green technologies when undertaking repair, renewal or refurbishment programmes. An example may be the renewal of traditional tungsten incandescent and fluorescent light fittings for more efficient LED alternatives, as part of maintenance planning. This could represent a positive way forward for meeting their organisational commitments to lowering carbon emissions whilst at the same time saving energy costs. Other environmental sustainability commitments could include strategies around reducing waste, lowering energy consumption, developing carbon management plans and adopting sustainable transport policies. The authors presented case studies in Chapter 11 where environmental strategies such as these were introduced.

When procuring maintenance projects and programmes, there should be a growing awareness of the importance of environmental considerations, as part of their organisations' social responsibilities. In this sense all responsible organisations should uphold policies, strategies and values linked to corporate social responsibility. This in turn will have far reaching benefits for their businesses in terms of public perception, marketing, operational efficiency, adherence to ever increasing regulations and maximising future government funding opportunities. Other initiatives may include devising and implementing credible recycling programmes and deploying state of the art technologies to reduce energy and water consumption. Use of renewable sources of heat and power are particularly appropriate especially where grant funding is available and having a minimum standard of BREEAM for all new buildings commissioned.

Reduction in single-use plastic waste is particularly important in this regard. Organisations that have shown responsibility in 'doing their bit' for the sustainability and environmental agenda have benefitted from positive media coverage and gaining environmental accreditation. Such accreditation could include ISO 14001 Certification; Environmental Management Systems and ISO 50001 Certification; Energy Management. Organisations should also look at their supply chains and have influence on them for responsible procurement of supplies and services. In addition, adopting sustainable transport policies can have many beneficial outcomes for businesses in reducing carbon emissions and can be linked to cycling schemes, car sharing and subsidies for using public transport. All the aforementioned measures, if implemented, should benefit organisations when they are tendering for works by being able to demonstrate that they are responsible and aware of the environmental sustainability agenda. This is particularly the case in the public sector where bid evaluation criteria linked to appointments of consultants and contractors are more frequently focusing on environmental sustainability aspects.

13.3.4 Improving health and safety in maintenance management

Health and safety in the UK's built environment have improved greatly over the last decades. Notwithstanding this premise the fatal injury rate is 1.62 per 100,000 workers which is around four times the all-industries rate. According to the HSE in 2020–2021 142 workers in the built environment were killed in the UK, and accordingly there is still room for improvement in health and safety management. Those ethically motivated construction organisations that are striving to reduce injuries and fatalities should consider not simply complying with health and safety legislation but exceeding regulations. In practice they can achieve this by introducing enhanced safety measures and initiatives whilst carrying out their maintenance programmes. Introducing near miss reporting with essential improvement measures for each reportable incident is one example of such an initiative and has proved very effective in the past in reducing hazards on sites. Other initiatives that companies could adopt include becoming members of the Construction Skills Certification Scheme and ensuring that all construction and maintenance staff are properly trained and gained accredited to the scheme.

13.3.5 Raising awareness of how successes, failures and key risks within organisations can be influenced by maintenance management practices

Organisations should identify how adopting policies and delivering strategies linked to maintenance management can assist them in gaining more efficiencies and achieving other successful outcomes for business owners, customers and employees. Accordingly, they should consider how asset management can benefit their businesses and bring advantages for all stakeholders.

As well as considering the benefits and successes brought about by good maintenance management practices, the potential risks associated with failures should also be taken into account. This is particularly important when considering the real estate value of organisational property portfolios and the financial consequences of deterioration through failures in maintaining their building assets. In addition, the legal risks associated with noncompliance with statutory regulations and breaches of health and safety legislation should not be underestimated. Accordingly, the book should be regarded as a standard reference for businesspeople in understanding how maintenance management affects organisational risk.

Ethical leadership in asset and maintenance management is another important factor for organisations to uphold. Accordingly, they should ensure that certain management measures are all in place which include adopting the right project controls, communication strategies, ethical codes of conduct and collaboration approaches. Furthermore, there is a strong case for identifying a set of clear guidelines, national or international, to support the adherence to and regulation around future maintenance management strategies.

13.3.6 *Promoting collaborative working and partnering principles in maintenance management*

The book has promoted organisations to use collaborative working and partnering principles, wherever possible, as part of their maintenance and asset management strategies. This could involve public sector organisations partnering with the private sector and therein allowing supply chains to engage and assist in maintenance-related decision-making. This could provide the way and means to encourage improved practices and behaviours in sharing expertise and knowledge, whilst improving business relationships. The extent and degree of collaboration has also been shown to underpin relationships of key stakeholders which in turn can lead to project success (Challender *et al.* 2019). The emergence of these arguments appears to affirm that trust is one of the most important factors to risk minimisation on projects. The use of collaborative procurement methods and improvements to relationships that they bring about, applying the principles of Egan (1998, 2002), can procure more successful project outcomes. Success in this way can be measured through significant improvements in cost predictability, safety and time predictability. For this reason, organisations as part of their maintenance management regimes should conduct strategies and continued professional development (CPD) linked to promoting collaboration.

13.3.7 *Using the surveyor's practical guide to form the basis of a practical toolkit for guidance and teaching of maintenance management in the built environment*

Chapter 9 of the book presented a practical guide or toolkit, which Professor Challender designed to assist students in the inspection and building survey processes associated with maintenance management. The authors hope that this teaching and learning guide can streamline future courses taught in universities and colleges, by helping students navigate their way through building surveying processes and methodologies.

13.4 Education, training and continued professional development (CPD) around maintenance management

From the interview conducted with professional estates practitioners, for the book, they reported that their organisations had not implemented enough training programmes linked to improving maintenance management practices. This clearly presents a deficiency in current practice and a dilemma for the industry. Reflecting on this, it is recommended that organisations should promote continued professional development in the workplace for the teaching and learning for improving maintenance management. In addition, professional bodies and construction regulatory bodies can adopt training programmes to remedy this educational dilemma, and in this way, measures can be embraced by their members. In this regard, development of skills, knowledge and awareness of considerations around asset management should all feature very highly in future training programmes. This is especially important within the context and environment of the built environment where things, regulations and developments are dynamic and constantly changing. The RICS approach this by imposing mandatory requirements for its members to submit written documented proof of at least 20 hours of CPD per calendar year, with penalties and sanctions applied for noncompliance. Some would argue that a 'carrot rather than the stick approach' would prove more successful and perhaps CPD participation in such could be better promoted through motivational measures rather than what could be regarded as old-fashioned penalties designed to punish. The implementation of CPD training may be best achieved through workshops, conferences, mentoring, lectures, webinars and by studying a new qualification. Other ways to disseminate CPD material could be through working forums, talks, discussions and seminars designed to endorse good maintenance management.

Best practices could be adopted in this way to improve old and outdated traditional practices and behaviours and achieve more successful outcomes for maintenance programmes. In addition, this could raise the reputation of those organisations that have failed in the past to take their maintenance responsibilities seriously. In this way, businesses can meet all their statutory requirements and avoid breaches and associated risks of prosecution. Notwithstanding this premise, only when maintenance management is commonplace and placed centre stage within organisations will professionalism be enhanced leading to improved business outcomes.

Investing in this type of training will ensure that staff are kept up to date on latest developments and changes in regulations. In addition, it has been suggested that colleges and universities should tailor more academic modules and courses to the subject of professional ethics to address this deficiency in knowledge and education. With the right degree of education, training and CPD, construction professionals can become more aware of how improved maintenance management strategies can benefit their estates portfolios whilst at the same time generating greater business efficiencies.

13.5 Summary

There have been assertions that estates and asset management in the built environment are inefficient, unsafe, wasteful, compromised on quality and cannot deliver to time and budgetary constraints. Implications and consequences of these dilemmas could include reputational damage and image considerations for organisations. To respond to this predicament there is a need to rethink the concept of maintenance management, especially given the many different challenges in maintaining buildings and facilities to keep them safe, compliant and conducive to user needs. Sometimes such challenges have resulted in failures where buildings have been rendered potentially unsafe leading to building closures and expensive remedial works. In the past silo mentalities have emerged around estates issues and maintenance of building assets and a clear disconnect from operational planning and core business activities. To avoid this in the future there is a growing need to elevate the importance of asset and maintenance management within organisations on a par with their core business activities. To encourage this shift in perception, businesses have become aware that providing good, well-maintained and legally compliant working environments for employees can improve their reputations and in turn increase their success in the marketplace.

The book has presented a strong case for fully integrating maintenance management into operational planning. Accordingly, decision making around business propositions and proposals should consider how the assets and estates of organisations can help them perform more successfully, efficiently and reduce disruption to business activities. In addition, when procuring maintenance projects and programmes, there should be a growing awareness of the importance of environmental considerations, as part of their organisations' social responsibilities. Initiatives may include devising and implementing credible recycling programmes and deploying state of the art technologies to reduce energy and water consumption.

Improving health and safety in maintenance management should also be considered wherever possible and in practice this can achieve this by introducing enhanced safety measures and initiatives whilst carrying out maintenance programmes. Furthermore, there is a growing need to raise awareness of how successes, failures and key risks within organisations can be influenced by maintenance management practices. Successes could include raising employee satisfaction levels through improvements in working environments whereas failures and risks could be associated with deterioration of buildings and breaches of regulations and statutory requirements. In addition, promoting collaborative working and partnering principles in maintenance management could provide the way and means to encourage improved practices and behaviours in sharing expertise and knowledge, whilst improving business relationships.

It is recommended that organisations should promote continued professional development in the workplace for the teaching and learning for improving maintenance management. Best practices could be adopted in this way to improve old and outdated traditional practices and behaviours and achieve more successful outcomes for maintenance programmes.

References

Challender, J., Farrell, P. and McDermott, P. (2019). *Building collaborative trust in construction procurement strategies*. Wiley.

Egan, J. (1998). *Rethinking construction. The report of the construction task force*. DETR.TSO, pp. 18–20.

Egan, J. (2002). *Accelerating change. Rethinking construction.* Strategic Forum for Construction.

Chapter 14

Summary of key points, reflections, overview and closing remarks

14.1 Introduction

This final section will summarise each chapter of the book and extrapolating the key findings and issued raised. Following on from what has been articulated and discussed, it will present some final reflections, an overview and some closing remarks for the future of maintenance and asset management in the built environment, taking account the inherent dilemmas and challenges that are faced.

14.2 Summary of the key issues raised throughout the book

14.2.1 Introduction to the book and the need to rethink building and facilities maintenance management

The UK government's Industry Strategy for Construction 2025 has caused a paradigm shift in the construction industry's view of the maintenance and upgrading of existing buildings. Accordingly, providing improved asset and maintenance management strategies has become a hot topic and one that businesses need to pay special attention to if they want to succeed. This book has responded to such needs brought about by the prominence of this area of practice, which has been neglected in previous years. *Building and Facilities Maintenance Management: An Innovative Operational Approach* is concerned with how we make improvements and positively contribute to existing estates management practices in terms of providing safe and sustainable work environments. The aim of the book is to provide the foundations for advance thinking and shape future practices in maintenance management and articulate how there can be a vast array of methods, motives and rationales for achieving more successful outcomes on both an organisational and industry-wide basis.

There is a need to rethink the concept of maintenance management. In recent years companies have realised that they need to focus on nonfinancial strategies alongside their economic goals. This involves behaving ethically in providing good, well-maintained and legally compliant working environments for employees which in turn can raise their reputations and in turn increase their success in the marketplace.

The book has examined how buildings and estates can be regarded as business enablers rather than expensive assets which burden companies with repairing liabilities. In this regard it made recommendations to overcome poor and outdated traditional working practices and behaviours in the pursuit of more successful business outcomes. The book is intended to assist students, academics, construction-related practitioners and clients in their awareness, breadth of

DOI: 10.1201/9781003358800-14

knowledge and comprehension of the issues around maintenance management considerations, with the overarching aim of delivering more successful outcomes.

The research for the book has been mostly derived from a combination of sources which include the authors' own experiences, and interviews with a wide range of construction professionals and literature. In addition, the authors have conducted a literature review of many of the aspects and subject matters covered in each chapter, and utilised various academic sources which are included in the references. Most chapters were supported by case studies from the authors' own experiences. As such it represents a practical guide for students, academics and practitioners that can develop into a common framework on how to initiate, procure and manage more collaborative maintenance management strategies and working philosophies. The research findings were presented to prioritise use by professional practitioners and therein provide a practical guide for adherence to best practices as a toolkit, for improvements in asset management in the built environment.

14.2.2 Significance and nature of building maintenance

There is no universal definition for maintenance, but clearly there are common strands which exist to articulate its meaning in practice. As well as natural deterioration processes, buildings when not maintained to the correct standard, can fail unexpectedly from physical or performance perspectives. This can lead to malfunctions of building services or emergence of physical building defects which can cause significant disruption to users and expensive repairs for building owners. To avoid such failures and ensure that buildings as valuable assets do not depreciate over time, it is important that maintenance regimes are put in place. Such regimes can be affected by use and type of buildings and examples were presented in this chapter to explain the different scenarios e.g. listed buildings.

It is essential to consider the standards by which building owners wish to maintain their estates. These standards can have a direct effect on the comfort and satisfaction of building occupants which in turn could affect the performance of businesses. There can be different approaches to setting building standards and some building owners may wish to set standards very high to present the right image of their organisations, and thereby attract new customers. Conversely others may regard built assets as merely accommodation for conducting their core business functions within and therefore seek to limit repairs, maintenance and any improvements to a minimum. In addition to the standards associated with state of repair and physical condition, there are also expectations for building accommodation to be fit for purpose in terms of space and functional suitability to meet the expectations of building occupants.

Essential maintenance includes compliance with regulations and statutory requirements and any breaches or infringements could be unlawful and possibly result in litigation and enforcement action. Moreover, desirable maintenance refers to works of a non-statutory nature including redecoration, window replacement, new finishes, and nonessential refurbishment programmes.

The Health and Safety at Work etc. Act 1974 requires employers to maintain the health and safety of their employees at all times. This is particularly important when carrying out maintenance programmes when arguably the level of risks to occupants and members of the public is greater. Accordingly, works needs to be carefully managed with use of method statements and risk assessment tools to ensure health and safety is maintained at all times. There have been concerted efforts to improve health and safety management when carrying out construction and maintenance programmes. Improvement measures have included 'near miss' reporting, accreditation such as the Construction Skills Certification Scheme and the introduction of new regulations including the Construction Design and Management Regulations (CDM).

Estates can be sophisticated and complicated and can present endless risks from unexpected events which if not managed correctly can be extremely disruptive or disastrous. The impacts from these risks and issues if not managed correctly could result in unacceptable costs being incurred, time delays, performance shortfalls or closure of facilities in some cases. Examples of risks were given in this chapter which included fires, floods, electrical failures, loss of heating and cooling, and information technology malfunctions e.g. cyberattacks. It is therefore important for risk assessments to be undertaken to identify potential estates issues, their severity and any mitigation measures to limit their impact. These risk assessments would normally become part of business continuity and disaster recovery planning.

Maintenance can be described as unscheduled or planned maintenance. Planned preventative maintenance programmes (PPMPs) can reduce the degree of 'reactive unplanned' maintenance. They incorporate regular checks, servicing and repairs to ensure that buildings remain in full working order and condition for as long as possible. Furthermore, they should predict potential defects or faults which could arise from condition survey information to plan, schedule and programme remedial works more effectively. This predictive approach is designed to avoid potential defects emerging and to minimise resultant damage and disruption associated with prolongation of any issues. Examples to assist readers have been provided in this chapter which have included defects associated with timber decay and concrete cancer. Benefits of PMPs can include the ability to group works into programmes e.g. roof replacement, which can bring financial cost savings from economies of scale.

Renewal as opposed to repair decisions may be made when buildings or some of their component elements are approaching obsolescence. There are four different types of obsolescence, and these include physical deterioration and obsolesce brought about by function, location and economic forces. A case study was provided in this chapter where a relatively new building procured by the lead author was adversely affected by unexpected economic forces leading to its demolition. Accordingly, predicting time horizons for the life of buildings, referred to as life cycle analysis, is something that can be daunting for maintenance teams.

The next chapter followed on from the concepts raised and introduced delivery strategies associated with maintenance policies, programmes and systems.

14.2.3 Maintenance and asset planning strategies, programmes, policies and systems

Strategic corporate business plans articulate the vision, mission, short- and long-term aims, goals and objectives of organisations but also should relate to building and estate assets. Accordingly building maintenance strategies and policies should include statements of how organisations seek to maintain their buildings and estates and keep them operationally fit for purpose whilst retaining their assets values. A clear set of objectives should be articulated together with the methodologies and approaches for how maintenance team plan to deliver them. In this way, maintenance strategies and policies should become frameworks for implementation of maintenance operations.

Maintenance programmes rely on compiling maintenance plans with associated allocated timescales for implementation, and quality assurance measures. This involves planning resources, budget setting and procurement of maintenance and works contracts. Furthermore, they will need to consider which maintenance works would be more cost-effective to carry out in house, with directly employed maintenance tradespersons, or to outsource through external contractors.

Asset management in property terms should ensure that buildings and land provide the higher level of benefits to organisations in meeting their business aims and objectives. It should also achieve the best possible utilisation of real estate assets and in doing so minimise the opportunity costs in tying up resources in buildings and land. Asset management should include areas such as maintenance management, environmental sustainability, energy efficiency, procurement of services, project management, information technology, security, waste management and cleaning.

In preparing asset management plans, it is necessary for organisation to undertake condition surveys of all their buildings and estates portfolios. These surveys should establish the state and condition of all elements of building structures, fabric, electrical and mechanical services, internal spaces, external areas, and equipment, fixtures and fittings. They should consider statutory compliance and involve checking certification on such areas as Legionella, gas and electrical certification. Furthermore, they should appraise and evaluate the current cost of running and maintaining estates assets including existing and predicted maintenance costs, rents, potential acquisitions, disposals, and building values and any capital depreciation. In balancing budgets, maintenance managers will need to consider prioritisation of certain works over others. There should be consideration of the implications of deferring certain works and any effects that this could have on business operations and continuity, occupier satisfaction and building deterioration.

The process by which maintenance requirements are established is based on expected levels of needs. Information sources to capture these needs emanates from building condition surveys, known building requirements for statutory compliance, reported defects, repairs and faults by building occupants and repairing obligations in leases. Building condition surveys will normally be carried out at intervals normally ranging from five years to ten years. They should record the physical condition of all elements of buildings including structure, fabric, finishes and building services. Building elements should be categorised to reflect their state of condition and the level of urgency for any recommended remedial works. Normally the repair works recommended in condition surveys will be categorised into minor or major works of renewal, repair, redecoration, refurbishment or adaption. Irrespective of which category they are placed within, each element of work should be fully quantified, prioritised and costed. Prioritisation is therefore an important aspect of maintenance planning and one which can sometimes require surveyors to use their knowledge of building pathology to predict rates of deterioration. However, it is not only linked to condition and consequential damage, as influenced by compliance with regulation and statutory requirements. In prioritising maintenance works it is preferable for minimum timescales to be designated for each classification of priority. Where works are related to statutory compliance and adherence with health and safety they should be deemed as urgent and unavoidable. Procurement of maintenance works, either in house or externally outsourced, is a major consideration for maintenance teams. Decision-making around this could be predicted on the size and nature of businesses. For instance, smaller companies who do not have the financial resources to invest in their own teams or feel that building maintenance goes far and beyond their core businesses.

Maintenance programmes can be regarded as enabling tools for facilitate maintenance teams in forward planning works into the future. They should be designed to avoid cases where buildings become deteriorated to the point where their condition adversely brings about failures of their structures, fabric or services. Backlog maintenance represents a programme of maintenance works or tasks which are deemed essential and have been scheduled or planned to be carried out. Backlog maintenance can normally be put in the context of value in how much expenditure needs to be invested to bring buildings or facilities back up to an acceptable state of repair and condition.

Normally in the first year following completion of a new building or refurbishment there is a *defects liability period* whereby the main contractor is responsible for faults or latent defects which may become apparent in the occupation period. This liability, however, does not cover accidental damage or wear and tear by building occupants or any building failures that are deemed to have emanated for poor building maintenance. Finally, single-source responsibility for defects and faults in this period, by using the same installation contractors and subcontractors for servicing and maintenance, can avoid issues and disputes over liabilities during this period.

14.2.4 Common maintenance issues and managing defects

The pathology of building maintenance is related to the school of applied sciences and revolves around the investigation, inspection and reporting of building defects with principle aims and objectives around the need to comprehend the processes of building degradation. Such investigation work should identify the source or sources of building problems through diagnosis of the underlying technical issues and then determine the extent and specification of the required remedial works. In the UK building pathologists are normally building surveyors but if a particular building defect is related to major complicated technical problems, then specialist engineers may be appointed. When building surveyors are determining remedial works, following diagnosis of defects, it is important for them to focus on the root problem (cause) of the problem rather than simply address the symptoms of the defect. Certain building failures result from poor design and quality of build during the construction stage or refurbishment whereas others may simply arise due to natural forces associated with weathering and normal use. The condition of a building is not always correlated with its age, on the basis that some newer buildings can have more defects and signs of wear and tear than older buildings, as a result of poor design, workmanship and quality of build. Another factor relating to building condition and whether they are preserved or deteriorate, will depend on the degree of maintenance they receive over their respective life cycles.

The diagnosis of building defects is not always straightforward as there may be more than one force at play in terms of multiple sources of different problems. In such cases this can sometimes complicate normal lines of investigation and enquiry and present difficulties in determining the cause and associated remedial works. When considering the typology of defects in buildings one should distinguish between different types of problems and these can be categorised by type and cause into i) settlement and subsidence, ii) condensation, dampness and water penetration, iii) loss of bond and adhesion, iv) decay and corrosion and v) movement and cracking. It is important to understand the nature and effect of agents that can lead to building defects and these include electromagnetic, mechanical, chemical, building users, thermal and biological forces and effects. The most common defects arise from poorly manufactured materials, components and elements alongside poor workmanship in the construction process and it is therefore important wherever possible to only use materials and building components that are adequately accredited for the standards or quality. Another common cause of premature failure and defects is a general understanding and lack of maintenance of buildings. Accordingly planned preventative maintenance systems should be instigated to address maintenance issues before they become more serious defects there are still those who will only action works on a reactive basis. Furthermore, design team consultants should be diligent in their designs and specifications to ensure buildings are as maintenance free as possible and allow for access for inspections and maintenance works.

The repair or replace question frequently arises when considering remedial works and how building components and elements and how they interact and work with each other. It should be considered whether a patch repair will be sufficient over time or merely storing up a bigger problem for a later time. Sometimes remedial works do not solely look to repair or renew a particular element back to its former state but to improve and adopt the design to make it function more effectively. Other considerations for the repair/renewal decision-making process could be granting funding for replacement works, disruption during the works, sustainability, life cycle costing and clearly affordability for building owners. Notwithstanding this premise, the options that surveyors and designers recommend should be arrived at through carefully considered mechanisms and consultation with their clients. Normally such recommendations would be accompanied by technical reports which clearly set out the different options, and the benefits, disadvantages and implications of each one. In a scenario that the repair or renewal costs are disproportionately high to the value of a building, clients may decide to demolish and redevelop, change the use of the building or simply sell the building and relocate elsewhere.

When considering remedial works to address defects and maintenance issues it is important to factor in the degree of complexity and wide-ranging specialist nature of the repair contract. For this reason, it is essential to only appoint suitably qualified and appropriately selected main contractors, subcontractors and specialist contractors who are suitably qualified for the nature of work they are required to carry out. When the agreed remedial works are agreed between the client and their supervising officer/surveyor overseeing the work, a brief outline of the works should then be prepared with budget costings and a proposed plan of implementation with timelines. Remedial works will be closely managed by the supervising officer, normally the surveyor who is overseeing the works, and this will involve frequent inspections of the works and certification at various stages of the project. It may be necessary to have reviews of the work if there are unforeseen areas of work that cannot be adequately specified owing to access difficulties and/or opening up works via destructive intrusive investigation. Sometimes trials and tests of areas of repair will be undertaken to evidence that a particular repair method is working correctly before being applied over a much larger area. During the remedial works it is essential to have levels of supervision to reflect the nature and complexity of the works. On specialist works a competent and specialist supervisor may have to be commissioned to undertake the inspections and to test the areas of repair where appropriate or deemed necessary. On completion of the remedial works, period inspections should be carried out to ensure that a repair system is performing adequately and possibly to detect any early signs of failure that can be addressed before further problems emerge.

Notwithstanding the array of information that is available for diagnosis and addressing building defects there are still wide-ranging degrees of building defects that do frequently occur. In some more severe cases such defects are left to develop unchecked without the necessary remedial works which subsequently causes more damage, e.g., dampness and timber decay. In a similar way to the health sector, best practice in surveying should continue to be disseminated around prevention of defects as well as cure. Certainly, avoidance of defects to mitigating measures and strategies are normally preferable and this can be achieved through regular building surveys as part of planned preventative maintenance policies and systems.

14.2.5 Maintenance and conservation of historic buildings

Restoration works on all classifications of historical and protected buildings should wherever possible involve using authentic natural materials, coupled with forging the balance between traditional construction techniques and modern advancements. Materials used for restoration projects do not always have to be on a like for like basis for those building elements being repaired. However, if alternative materials are being considered for historic buildings they must be proven to be tried and tested.

Historic England holds a register with specialist contractors competently qualified with the necessary expertise to undertake the most complicated restoration project. They also have a suppliers list that can provide a wide range of different materials used commonly in restoration programmes. Whichever materials are selected they must be highly compatible and sympathetic to original materials used within the structure and fabric of buildings.

Regular inspection and a robust maintenance regime should be applied to all heritage buildings especially those that are in conservation areas and/or are listed. This should involve building owners having robust maintenance policies and systems in place coupled with the formulation and utilisation of 'maintenance plans'. These should consider individual building requirements and accordingly should be tailored to properties depending on their heritage value and condition.

Conservation areas are designated places of historical or architectural interest whereby there is a desire to enhance and preserve the appearance and character of such places. One can identify which areas have been designated conservation areas by contacting the planning authority within the local authority. Planning restrictions in conservation areas are normally concerned with the implications that any developments or alterations have on the external appearance of properties. The principle of conservation areas, as a rule, is that permission must be sought and granted for any alterations to the exterior of buildings, unless works come under within the parameters of 'permitted development'.

In the UK a building, monument or structure can become listed when it is deemed to have special historic and architectural interest and considered to have national importance and worthy of protection. There are three categories of listed building, and these relate to Grade I (buildings of exceptional interest), Grade II * (buildings of more than special interest) and Grade II (buildings of special interest). Whilst conservation areas are focused on external alterations to properties, buildings designated as listed cannot have their exterior or interior altered without permission. Examples and photographs of UK listed buildings have been given in the chapter. Building owners are legally obliged in some cases to maintain and repair their listed buildings and may face prosecution in criminal courts if they allow them to fall into disrepair. There are numerous interventions to listed buildings that applicants are unlikely to gain consent for and these include the removal or alterations to historic or architectural features.

Case studies have been described in this chapter to articulate some of the many issues and challenges faced when embarking on restoration projects. These related to vulnerable listed buildings in Greater Manchester which the author was personally involved in. The first case study related to the former Stockport Union workhouse, Cheshire under the project direction of the author and the second Gorton Monastery, East Manchester. In addition, a third case study relating to a restoration project carried out by the author examined appropriate building technologies and materials for restoration works.

14.2.6 Common internal maintenance issues, defects and remedial works

The categories of the internal different defects are:

Rising dampness
Condensation
Water penetration through buildings
Timber decay. Wood-boring insect
Presence of asbestos and other hazardous materials
Wall, floor and roof defects
Defective electrical, gas and water installations

When considering the types of internal maintenance issues, one needs to closely look at the construction of the buildings and focus particular emphasis on any elements of them which could give rise to defects and remedial works. For instance, if it's an older building, elements of it may not be constructed from conventional materials.

Rising dampness can be described as prevalent where there is either no damp-proof course (DPC) installed in any wall structure or an existing DPC has become damaged. Damp-proof courses can over time become damaged or compromised allowing moisture 'bridging' to take place and rise up the wall by capillary action. Dampness can cause internal damage to finishes including plasterwork and can lead to timber decay of floors, architraves and skirtings if it is prevalent for long periods of time. The noticeable signs of rising dampness can be water staining, possibly tide marks on wall finishes and mould growth usually no more than one metre from ground level. There are many different visual indicators that a building is suffering from the effects of rising dampness, and these are normally present on walls no more than 1 m above ground floor level. These include dark staining, efflorescence and paint flaking and blistering. In addition to these visible signs there may also be a damp odour in rooms affected. Rising damp can sometimes be misdiagnosed as condensation or water penetration and it is therefore imperative for surveyors to carefully consider all the symptoms and not initially jump to conclusions. Remedial works to address rising dampness may vary and can include isolated repairs to damp-proof courses (DPC), or complete reinstallation of a new DPC system.

Condensation can happen in buildings when moist warm air comes into contact with surfaces which are much cooler and below the dew point. Window surfaces are a good example of such surfaces where condemnation is prevalent for this reason. Surface condensation will normally occur in buildings where moisture within the air in the form of water vapour meets surfaces which are colder and below dew point temperatures. For this reason, cold internal surfaces of external walls, where there is insufficient insulative qualities, can be particularly vulnerable. There are numerous visible signs and problems caused by condensation, and these could include random staining on walls and ceilings and buildup of mildew on surfaces. Other visible signs of condensation could include evidence of speckled black mould growth across the internal surfaces of walls, and particularly prevalent in the corners of external walls where surface temperatures can be colder. However interstitial condensation can occur within the fabric of buildings which may be hidden and therein more difficult to diagnose. Some of the measures that can be adopted to control condensation can include ways and means to reduce the humidity whilst increasing air temperature within rooms. This can be achieved through increased ventilation and other dehumidifying measures whilst increasing heating. If the source of the humidity is cold bridging or dampness, then measures to address these defects should be instigated. Remedial works to address cold elements of the building fabric such as walls can be complex but could include the installation of proprietary insulation systems to the external face of walls followed by external cladding or rendering.

Water penetration, otherwise known as water ingress, occurs when water transgresses laterally from the outside to the inside of a building through its fabric. Perhaps the most common source of water penetration occurs when roofs are not fully wind and watertight. This occurrence can happen in a variety of ways and for different reasons. These include defective flat or pitched roofs causing leaking, wrongly installed cavity insulation, blocked or damaged rainwater goods or poorly installed cavity insulation. However, in a lot of cases it can occur from a general lack of maintenance of roof finishes and external wall e.g. damaged or missing pointing or render finishes. Sometimes diagnosing the cause of any dampness is problematic and may require specialist surveys and equipment which could include thermal imaging cameras or electrical resistance meters. Remedial measures and remedial works for water penetration can vary but will normally involve overcoming external defects associated with leakages through walls and roofs e.g. missing or damaged roof coverings.

Timber decay can include dry rot, wet rot or wood-boring beetle attack. Dry rot can quickly transition and spread throughout buildings and through other surface other than timber e.g. brickwork. This makes it particularly harmful and can cause widespread structural damage because it digests those part which give the wood its integrity and strength. The optimum conditions for growth occur at between 30–40%, which is normally associated with dampness through the building structure or from internal leakage e.g. baths and other water installations. Identifying and therein diagnosing dry rot can be difficult in certain circumstances as it traditionally develops in areas which are damp but not exposed to light. Therefore, it can be hidden at early stages of its development although in later stages it can become visible as white mycelium strands or fruiting bodies resembling mushrooms. It can also become visible as cuboidal cracking in the damage it can inflect on timber. As well as these visible signs it can be identified from a musty pungent odour in those building affected.

Wet rot tends to be more common than dry rot. It can be in the form of white and brown rot. The signs of wet rot can be a discoloured spongelike consistency and surface cracking of timber. Corners of windows and door frames where joints occur are particularly vulnerable where water ingress has occurred over long periods of time. Timber affected by wet rot tends to become spongy to the touch and easily penetrated by a knife or screwdriver. Wet rot normally occurs when there are persistent damp conditions pertaining to moisture contents of between 50% and 60% which is clearly much higher than dry rot. In most cases, remedial works will involve cutting out affected areas and timber for replacement. However, in some cases where damage has significantly affected a building e.g. wet rot in windows, it may render complete replacement of the affected component.

A wood-boring insect is a term which covers a multitude of different beetle species where their larvae eat and destroy timber. Beetle attack is more commonly referred to as 'woodworm'. Wood-boring beetles, like dry and wet rot, are particularly suited to conditions which are damp. The most common type of beetle in the UK is the common furniture beetle, although the deathwatch beetle and the house longhorn beetle can generally be more destructive. If the damage afflicted to timbers is considered significant then those affected timbers especially if structural members may be beyond repair and will require replacement. If the damage is less severe, then it may be possible to adopt a programme of specialist treatment which includes targeting the insects located within the timbers.

Wall, floor and roof defects are commonly linked to imperfections present in the structure and fabric of buildings. Such defects may have resulted from penetrations and damage to these structures caused by inadequate design or poor workmanship during the initial construction. They can be associated with movement cracking, poor maintenance, damage from external forces including wind, snow and rain and may present other contributory factors. Defects can be very visible internally; they are in most cases caused or at least linked with what is happening externally. Visible symptoms can include wall cracking to ceilings walls and floors affected by movement of buildings, or in some cases bulging of walls and floors and sagging of ceilings. Walls and floors in basements, especially in older buildings e.g. Victorian cellars are particularly prone to dampness. Where basements are to become habitable spaces then a proprietary 'tanking' solution may be required to render the spaces watertight.

Asbestos is a naturally occurring fibrous material which has properties which are resistant to corrosion chemical attack and heat. Asbestos is part of a group of six naturally occurring minerals composed over very then microscopic fibres and these include Actinolite, Amosite, Anthophyllite, Chrysotile, Crocidolite and Tremolite. It represents an inexpensive and strong material which made it a popular choice as an integral component of buildings constructed before the year 2000. It presents a major health risk for individuals who have been exposed to it. Notwithstanding this assertion, the presence of asbestos can sometimes be difficult to detect as it was used in many different building products over many years which include insulation, vinyl floor tiles, cement and finishes. All business premises in the UK are required by law to keep an asbestos register of their premises to avoid any asbestos being disturbed. In addition, it must be managed under the Control of Asbestos Regulations 2012.

Mechanical and electric installations, otherwise known as building services, can pose a real and imminent danger for building occupants or alternatively in the case of unoccupied buildings, surveyors and other visitors that enter into them. This chapter has identified hazards associated with electric, gas and water installations. Electrical hazards can include electric shocks, fire or explosion from unsuitable electrical installations and equipment. For this reason, an Electrical Safety Certificate, sometimes referred to as an Electrical Installation Condition Report (EICR) or a Fixed Wiring Certificate, is required under law in the UK for all commercial property. Gas services and equipment must be installed and maintained correctly by law to prevent hazards associated with gas or carbon monoxide release, possibly leading to asphyxiation, fire and explosion. By law in the UK all gas services and equipment can only be installed by a competent and qualified engineer registered under the Gas Safe certified scheme.

The main risk to health and safety for water installations in buildings is from Legionella. Water systems which are old and poorly maintained present an increased risk for exposure to Legionella and there should be careful measures implemented to reduce the hazards. These include draining down water systems before demolition or during refurbishment works. In this way it can prevent the stagnation of water. Another alternative hazard for water installations in properties is to maintain clean and safe drinking water free from contaminants for occupants and visitors.

14.2.7 Common external maintenance issues, defects and remedial works

The categories of the different external defects include the following:

Brickwork defects, pointing, cracking, dampness
Differential settlement and subsidence

Window and door defects
Flat and pitched roof defects
Defective rainwater goods
Presence of hazardous and potentially flammable materials
Defective external areas and hardstandings

When considering the types of external maintenance issues, one needs to closely look at the construction, age and use of buildings. An older industrial unit for instance may exhibit very different signs of disrepair when compared to a modern dwelling.

The use of brickwork and its popularity can relate to its longevity as if built correctly it can last for many years with relatively little in the way of maintenance. However, poor quality control during construction could affect external walls and mortar pointing between brick courses leading to defects. Examples given in this chapter have included mortar not being mixed with the correct ratios of sand and cement and new brickwork being exposed to extremes of weather during construction, without protection. Notwithstanding this premise even when the quality of materials and workmanship is good, given external brickwork is exposed to the elements and extreme conditions, this can affect the best masonry walling over time. Other problems for brickwork could include frost damage, shrinkage, efflorescence, contamination from air pollution, sulphate attack, rusting cavity ties and damaged render. Cavity tie corrosion can be a particular problem causing horizontal cracking of brickwork and bulging where the outer leaf of brickwork moves away from the inner leaf. In some cases, cavity ties can be carefully removed and new one retrospectively installed but this is an expensive and time-consuming exercise.

Dampness can affect brick external walls, and this can be caused through water penetration or rising dampness. Both these sources of dampness can damage brickwork, causing joints to become cracked and perished.

Differential settlement and subsidence can cause major problems for buildings. Differential settlement occurs when parts of a building, for instance one elevation, is settling at a different rate than another. Differential settlement could occur where there are weak pockets of land under buildings, sometimes referred to as hot spots which could allow local downward movement in these areas. Structural tensile stresses in buildings are generated in these instances causing movement and resultant cracking. Subsidence can in most cases be more serious than differential settlement and refers to the gradual caving in or sinking of an area of land. It may occur where land has been infilled in the past or where groundwater problems are causing land to be washed away. For buildings constructed on land prone to subsidence can be devastating and may render them dangerous structures with demolition the only alternative. Visible signs of differential settlement and subsidence could include cracking to external and internal wall, floors becoming unlevel, bulging walls and noticeable leaning of walls and properties. Remedial works can include underpinning to create new deeper foundations or proprietary injection systems with cement grout or geo-polymer materials to stabilise and compact the ground.

Window and door defects could include timber decay, warping of frames, peeling or deterioration of paint finishes or defective double-glazing panels. The most predominant type of timber decay for external windows and doors is wet rot. Visible signs to look out for include missing, blistered or damaged paint and darkening of timber especially on corners of frames. On prodding such area timber affected by wet rot is found to be soft and spongy in nature and easily penetrable with a knife or screwdriver.

Flat roof defects can be synonymous with inadequate design and construction issues. These could include constructing flat roofs without adequate gradient to discharge rainwater. In addition, incorrectly installed insulation or a general lack of ventilation can cause problems associated with condensation in roof structures. Damage to roof coverings can emanate from the effects of wind, snow, sun and rain. Blockage to outlets can also result in ponding on roof surfaces which could lead to differential movement causing cacking and tearing of roof surfaces. There are many types of pitched roof, but gabled roofs are the most common. They can be formed by trussed drafters or from traditional roof rafters and purlins. Defects can include broken or missing tiles and slates, damaged sarking felt, defective ridge tiles and timber decay in roof spaces. They can also be prone to a heavy buildup of vegetation growth particularly moss and algae. Flat and pitched roofs can be particularly vulnerable where they meet abutments such as chimneys, parapet walls or pipework. The intersections in each case must be carefully designed and constructed with flashings or collars, in the case of pipework, to protect the roof from water penetration at these vulnerable points. Defective rainwater goods serving roofs can be prone to blockages, cracking and splitting, deflection and leaking joints. Any leakage from rainwater goods could have an adverse effect on brickwork where walls become exposed to dampness.

When conducting building inspections, surveyors should be looking for signs of any materials which could be hazardous or be deemed dangerous in terms of flammability. The presence of asbestos on older buildings is certainly one to be aware of as this was used for many years in walling and roofing products. Aluminium composite cladding (ACM) which was a cause of the Grenfell Tower disaster should be identified wherever possible although this may be subject to separate specialist surveys.

Concrete cancer can be prevalent for a lot of older buildings in the UK and Europe as it was used for commercial buildings and tower blocks especially in the 1960s. Deterioration associated with this condition can significantly weaken concrete. Furthermore, it can become an obvious hazard to the health and safety of individuals if concrete spalling from the external facades of buildings becomes evident. It can be identified by cracking and displacement of concrete and can be exacerbated by climates linked to high humidity and heavy rain. Concrete cancer can be repaired using polymer modified or electrochemical treatment, but in most cases repair systems rely on the replacement of affected concrete and steel.

Inspection of external areas and hardstanding should be part of most building surveys. Boundary walls and retaining walls can be prone to excessive dampness which can lead to frost action damage to brick faces from frost action. Mortar joints can also become cracked and receded if dampness in external walling persists and repointing in such cases should be instigated. Where retaining walls can be exposed to lateral forces of higher ground which they are supporting and hydrostatic pressures from the underground water. This may cause them to move under these forces and lean outwards, become structurally unstable and in extreme cases require them to be completely rebuilt.

14.2.8 Surveying existing buildings

Conducting building surveys call for the forensic identification, collection and recording of evidence to support a data-led approach to defect diagnosis and a comprehensive knowledge of building typologies. In this pursuit it is vitally important for building surveyors to assess all the possible causes of defects and not simply jump to assumptions as relatively straightforward building problems can sometimes be misconstrued for more complicated and expensive building defects and vice versa. If a building defect diagnosis is incorrect it is almost certain that remedial works to address the problem will also be wrong, resulting in potentially unnecessary expensive repairs which will have little or no benefit in rectifying the problem.

There are many different types of surveys that can be carried out on buildings and each one will very much depend on the purpose of what is required from the survey. If clients are not clear which type of survey, they require then surveyors should outline their options, the respective approximate costs of each and make recommendations where necessary. Before commissioning building surveys, it is important to consider the purpose for which they are required, especially as they are designed as health checks for different types of buildings and should be tailored to the individual needs of clients. This is especially pertinent considering the cost of any type of survey and ensuring their remit covers the brief for which client require in a particular instance.

In the UK most surveys are normally governed by the Royal Institution of Chartered Surveyors (RICS) as the professional body and many codes of conduct regulate for members as to how they should be conducted. Violation of such codes of conduct can result in members of the RICS facing disciplinary action from the professional body sometimes resulting in fines or in extreme cases their dismissal from the institution.

Mortgage valuation surveys are normally organised by the institution providing the mortgage for a particular property. In this sense it is important for prospective purchasers to note that these types of survey are carried out for the lender's purposes only and not for the benefit of the customer. These surveys would confirm the value of the property, based on previous historical sales data in the area, the surveyor does not always have to inspect the property to undertake the valuation, and in a lot of cases the survey reports can be prepared 'desktop'. For this reason, mortgage valuation surveys can be prepared cheaply and quickly but as they don't consider condition are a riskier alternative to Homebuyers surveys. According to the RICS there are three types of Homebuyer surveys categorised as Level 1, 2 and 3. Level 1 is the most basic whereas Level 3 (formerly known as full structurer surveys) contain detailed information on all building elements. Level 2 is the most common form of Homebuyer and includes a reasonable amount of information on the condition of building and defects pertaining to a property.

A schedule of condition schedules includes details pertaining to the condition of a given property to be retained as a record for some time in the future and normally form part of an appendix to a lease. A well-constructed schedule of conditions should benefit both landlords and tenants as terms of reference at the end of leases, and thereby protect their respective interests.

A schedule of dilapidations contains a description and record of defects and disrepair pertaining to a property normally towards the end of a lease. They can be categorised into interim, terminal and final schedules. Interim schedules are served within the lease period whereas terminal served towards the end of the lease and final after the lease has expired.

It is important for a contract to be entered into on the basis of this fee quotation prior to surveys being carried out and for surveyors to agree with their clients as to the extent of opening up and making good access into and around the premises in order to conduct the inspections. Most building survey reports will exclude certain items which are considered as either too specialist for the type of survey being carried out and therein beyond the scope of the survey. This could include those areas that cannot be readily inspected, e.g., underground drainage or those elements which require further specialist advice, e.g., building services installations.

Building surveys may also be linked with maintenance and preparing a planned preventative maintenance programme. Such programmes will predict what level of maintenance needs to be undertaken on a year-by-year basis to keep a building in good repair. Programmes of this nature seek to ensure that building issues do not evolve into building defects and cause ancillary damage to the structure and fabric of buildings.

Undertaking any type of building surveys calls for a forensic, methodical and systematic approach to the inspection process. Furthermore, desktop investigations prior to arriving at the property can sometimes be useful in identifying local problems in a particular area and this could include flooding and any underground mines. Other desktop information that may be useful could include previous building surveys of properties. Sometimes building surveys are undertaken in two stages, with the first stage comprising of an initial inspection to get an overall high-level perspective on a particular defect, and to assess further investigatory works. The second stage would be more in depth and could include carrying out specialist investigatory works especially those hard-to-access areas or those unable to be viewed normally.

On arriving at a property, every surveyor initially walks around to assess any difficulties in carrying out the survey such as access problems associated with heights, depths and openings, confined areas or unsafe areas of the building. For health and safety reasons and to ensure that the employers insurance cover is not being compromised it is usual to carry out a very brief risk assessment and survey plan to confirm that inspecting the property is not putting the surveyor in any imminent dangers. The exact way surveys chose to conduct their surveys and record their inspection findings varies from one individual to another. Some surveyors chose to use a checklist of internal and external elements and write down their findings in preparation for typing their reports at a later stage.

There are many challenges and obstacles when undertaking building surveys. Carrying out inspections in premises which are occupied and accessing all areas around live operations whilst trying to avoid disruption to end users can be particularly challenging and requires a careful degree of planning and the help of the owners or occupiers to negotiate around certain conditions and constraints. All types of property and categories of survey can present specific problems where intrusive and/or destructive investigations are required. This could involve opening up walls or floors to gain visibility or access for further investigations. Clearly such works in their own right will create disruption so alongside and all making good works, best planned out of normal working hours if at all possible. When carrying out building inspections it is important wherever possible for surveyors to define the severity of defects against predetermined metrics to assist them in making their assessment. This is to avoid cases where one surveyor regards a particular building problem as moderately severe, and another classifies the same problem as very severe. Opening up works may be required to inspect further into the structure and fabric of buildings, and this can sometimes include taking up floor boards, uncovering roof timbers, removing wall panelling and exposing ceiling and roof voids.

When surveyors are undertaking building surveys it is essential for them to adopt logical and pathological methodologies and this can include the sequence of the inspections and the mode by which they record their findings. In undertaking diagnosis of building defects each potential defect should be considered carefully to forensically assess all the symptoms that have been presented and to examine all the possible or probable sources and causes of the problems. This process may lead surveyors to not just one possible cause but to a narrowing down of potential causes for further reflection. Thereafter, the successful diagnosis will ultimately allow a technical report to be prepared which will detail the extent of the problem, how the underlying causes have been arrived at and most importantly recommendations of all the necessary remedial works. Sometimes options for different remedial works will be given. These can relate to temporary repairs to monitor the problems or the availability of clients' funds or the 'full and final' solution which will seek to cure the problem permanently, albeit at a higher level of cost.

14.2.9 An innovative project to assist building surveying students: designing a handbook as a practical teaching and learning guide for surveying buildings

A handbook was designed as a useful, simple and innovative teaching and learning aid for worthy inclusion within the curriculum. It was felt that it could offer students a useful tool as relative beginners to the task of carrying out complex building surveys and hopefully narrows the gap between traditional classroom teaching and those skills required by professional practice. From experience building surveys of this kind can be bewildering when undertaking them for the first time and it is believed that the handbook will hopefully make the process more comprehensive for students in the future.

It has future development opportunities for professional practice as much for education in that it presents a more effective and efficient means of collating building information on site. It has the potential to become integrated by way of palm held computers or SMART phones for instance to electronically input details on site rather than rely on hard copies. This technology could be networked and downloaded on to a software system that could effectively prepare a report automatically. This would make the whole process so much more efficient and greatly reduce the resources and time currently expended by surveyors and administration staff in preparing the completed report. This is particularly relevant in the current economic climate, in construction and surveying especially, where companies are seeking alternative ways to reduce costs and become more competitive accordingly.

In terms of developing the handbook further, feedback from the pilot study and constructive criticism from participants was sought. Several recommendations emerged from this feedback for future development including:

1. Reformat to include more technical details which would normally be required as part of an RICS Homebuyers Report particularly including building services and external areas which were excluded from the Building Surveyors handbook.
2. Introduce degrees of responses to certain questions.
3. Include a possible a traffic light system to deal with the severity of some defects.

14.2.10 The relationship between maintenance management and environmental sustainability

The agenda for change to reduce carbon emissions and attain global targets on such initiatives as net-zero carbon cannot be disentangled from future maintenance management strategies. Accordingly, environmental sustainability encapsulating energy management should be at the heart of any future estates strategies if they are to be successful. This is especially important for organisations which have been facing for many years of inflationary rises in utility, construction and maintenance costs, affecting their profitability. This chapter responded to this challenge and presented many ideas and recommendations hopefully to create the impetus and catalyst for rethinking current practices and a move to more environmentally sustainable ways of working.

The need to create more energy efficiency and improve environmental sustainability to reduce emissions is clearly part of a wider global agenda to reduce carbon generation. Certainly, in the UK there has been for many years a noticeable increase in participation in accreditation schemes linked to energy efficiency and sustainability and these have included Passivhaus and BREEAM. Such schemes have sought many ways and measures to achieve significant carbon reductions, mostly through good building design. Measures have included passive design, use of renewable technologies and responsible specification of building materials to reduce waste and increase recycling.

When considering maintenance management, it is imperative to consider energy efficiency as part of any policy, strategy or maintenance regime. This chapter has suggested that a major step change is required, especially in the case of the UK, given that the government has committed to a new zero carbon pledge by 2050. Energy management should therefore represent a key objective in the pursuit of environmental sustainability for any organisation. It should influence the way buildings are designed, constructed and used throughout their occupancy. Potential environmental initiatives could include new methods of construction, control systems, energy-efficient LED lighting and technologies for harnessing wind and solar power. However, organisations need the tools to be able to accurately calculate, monitor and predict their energy consumption as a measure of performance, and building management systems are one way of enabling this. Notwithstanding this premise, on all new systems introduced it is important for inductions and training on new energy management systems for estates and maintenance teams.

A case study of Energy House 2.0 was presented in this chapter which the lead author was heavily involved with. The facility represents the largest research centre of its type in the UK in providing controlled environments for testing different types of buildings, structures and technologies. It is hoped that this award-winning project can spearhead the way forward for more energy-efficient technologies to be introduced in construction and the maintenance of existing estates.

Education and training for staff is important especially for estates and maintenance teams and building users who should be influenced to preserve energy in their workplaces. However, sometimes cultural barriers associated with old traditional ways of working may be difficult to break through in this regard, with some staff feeling uncomfortable and threatened by new initiatives and technologies. To respond to this challenge a programme of continual professional development may be required within organisations.

The life cycle of buildings relates to the period from their inception at the start of the conceptual design process to their demolition at the end of life. Life cycles are affected by the extent and quality of repairs and maintenance that estates and buildings will receive which can affect their condition and ultimately influence their deterioration and longevity. The quality of their design and construction will also play a major part in how they perform and the duration of their occupation in use prior to demolition. The degree to which buildings are flexible and versatile enough to deal with changing requirements made of them throughout their life cycle is very important as any creative reuse could improve their longevity. Estates team should therefore be mindful to plan for future proofing buildings at early design stages.

The construction industry has made concerted efforts in recent years to curb the consumption of raw materials, reduce their 'embodied energy' in the construction process and the energy used during their occupation periods. This chapter has an example of a system, aptly named CarboniCa, which is a digital carbon reduction tool and can measure carbon emissions across the whole life of buildings. In this way designers can choose those materials and forms of construction which yield the lowest embodied energy levels. There should also be close consideration of operation and maintenance costs, commonly known as OPEX costs, alongside capital costs, as these can amount to most of the costs incurred, across the lifetime of building assets. These will include all ruining costs such as maintenance, energy, water, remodelling and eventually demolition.

In estates and maintenance management there will be instances where the environmental effects on decision-making need to be understood and considered and an environmental impact assessment (EIA) is a tool used to assess the significant effects of a project or development proposal on the environment. The different stages, scope and characteristics of such assessments were described in this chapter which will allow estates policies to consider the overall effects and implications of developments and maintenance strategies at early design stages.

Organisations should be constantly looking to improve their estates through good environmentally sustainable design *to* include energy and resource efficiency *and* pollution prevention. *Accordingly*, the inclusion of many renewable energy and sustainability technologies should be considered wherever possible, including photovoltaics, green roofs, wind turbines, rainwater harvesting, ground and air source heat pumps and active solar panels. In addition, the design of building envelopes represents a major factor in determining the amount of energy buildings will use in their operation. They should consider such factors as passive solar design, window positioning, thermal massing, daylighting and internal environments.

Estates and maintenance management teams in organisation should strive for policies to reduce their waste production as part of their environmental sustainability strategies and programmes. These policies should consider avoidance, reduction, reuse, recycling, energy recovery, treatment or disposal.

When redevelopment of an estate is being planned, organisations at early stages in the master planning and design processes should consider wherever possible the introduction of intelligent buildings which can incorporate some of the environmentally sustainable technologies referred to in this chapter. These can include passive design, building automation/management systems, renewable resources and technologies, smart building materials and intelligent envelopes. They can save energy and lower wastage, leading to lower carbon emissions by managing occupied space. Other benefits include facilitating real-time responses, providing early warning systems, the ability to anticipate potential failures in advance and enabling smart maintenance of equipment, plants and machinery.

Living buildings can be described as regenerative facilities which provide occupants with connections to nature, air, light, food and community. They have been proven to be able to create positive impacts on both natural and human systems. Benefits include improved health and well-being of all occupants, reduction of carbon emissions, no pollution and generation of no waste.

Finally, landscaping has other benefits associated with environmental sustainability in improving biodiversity, ecology and providing a natural habitat for winds and wildlife. Furthermore, it can also provide health and well-being benefits for people and reduce energy consumption if designed to avoid buildings overheating by providing cooling and shade.

14.2.11 Case studies related to embedding environmental sustainability into estates practices

Two case studies were presented relating to a college and university in the North West of England, where the lead author was responsible for devising and implementing new environmental sustainability strategies. These provided an important piece of action research for projects linked to environmental ethics and in the second example used a ‘collaboration toolkit’. They articulated the benefits of developing successful sustainability outcomes with environmental ethics at the heart of decision-making. In this regard there have been many policies and initiatives developed which have assisted the two case study organisations in meeting their environmental sustainability aims and objectives and to be able to constantly monitor and track their progress against predetermined goals and targets. Accordingly, it has created the partnering and collaboration ethos which is considered vitally important for delivery strategies and innovation, linked to engineering processes and projects around environmentally sustainable construction. Without this it could otherwise have been problematic and challenging in attaining the successful environmental sustainability outcomes. It is hoped that the toolkit can provide the right platform for construction and engineering projects in globally and therein achieve similar environmentally beneficial outcomes for many other organisations.

14.2.12 *The role of information technology and digital twins in asset and maintenance management*

Having information, which is accessible, comprehensive and accurate is regarded as essential for the strategic and operational management of property estates. However, any information system will need to be user friendly and flexible enough to cater for future changes. Such information can be used to form estates management systems which seek to assist with asset planning. Examples of such information could include floor plans, certificates, guarantees, planned preventative maintenance programmes, building occupancy data, types of building construction and health and safety risk assessments.

The built environment sector is now very much a data management business, where concepts like building information modelling (BIM) are now essential to driving information-led productivity in estates projects and programmes. The digitalisation of the built environment is an opportunity to tackle industry's challenges by taking advantage of the general availability of best practices from other sectors and of advanced techniques and tools, digital workflows and technological expertise to shift to a higher level of efficiency and to become a digital industry.

Digital transformation in the built environment is the practice of how digital technologies can allow the sector to develop what could not be created before, to construct new forms of building, to transform estates and respond to growing asset management and sustainability challenges. Accordingly, built environments need to use innovative technologies to allow them to develop, sustain growth and exploit high-tech products to improve their estates.

Digital twins can be utilised for streamlining the operation of building assets. They can harness data to prepare the future digital operational life of built assets. In this way they can create uniformed smart built asset databases, saving money and time by not having to recreate these applications. The use of digital twins also allows asset owners and operators to have greater leverage over their buildings and service networks, with the capacity to enforce and test against specific targets.

Digital twins have the power for building automation operations and assisting with prediction of future maintenance requirements. Furthermore, remote assistance technology enables estates departments to remotely monitor and control built assets, plants and machinery. Cloud computing and data analytics integrated into the digital twin offer insights into building assets and their performance, enabling estates teams to take timely and well-informed action, promoting feedback mechanisms that lead to product and process improvements.

Finally, smart buildings are an excellent example of applications that benefit from digital twin machine learning capabilities. The University of Salford's Energy House 2.0, referred to in Chapter 10 of the book, is one example of research and development into smart buildings and one which should be promoted within the built environment.

14.2.13 *Implications in practice for maintenance management in the built environment*

There have been assertions that estates and asset management in the built environment are inefficient, unsafe, wasteful, compromised on quality and cannot deliver to time and budgetary constraints. Implications and consequences of these dilemmas could include reputational damage and image considerations for organisations. To respond to this predicament there is a need to rethink the concept of maintenance management, especially given the many different challenges in maintaining buildings and facilities to keep them safe, compliant and conducive to user needs. Sometimes such challenges have resulted in failures where buildings have been rendered potentially unsafe leading to building closures and expensive remedial works. In the past silo mentalities have emerged around estates issues and maintenance of building assets and a clear disconnect from operational planning and core business activities. To avoid this in the future there is a growing need to elevate the importance of asset and maintenance management within organisations on a par with their core business activities. To encourage this shift in perception, businesses have become aware that providing good, well-maintained and legally compliant working environments for employees can improve their reputations and in turn increase their success in the marketplace.

The book has presented a strong case for fully integrating maintenance management into operational planning. Accordingly, decision-making around business propositions and proposals should consider how the assets and estates of organisations can help them perform more successfully, efficiently and reduce disruption to business activities. In addition, when procuring maintenance projects and programmes, there should be a growing awareness of the importance of environmental considerations, as part of their organisations' social responsibilities. Initiatives may include devising and implementing credible recycling programmes and deploying state of the art technologies to reduce energy and water consumption.

Improving health and safety in maintenance management should also be considered wherever possible and in practice this can be achieved by introducing enhanced safety measures and initiatives whilst carrying out maintenance programmes. Furthermore, there is a growing need to raise awareness of how successes, failures and key risks within organisations can be influenced by maintenance management practices. Successes could include raising employee satisfaction levels through improvements in working environments whereas failures and risks could be associated with deterioration of buildings and breaches of regulations and statutory requirements. In addition, promoting collaborative working and partnering principles in maintenance management could provide the way and means to encourage improved practices and behaviours in sharing expertise and knowledge, whilst improving business relationships.

It is recommended that organisations should promote continued professional development in the workplace for the teaching and learning for improving maintenance management. Best practices could be adopted in this way to improve old and outdated traditional practices and behaviours and achieve more successful outcomes for maintenance programmes.

14.3 Final reflections, overview and closing remarks

This book has discussed many different aspects and issues that influence and affect maintenance and asset management in an organisational context. Measures to improve the practice around maintenance in the built environment from education and training have gone some way to improve the way the industry works but there are still far too many cases emerging of poor practices that are still blighting the sector. Although arguably these practices are emerging from a small minority of the sector, they are creating a bad press for the whole industry and further measures should be instigated to address this dilemma. Traditional maintenance responses in the past, at an institutional level, have been shortsighted in only looking inwards at property-related issues linked to building deterioration and capital costs. Hopefully this book has highlighted that a more proactive and innovative approach is required, by looking at benefits for business operations to address some of the challenges in promoting good asset management. Accordingly, leaders of organisations and professional institutions should be leading the way for a cultural change in the built environment to train, educate and motivate construction individuals and organisations in more innovative proactive ways, measures to ensure compliance and the benefits that they can bring for the sector. The book has identified and recommended many different measures, in this regard, which include motivational management, leadership, organisational goals linked to environmental sustainability, health and safety and corporate social responsibility. These could be achieved through more focus on further education and higher education course modules linked to maintenance and asset management and continued professional development (CPD) through workshops and training events in the workplace. These measures will hopefully contribute to providing a more proactive position for maintenance and asset management in the built environment. It is accepted, however, that to bring about these cultural changes will take resources, conviction, forward thinking and in some cases courage not to engage in established and outdated traditional practices. These improvements once ingrained within the built environment could then reap massive rewards in providing a safer, comfortable and more enjoyable working environment for all, whilst promoting better operational performance for businesses.

Appendix 1A

Five-year capital programme

HOUSING (HRA) PORTFOLIO
CAPITAL PROGRAMME
2009/10 to 2013/14 5-YEAR FORECAST EXCLUDING CARRY FORWARDS

	2009/10 Forecast £000	*2010/11 Forecast £000*	*2011/12 Forecast £000*	*2012/13 Forecast £000*	*2013/14 Forecast £000*	*5 Year Total £000*
Housing Revenue Account						
Springfields, *Waltham Abbey (Works & Fees)*	0	0	0	0	0	0
Norway House Improvements	50	50	50	50	50	250
Communal TV Upgrade/Other	0	0	0	0	0	0
Total Planned Maintenance	**50**	**50**	**50**	**50**	**50**	**250**
Boiler Replacements	300	300	300	300	300	1,500
MVHR/Ventilation Installations	20	40	40	40	40	180
New Heating Upgrades	450	400	400	400	400	2,050
Rewiring – Kitchens & Bathrooms	150	150	150	150	150	750
Rewiring – Heating	150	150	150	150	150	750
Central Heating/Rewiring Subtotal	**1,070**	**1,040**	**1,040**	**1,040**	**1,040**	**5,260**
PVCu Double Glazing/Door Replacement	150	150	150	150	150	750
Roofing	500	600	600	600	600	2,900
Drainage Works	100	100	100	100	100	500
Asbestos Removal	95	115	115	115	115	555
Door Entry	143	80	80	80	80	463
Energy Efficiency Works	200	200	200	200	200	1,000
Communal Water Tank Renewals	165	75	75	75	75	465
Total Other Planned Maintenance	**2,423**	**2,360**	**2,360**	**2,360**	**2,360**	**11,863**
Balcony Resurfacing	25	25	25	25	25	125
Structural Watercourse Improvements	10	10	10	10	10	50
Parsonage Court – Conversion of Warden Accom.	120	0	0	0	0	120
Miscellaneous Structural Works	250	600	600	600	600	2,650

(*Continued*)

(Continued)

	2009/10 Forecast £000	*2010/11 Forecast £000*	*2011/12 Forecast £000*	*2012/13 Forecast £000*	*2013/14 Forecast £000*	*5 Year Total £000*
Total Structural Schemes	**405**	**635**	**635**	**635**	**635**	**2,945**
External Lighting Schemes	23	5	5	5	5	43
Total Cyclical Maintenance	**23**	**5**	**5**	**5**	**5**	**43**
Small Capital Repairs	**400**	**350**	**350**	**350**	**350**	**1,800**
Kitchen & Bathroom Replacements	2,067	878	878	878	878	5,579
Total Cost Reflective Repairs	**2,067**	**878**	**878**	**878**	**878**	**5,579**
Fencing	15	15	15	15	15	75
Environmental Improvements – Shops	312	120	120	120	120	792
Off Street Parking	460	341	341	43	43	1,228
Estate Environment	102	102	102	102	100	508
CCTV	30	50	50	50	50	230
Total Non-Cost Reflective Repairs	**919**	**628**	**628**	**330**	**328**	**2,833**
Welfare Heating	50	50	50	50	50	250
Other Disabled	380	400	400	400	400	1,980
Disabled Adaptations	**430**	**450**	**450**	**450**	**450**	**2,230**
Norway House Fire Alarm	0	0	0	0	0	0
Careline Equipment Upgrade	50	0	0	0	0	50
Other (New Roof Covering Loughton Way)	8	0	0	0	0	8
Other Repairs and Maintenance	**58**	**0**	**0**	**0**	**0**	**58**
Feasibilities	**15**	**15**	**15**	**15**	**15**	**75**
Housing DLO vehicles	**50**	**50**	**50**	**50**	**50**	**250**
TOTAL HRA	**6,840**	**5,421**	**5,421**	**5,123**	**5,121**	**27,926**

Appendix 1B

Thirty-year capital programme

HOUSING (HRA) PORTFOLIO
30 YEAR CAPITAL PROGRAMME

		Years 1–5 £000	*Years 6–10 £000*	*Years 11–15 £000*	*Years 16–20 £003*	*Years 21–25 £000*	*Years 26–30 £000*
1	**Future Planned Maintenance**	**15,401**	**13,038**	**12,273**	**11,286**	**11,483**	**11,063**
	Roofing	2,750	4,151	3,778	2,544	1,012	1,778
	Rewiring	1,500	1,498	768	996	1,482	1,888
	Window and Door Replacement	750	900	748	2,293	3,779	2,274
	Heating and Boiler Replacement	3,570	2,141	2,729	1,228	*985*	*1,398*
	Asbestos Removal	575	575	575	575	575	575
	Drainage	500	500	500	500	500	500
	Energy Efficiency	1,000	1,500	1,500	1,500	1,500	1,003
	General Improvements	3,723	1,000	1,000	1,000	1,000	1,000
	Door Entry Security	438	148	125	100	100	100
	Ventilation	220	250	250	250	250	250
	Communal Water Tank Replacement	375	375	300	300	300	300
2	**Structural Schemes**	**2,475**	**500**	**500**	**500**	**500**	**500**
3	**Cyclical Maintenance**	**35**	**75**	**75**	**75**	**75**	**75**
4	**Small Capital Works**	**1,850**	**1,750**	**1,750**	**1,750**	**1,750**	**1,750**
5	**Cost-reflective Improvements**	**5,586**	**1,800**	**2,250**	**3,550**	**2,020**	**1,650**
	Kitchen and Bathroom Renewals	5,586	1,800	2,250	3,550	2,020	1,650
6	**Non-cost-reflective Repairs**	**1,693**	**1,500**	**1,500**	**1,250**	**1,000**	**1,504**
7	**Disabled Adaptations**	**2,190**	**2,321**	**2,962**	**3,780**	**4,825**	**6,158**
8	**Other Repairs and Maintenance**	**0**	**500**	**500**	**500**	**500**	**500**
9	**Feasibilities**	**75**	**75**	**75**	**75**	**75**	**75**
10	**Housing DLO Vehicles**	**250**	**290**	**290**	**290**	**290**	**290**
11	**Less Income**						
	TOTAL EXPENDITURE	**29,555**	**21,849**	**22,175**	**23,056**	**22,518**	**23,565**

Appendix 1C

Decent homes summary

Fails Decent Homes	*Base Year Failures 2001 (Business Plan)*		*Current Failures 2008/09*		*All Current and Potential Failures 2010*	
	1627		*237*		*703*	
Stock	7088		6624		6624	
Non-decent Homes as a % of Total Stock	22.95%		3.58%		10.61%	
	No.	£	No.	£	No.	£
Min. Fitness Standard	8	8.000	0	-	0	-
Key Building Components	**688**	**1,140,100**	**340**	**925,700**	**1042**	**2,818,600**
Walls	6	1,500	37	92,500	85	£212,500
Roof Cover	28	56,000	38	76,000	146	£292,000
Chimneys	1	400	6	2,400	8	£3,200
Heating	210	462,000	74	199,800	327	£882,900
Electrics	443	620,200	185	555,000	476	£1.428,000
Modern Facilities	**1180**	**1,657,100**	**147**	**370,000**	**414**	**908,000**
Kitchens	513	923,400	38	152,000	40	£160,000
Bathrooms	667	733,700	109	218,000	374	£748,000
Thermal Comfort	**459**	**573,750**	**38**	**38,000**	**38**	**38,000**
TOTALS	**2,327**	**3,378,950**	**525**	**1,333,700**	**1.494**	**3,764.600**

Appendix 2

RICS home surveys

coseyhomes

RICS

RICS Home Surveys

Survey level two: HomeBuyer report - Survey & Valuation

Property address

XXX

Client's name

XXX

Date of inspection

29th August 2017

rics.org

Contents

A

Introduction to the report

This HomeBuyer Report is produced by a RICS surveyor who is a member of the RICS Valuer Registration Scheme. The surveyor has written this report for you to use. If you decide not to act on the advice in this report, you do this at your own risk.

The HomeBuyer Report (Survey & Valuation) aims to help you:

- make a reasoned and informed decision on whether to go ahead with buying the property
- make an informed decision on what is a reasonable price to pay for the property
- take account of any repairs or replacements the property needs; and
- consider what further advice you should take before committing to purchase the property.

Any extra services we provide that are not covered by the terms and conditions of this report must be covered by a separate contract.

If you want to complain about the service, please refer to the complaints handling procedure in the 'Description of the RICS HomeBuyer (Survey & Valuation) Service' at the back of this report.

Property address

XXX

3 About the inspection

Surveyor's name

Surveyor's RICS number

Company name

Cosey Homes Chartered Surveyors

Date of the inspection

Tuesday 29th August 2017

Report reference number

SAMPLE HOMEBUYERS 2017

Related party disclosure

This report was carried out by Philip Marsh MRICS BSc Building Surveying. I am an independent Chartered Building Surveyor. I can confirm that there is no conflict of interest in relation to the report.

This report has been checked by quality checked by Mike Cosy MRICS Civil Eng. HND. BA. DipSurv FCABE.

Full address and postcode of the property

XXX

Weather conditions when the inspection took place

When I inspected the property, the weather was hot and dry.

The status of the property when the inspection took place

The property was unoccupied and furnished.

Property address

XXX

B

About the inspection (continued)

We inspect the inside and outside of the main building and all permanent outbuildings, but we do not force or open up the fabric. We also inspect the parts of the electricity, gas/oil, water, heating and drainage services that can be seen, but we do not test them.

To help describe the condition of the home, we give condition ratings to the main parts (the 'elements') of the building, garage and some parts outside. Some elements can be made up of several different parts.

In the element boxes in parts E, F, G and H, we describe the part that has the worst condition rating 1st and then briefly outline the condition of the other parts. The condition ratings are described as follows.

3	Defects that are serious and/or need to be repaired, replaced or investigated urgently.
2	Defects that need repairing or replacing but are not considered to be either serious or urgent. The property must be maintained in the normal way.
1	No repair is currently needed. The property must be maintained in the normal way.
NI	Not inspected (see 'Important note' below).

The report covers matters that, in the surveyor's opinion, need to be dealt with or may affect the value of the property.

Important note: We carry out only a visual inspection. This means that we do not take up carpets, floor coverings or floorboards, move furniture or remove the contents of cupboards. Also, we do not remove secured panels or undo electrical fittings.

We inspect roofs, chimneys and other surfaces on the outside of the building from ground level and, if necessary, from neighbouring public property and with the help of binoculars.

We inspect the roof structure from inside the roof space if there is access (although we do not move or lift insulation material, stored goods or other contents). We examine floor surfaces and under-floor spaces so far as there is safe access to these (although we do not move or lift furniture, floor coverings or other contents). We are not able to assess the condition of the inside of any chimney, boiler or other flues.

We note in our report if we are not able to check any parts of the property that the inspection would normally cover. If we are concerned about these parts, the report will tell you about any further investigations that are needed.

We do not report on the cost of any work to put right defects or make recommendations on how these repairs should be carried out. Some maintenance and repairs we suggest may be expensive.

! Please read the 'Description of the RICS HomeBuyer Report Service' (at the back of this report) for details of what is, and is not, inspected.

Property address

XXX

C

Overall opinion and summary of the condition ratings

This section provides our overall opinion of the property, and summarises the condition ratings of the different elements of the property.

If an element is made up of a number of different parts (for example, a pitched roof to the main building and a flat roof to an extension), only the part in the worst condition is shown here.

To make sure you get a balanced impression of the property, we strongly recommend that you read all sections of the report, in particular the 'What to do now' section.

Our overall opinion of the property

The property is considered to be a reasonable purchase although there are a number of defects which require immediate attention and which will require some expenditure at the outset. I would not expect any particular difficulty on resale in normal market conditions, provided that the necessary works are carried out to a satisfactory standard.

You should investigate the cost of these works prior to commitment to purchase. Once known, you may wish to re-negotiate the purchase price to reflect them.

It is very important that you read this report as a whole. In the main body of the report, I have given elements a Condition Rating of 2 or 3, I particularly refer you to the section at the end of the report entitled 'what to do now'. You must make sure that you have all of the repairs needed investigated by reputable contractors so that you are fully aware of their scope and financial implications before you purchase.

This report should be construed as a comment upon the overall condition of the property and is not an inventory of every single defect. The report is based on the condition of the property at the time of my inspection and no liability can be accepted for any deterioration in its condition after that date.

	Section of the report	Element number	Element name
3	E: Outside the property	E2 E6	Roof coverings Outside doors (including patio doors)
	F: Inside the property	F7 F9	Woodwork (for example, staircase and joinery) Other
	G: Services	G1 G2 G4 G5 G6	Electricity Gas/oil Heating Water heating Drainage
	H: Grounds (part)	H2	Other

Property address

XXX

C

Overall opinion and summary of the condition ratings (continued)

2

Section of the report	Element number	Element name
E: Outside the property	E3 E4 E5 E8	Rainwater pipes and gutters Main walls Windows Other joinery and finishes
F: Inside the property	F1 F2 F3 F4 F6 F8	Roof structure Ceilings Walls and partitions Floors Built-in fittings (built-in kitchen and other fittings, not including appliances) Bathroom fittings
G: Services	G3	Water
H: Grounds (part)	-	

1

Section of the report	Element number	Element name
E: Outside the property	-	
F: Inside the property	-	
G: Services	-	
H: Grounds (part)	-	

Property address

XXX

About the Property

Type of property

Mid mews house The property has 3 storeys.

Approximate year the property was built

1960s

Approximate year the property was extended

The property has not been extended. However, it is believed a fourth bedroom was added at ground floor level.

Approximate year the property was converted

The property has not been converted. However, it appears to have been formed into a house with unrelated occupants in each room that share common facilities.

Information relevant to flats and maisonettes

N/A

Accommodation

Floor	Living rooms	Bed rooms	Bath or shower	Separate toilet	Kitchen	Utility Room	Conser-vatory	Other	Name of other
Lower Ground									
Ground	0	1	0	1	0	1	0	1	Common room
First	1	0	0	0	1	0	0	0	N/A
Second	0	3	1	0	0	0	0	0	N/A
Third									
Other									
Roof Space									

Construction

The properties construction is cross wall construction with UPVC between the brick piers. The walls may be cavity with some form of timber stud construction which are plaster boarded internally and covered with the UPVC externally. The level of insulation is likely to be poor due to the size of the cavity if any. The ground floor

Property address

XXX

D

About the Property (continued)

is solid concrete whilst the upper floors are suspended timber. The roof covering could not be seen but based on similar properties in the area it is believed to be felt based and is supported on an early trussed roof system.

Property address

XXX

About the Property (continued)

Energy

We have not prepared the Energy Performance Certificate (EPC). If we have seen the EPC, then we will present the ratings here. We have not checked these ratings and so cannot comment on their accuracy.

We are advised that the property's current energy performance, as recorded in the EPC, is:

Energy Efficiency Rating

62 see EPC later in report

Environmental impact rating

55 see EPC later in report

Mains service

The marked boxes show that the mains services are present.

Gas	Electric	Water	Drainage
✔	✔	✔	✔

Central heating

Gas	Electric	Solid fuel	Oil	None
✔	✔	☐	☐	☐

Other services or energy sources (including feed-in tariffs)

N/A

Grounds

There is no garage. There are front and rear gardens with the rear containing a timber shed. There is street parking to the rear.

Property address

XXX

D

About the Property (continued)

Location

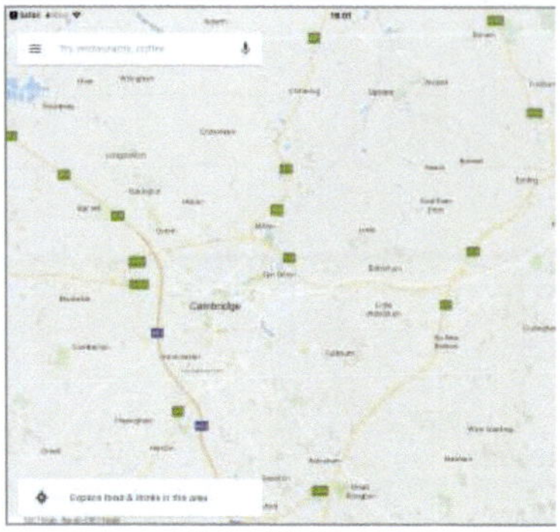

Notes Photo - 1

Facilities

Local environment

Property address

XXX

Outside the property

Limitations to inspection

For the purpose of this report, only significant defects and deficiencies readily apparent from a visual inspection are reported. Services can only be fully assessed by a specialist contractor.

Building standards are continually being upgraded and older properties can become increasingly out of date due to the passage of time, leading to a requirement for improved efficiency. It is inevitable, therefore, that these homes will probably have higher running costs compared to newly built properties.

We have not exposed the foundations of the property and without doing so, you must accept the risk of unseen defects.

We have not carried out any geological survey or site investigation and cannot confirm the nature or characteristics of the soil with regards to fill or possible contamination. Normal legal searches should confirm the past use of the site and if instructed, we will advise further. Our examination of the roof covering was confined to an inspection from ground level.

All accessible areas were inspected. No services were tested.

1 2 3 NI

E1 Chimney stacks

N/A

E2 Roof coverings

The roof covering could not be seen from the ground and as such requires further investigation. A similar block of properties across the road to the rear has felt type roofs. **3**

E3 Rainwater pipes and gutters

The rainwater goods are UPVC and terminate below ground. They appear to be in fair condition however, the vegetation requires removal. **2**

We cannot comment upon the serviceability of the system and confirm the water tightness of the joints unions or connections. These items should be regularly maintained and checked. Gutters can easily get blocked by leaves and debris and cause overflow resulting in damp walls. The stop ends are particularly vulnerable to leakage.

Property address

XXX

Outside the property (continued)

Photo - 2 Remove vegetation from guttering

E4 Main walls

The wall construction is believed to be timber frame construction between the brick piers. As such the majority of the walls were covered with UPVC cladding from first floor level upwards, only the ground floor was brickwork other than the piers which ran full height. As previously noted the cavity in the timber frame is unlikely to offer much opportunity for any real thermal improvement due to the constructions overall thickness.

To the rear the central pier requires making good and to the front there is evidence of damp staining from overflow pipes. It should be checked that this issue has been fixed.

2

Photo - 3

Photo - 4 Staining to front pier

Property address

XXX

Outside the property (continued)

Photo - 5

Photo - 6 Remove ivy

Photo - 7

Photo - 8 Repoint rear pier

E5 Windows

The windows were UPVC double glazed and appeared to be in fair condition. The cills were UPVC other than the ground floor rear bedroom which was timber and in need of maintenance.

2

In the bathroom is a high level timber frame roof light in need of maintenance. Access to this for opening will be difficult and as such it is likely to be a permanently shut feature.

Property address

XXX

E

Outside the property (continued)

Photo - 9 Maintain/replace roof light

Photo - 10 Maintain timber frame

E6 Outside doors (including patio doors)

The front and rear doors were UPVC double glazed as was the front side panel. The rear door would not open with any of the keys provided to me by the estate agent and the central side panel pane needs to be confirmed as being safety glazing due to its critical location.

3

At first floor level were French UPVC double glazed and safety marked doors leading immediately to the Juliet balcony.

Photo - 11

Photo - 12 Middle side pane to be confirmed as safety glazing. Back door key to be located and door tested to see if it opens.

Property address

XXX

Outside the property (continued)

Photo - 13

E7 Conservatory and porches

N/A

E8 Other joinery and finishes

The soffits and fascias were UPVC and appeared to be in fair condition.

2

At first floor level there was a Juliet balcony, the metal guarding requires maintenance and the height and gaps between the vertical members were not compliant with current standards. Current standards require a height of 1100mm and gaps which will not allow a 100mm sphere to pass through them. Whereas the gaps were in excess of 100mm and the height just under 1000mm. It is likely that this detail was built with the property under different standards as such it is only recommended that the balcony guarding be changed.

Property address

XXX

E Outside the property (continued)

Photo - 14 Maintenance required

Photo - 15

E9 Other

EPC and comparables

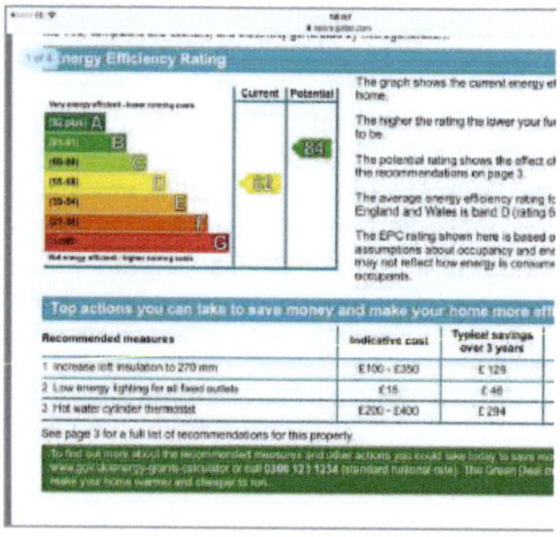

Energy Efficiency Rating

	Current	Potential
A		
B		84
C		
D	62	
E		
F		
G		

The graph shows the current energy ef
home.

The higher the rating the lower your fu
to be.

The potential rating shows the effect of
the recommendations on page 3.

The average energy efficiency rating fo
England and Wales is band D (rating 6

The EPC rating shown here is based o
assumptions about occupancy and env
may not reflect how energy is consume
occupants.

Top actions you can take to save money and make your home more eff

Recommended measures	Indicative cost	Typical savings over 3 years
1 Increase loft insulation to 270 mm	£100 - £350	£ 129
2 Low energy lighting for all fixed outlets	£15	£ 48
3 Hot water cylinder thermostat	£200 - £400	£ 294

See page 3 for a full list of recommendations for this property.

Photo - 16

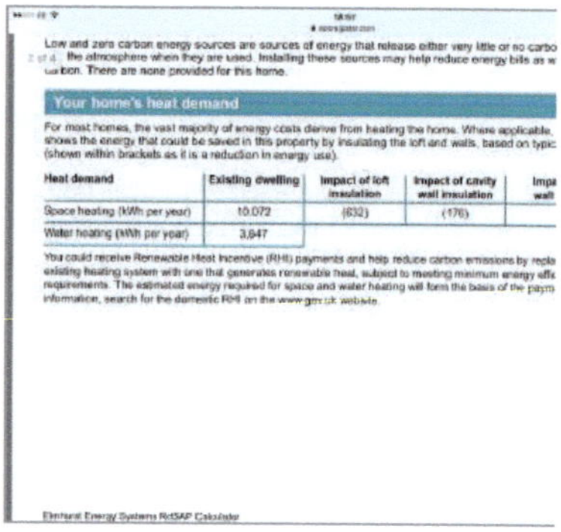

Low and zero carbon energy sources are sources of energy that release either very little or no carbo
the atmosphere when they are used. Installing these sources may help reduce energy bills as w
carbon. There are none provided for this home.

Your home's heat demand

For most homes, the vast majority of energy costs derive from heating the home. Where applicable,
shows the energy that could be saved in this property by insulating the loft and walls, based on typic
(shown within brackets as it is a reduction in energy use).

Heat demand	Existing dwelling	Impact of loft insulation	Impact of cavity wall insulation	Impa wall
Space heating (kWh per year)	10,072	(632)	(176)	
Water heating (kWh per year)	3,947			

You could receive Renewable Heat Incentive (RHI) payments and help reduce carbon emissions by repla
existing heating system with one that generates renewable heat, subject to meeting minimum energy effi
requirements. The estimated energy required for space and water heating will form the basis of the paym
information, search for the domestic RHI on the www.gov.uk website.

Elmhurst Energy Systems RdSAP Calculator

Photo - 17

Property address

XXX

Outside the property (continued)

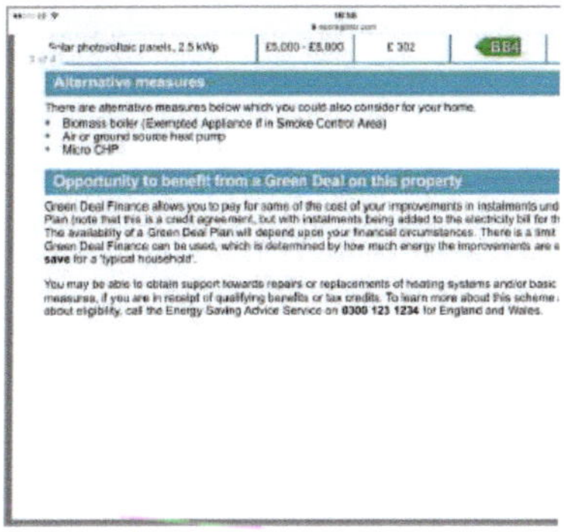

Solar photovoltaic panels, 2.5 kWp	£5,000 - £8,000	£ 302	B84

Alternative measures

There are alternative measures below which you could also consider for your home.

- Biomass boiler (Exempted Appliance if in Smoke Control Area)
- Air or ground source heat pump
- Micro CHP

Opportunity to benefit from a Green Deal on this property

Green Deal Finance allows you to pay for some of the cost of your improvements in instalments und Plan (note that this is a credit agreement, but with instalments being added to the electricity bill for th The availability of a Green Deal Plan will depend upon your financial circumstances. There is a limit Green Deal Finance can be used, which is determined by how much energy the improvements are **save** for a 'typical household'.

You may be able to obtain support towards repairs or replacements of heating systems and/or basic measures, if you are in receipt of qualifying benefits or tax credits. To learn more about this scheme about eligibility, call the Energy Saving Advice Service on **0300 123 1234** for England and Wales.

Photo - 18

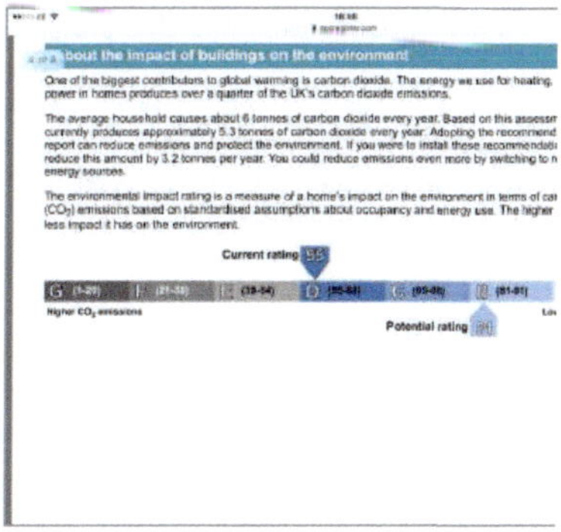

About the impact of buildings on the environment

One of the biggest contributors to global warming is carbon dioxide. The energy we use for heating, power in homes produces over a quarter of the UK's carbon dioxide emissions.

The average household causes about 6 tonnes of carbon dioxide every year. Based on this assessm currently produces approximately 5.3 tonnes of carbon dioxide every year. Adopting the recommend report can reduce emissions and protect the environment. If you were to install these recommendati reduce this amount by 3.2 tonnes per year. You could reduce emissions even more by switching to n energy sources.

The environmental impact rating is a measure of a home's impact on the environment in terms of ca (CO_2) emissions based on standardised assumptions about occupancy and energy use. The higher less impact it has on the environment.

Photo - 19

Property address

XXX

F

Inside the property

Limitations to inspection

Comment cannot be made on areas that are covered and concealed or not otherwise readily available. There may be detectable signs of concealed defects, in which case recommendations are made. If greater assurance is required on the matter, it would be necessary to carry out exposure works. Unless these are carried out prior to legal commitment to purchase, there is a risk that additional defects and consequently repair work will be discovered at a date.

It should be appreciated that original parts of the property are of period in nature. Accordingly such parts of the structure and fabric should not be expected as new and regard should be given to the natural deterioration of older products. It is possible that defects could occur between the date of survey and the date of which you take occupation.

Roof space was inspected from shoulder height position standing on a ladder, this inspection has its limitations.

1 2 3 NI

F1 Roof structure

2

The roof structure appears to be an early form of truss and supports a felt which shows signs of damage in places. It is not known whether the felt is asbestos based as such further investigation is required. In the roof space there is some boarding to joists and a very thin amount of insulation visible in other areas.

The roof construction needs further investigation as it would be expected that there would be cross flow ventilation provision to prevent condensation and roof timber decay. There was no evidence of this.

Photo - 20 Repair felt and find out what the material is before doing so.

Photo - 21

Property address

XXX

Inside the property (continued)

Photo - 22

Photo - 23 Water tank

Photo - 24 Felt could be asbestos based?

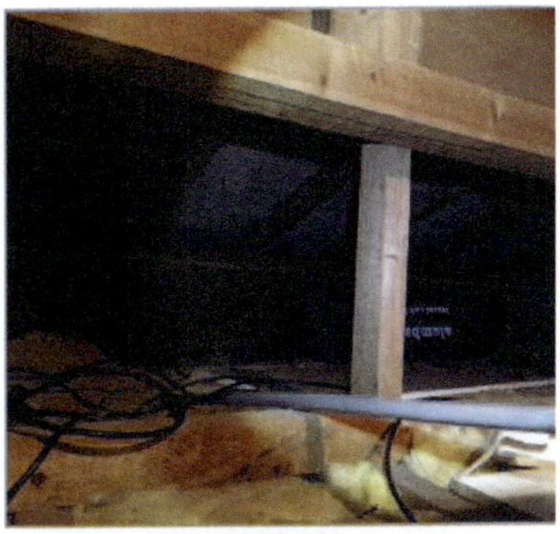

Photo - 25

F2 Ceilings

A textured decorative coating covers the ceilings in the utility room, cloakroom, living room and second floor. This may contain small amounts of asbestos fibres and if disturbed, it could be a safety hazard (see section J3 Risks). If the ceiling needs decorating or repairing, you will have to use a contractor experienced in this type of work or an asbestos specialist.

2

Other ceilings appear to be plaster skim with areas of making good required and poor patch up works in the past having been done. In the front second floor bedroom stickers are also stuck to the ceiling which when removed could cause damage.

Property address

XXX

F

Inside the property (continued)

Photo - 26 Could contain asbestos

Photo - 27 Poor plaster skim patch up

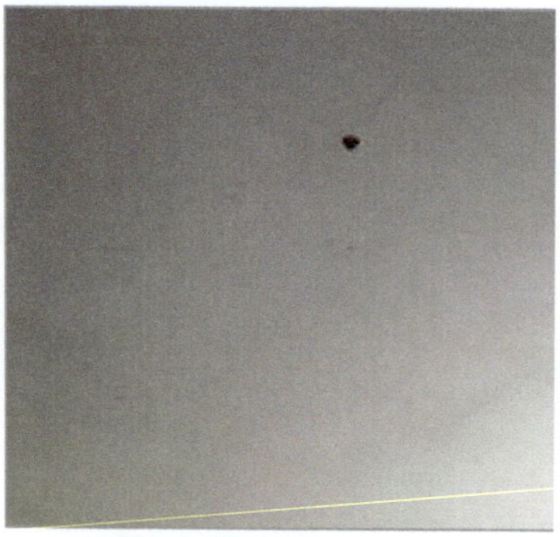

Photo - 28 Mechanical damage

Photo - 29

Property address

XXX

Inside the property (continued)

Photo - 30

Photo - 31 Dirt/mould

F3 Walls and partitions

2

The walls are a mixture of plaster skim, tiles, wall paper and timber panels in the bathroom. There are areas of making good required and what appears to be poor patching up. At second floor level on the stair is an almost vertical hairline crack running from the timber rail to the ceiling. There is a shorter crack at ground floor from the stair soffit down part of the wall. It should be ensured that the cracks do not extend through to the structure of the walls. They are believed to be vibration but should still be assessed by a structural engineer.

To the walls internally to the UPVC cladding, there appears to be expansion and contraction cracks between board joints and where radiators have been fixed.

The rear ground floor bedroom has poor detailing at its bottom with foam filling of gaps.

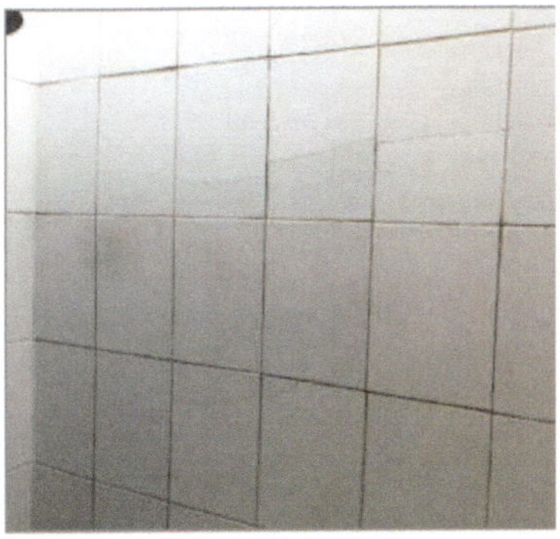

Photo - 32 Cracked tiles

Photo - 33 Timber panelling is a fire risk

Property address

XXX

F Inside the property (continued)

Photo - 34 Walls require making good, cleaning and redecoration throughout.

Photo - 35 Crack on second floor stair

Photo - 36

Photo - 37

Property address

XXX

Inside the property (continued)

Photo - 38 Crack to ground floor wall

Photo - 39 Expansion contraction cracking

Photo - 40

F4 Floors

The ground floor is solid whilst the upper floors are suspended timber. The condition of the floor coverings is poor with bad detailing at edges, along with general damage and wear. The floors are a mixture of carpet, vinyl and laminate.

2

Property address

XXX

F

Inside the property (continued)

Photo - 41

Photo - 42 Damage to cloakroom floor

Photo - 43

Photo - 44

Property address

XXX

Inside the property (continued)

Photo - 45

Photo - 46 Poor detailing

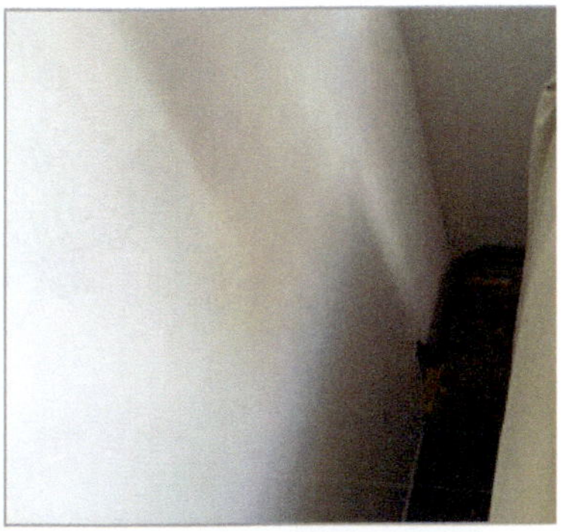

Photo - 47 Foam at bottom of wall

F5 Fireplaces, chimney breast and flues

N/A

F6 Built-in fittings (built-in kitchen and other fittings, not including appliances)

The kitchen appears to be ok but in need of repair works and sealant renewal at the edge of the worktop where it meets the wall. The floor of the under sink cupboard is broken. 2

Two bedrooms have built in wardrobes whilst on the landing and under the stair are store areas. General making good is required.

Property address

XXX

Inside the property (continued)

Photo - 48

Photo - 49

Photo - 50

Photo - 51 Damage to cupboard

Property address

XXX

Inside the property (continued)

Photo - 52

Photo - 53 Sealant required

F7 Woodwork (for example, staircase and joinery)

3

General cleaning, making good and redecoration are require to the ground floor timber door steps, the door frames, doors and stair guarding.

Of concern is that the means of escape is unsatisfactory in this property. A property with a second floor cannot rely on bedroom windows as being suitable for escape. As such a protected stair is required which means 30 minute fire doors with intumescent strips to all rooms off the stair at all levels, the floors provided with 30 minutes fire resistance and the stair protected with 30 minutes fire resistance. In addition mains interlinked fire detection with battery back up installed at each level would be required. The stair passes through the first floor rooms and as such is not a protected stair, this also makes the second floor rooms inner rooms to the kitchen and living room.

The property has glazed doors on the stair which are Georgian wired which were historically acceptable as fire doors but not currently. The glazing above the doors would need to be fire resisting and the doors to all rooms changed accordingly as they were not fire doors. The strips to the old Georgian wired doors are old and defective.

The boarding in the ground floor store to the under side of the stair is damaged and may be asbestos based. This should have 30minutes fire resistance.

It is possible that the doors at first floor could negate the protected stair requirement in its strictest form however, specialist advice should be sought from a fire engineer.

The guarding to the stair at first floor level is missing and as such is a fall risk, the gaps between guarding members at second floor level is in excess of current standards and the edge of the floor at second floor on the stair is exposed.

The timber internal threshold registered on the damp meter as 9.5% to the utility room and 11.5% to the rear. Insect and d rot are believed to set in with damp levels of 15-25%. There was no evidence of any damp related damage to these areas.

Property address

XXX

F Inside the property (continued)

The Georgian wired glazing is located in critical locations and as such should be safety glazing to reduce the risk of serious injury on breakage.

Over time, the wood to staircases can shrink and split loosening the various joints causing the stairs to creak when used. This is not a safety hazard, but you should overhaul and repair the stairs soon.

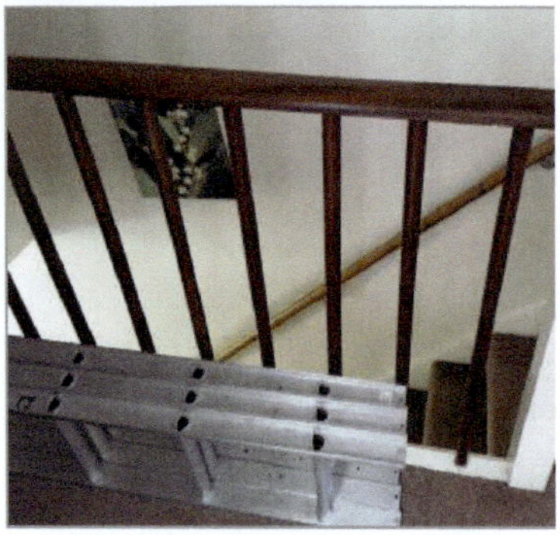

Photo - 54

Photo - 55 Split to door frame lining

Photo - 56 No guarding

Photo - 57

Property address

XXX

Inside the property (continued)

Photo - 58

Photo - 59 Georgian wired door should be safety glazing

Photo - 60 Cat flap breaches door integrity. It should not be cited in a fire door

Photo - 61 No fire protection to stair under side

Property address

XXX

F

Inside the property (continued)

Photo - 62

Photo - 63

Photo - 64 Crack to door glazing to bedroom. It should be fire rated if on a protected stair

Photo - 65 Damage to under stair cupboard door

Property address

XXX

Inside the property (continued)

Photo - 66 Defective fire door strips

F8 Bathroom fittings

The bathroom fittings are old and will likely require replacement in due course.

2

Photo - 67

Photo - 68

Property address

XXX

F

Inside the property (continued)

Photo - 69

Photo - 70

F9 Other

The cloakroom appears to be in fair condition but has a foul smell within it. The utility room requires general maintenance.

3

There is no mechanical extraction to outside air in the utility room, kitchen, cloakroom or bathroom. All have openable windows except the cloakroom and bathroom, as the location means the roof light would probably not be used. The cloakroom should under current standards have extraction at the very least. Fitting extraction to all aforementioned areas is recommended as this helps remove foul smelling and warm moist air which could condense on cooler services leading to mould and internal finish damage from damp.

Photo - 71

Photo - 72

Property address

XXX

G

Services

Services are generally hidden within the construction of the property. This means that we can only inspect the visible parts of the available services, and we do not carry out specialist tests. The visual inspection cannot assess the services to make sure they work efficiently and safely, or meet modern standards.

Limitations to inspection

No services were tested during this inspection. All services and associated fittings and constructions should be inspected and tested by the relevant contractor registered under a competent person scheme listed on the governments competent person scheme website. All accessible areas were inspected.

G1 Electricity *Safety warning: The Electrical Safety Council recommends that you should get a registered electrician to check the property and its electrical fittings and that a periodic inspection and testing is carried out at the following times: for tenanted properties every 5 years or at each change of occupancy, whichever is sooner; at least every 10 years for an owner-occupied home. All electrical installation work undertaken after 1 January 2005 should have appropriate certification. For more advice contact the Electrical Safety Council.*

1 2 3 NI

Consumer board is fitted with circuit breakers and appears in good condition. It is recommended that at any change of ownership the electrical wiring should be checked by an NICEIC qualified electrician. Earth bonding was not checked. Earthing is used to protect people from the risk of electric shock. If the earthing arrangements within your electrical installation were defective or inadequate, you could receive an electric shock from the equipment or appliance metal casing. The purpose of earthing is to provide a path for electric fault current to flow safely to earth to enable the circuit breaker or fuse to operate. Bonding is the connection of the incoming metal gas and water pipes to and is vital for your protection from electric shock. In a correctly earthed installation, any appliance or equipment developing a fault to the metal casing will be quickly disconnected by the operation of the circuit fuse or circuit breaker. In the absence of electrical certification, the surveyor must give designation 3 for this item.

All electrical wiring, installations, fixtures and fittings should be inspected and tested by an electrical engineer registered under a relevant competent person scheme listed on governments competent person scheme website.

3

G2 Gas/oil *Safety warning: All gas and oil appliances and equipment should regularly be inspected, tested, maintained and serviced by a appropriately qualified Gas Safe Engineer or Registered Heating Engineer and in line with the manufacturer's instructions. For tenanted properties by law a 12 monthly gas safety check must be carried out on every gas appliance/flue. A gas safety check will make sure gas fittings and appliances are safe to use. This is important to make sure that the equipment is working correctly, to limit the risk of fire and carbon monoxide poisoning and to prevent carbon dioxide and other greenhouse gases from leaking into the air. For more advice contact the Gas Safe Register for gas installations, and OFTEC for oil installations.*

The boiler is gas fired with what appears to be an asbestos cement flue extending vertically through the house, the termination of which could not be seen. The meter is located under the stair.

All gas installations and associated fittings should be inspected and tested by a Gas Safe registered engineer.
In the absence of certification, the surveyor must give designation 3 for this item.

3

Property address

XXX

G

Services (continued)

Photo - 73

Photo - 74

Photo - 75

G3 Water

The stop cock is located in the utility room, the Incoming water supply pipe may be of copper, Alkathene (plastic) or lead. Lead pipes are vulnerable to leakage and also considered as unhealthy. It is not always possible to see the incoming water main and therefore identify the type. Bringing a new supply into the house could be expensive depending where the main supply is. If this was necessary then there are certain regulations that apply.

Based on the property age the pipe could be lead as such further investigation is required.

In the roof space is a water tank.

2

Property address

XXX

G

Services (continued)

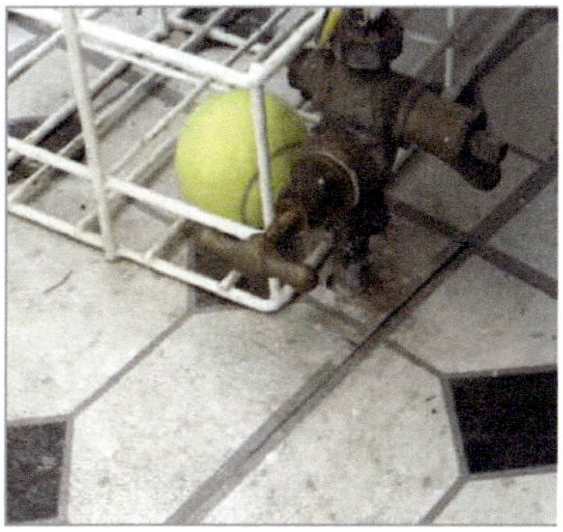

Photo - 76

G4 Heating

The heating system consists of the gas fired boiler and radiators throughout. There was evidence of electric storage heaters in bedrooms suggesting poor thermal insulation in walls.
All heating system components should be inspected and tested by an engineer registered under a competent person scheme listed on the governments competent person scheme website.
In the absence of electrical certification, the surveyor must give designation 3 for this item.

3

Photo - 77

G5 Water heating

In the second floor built in fitting off of the landing is a water cylinder. This should be tested and inspected by an engineer registered under a competent person scheme listed on the governments competent person scheme website as should the water tank.

3

Property address

XXX

Services (continued)

Photo - 78

Photo - 79

Photo - 80

G6 Drainage

3

The soil and vent pipe is UPVC internal and visible in the utility room, it is believed to be boxed in with the boiler flue where it passes through the floors above. The termination could not be seen, as such further investigation is required. The penetrations of both the flue and drainage pipe should be boxed in for there full vertical height by 30 minute fire resisting construction or fitted with pipe collars at compartmentation lines to prevent fire spread through floors.

The kitchen waste discharges via UPVC pipework to a hopper and then to a front gully into which the utility room drainage also discharges. The gully should be cleared out a grate fitted and the UPVC pipework checked to be suitable for foul waste.

The inspection chamber to the front was inspected, the lid appears to be rusting and replacement should be considered. The chamber construction condition was fair.

Property address

XXX

G

Services (continued)

Photo - 81 Services should be boxed in in fire rated construction or fitted with fire collar to soil and vent pipe

Photo - 82

Photo - 83 Front gully

Photo - 84

Property address

XXX

G Services (continued)

Photo - 85

Photo - 86

G7 Common services

N/A

Property address

XXX

H

Grounds (including shared areas for flats)

Limitations to inspection

We have not consulted any Geological or Ordnance Survey Maps and have been unable to establish any details as to the previous use of the site. We are unable to comment within the terms of this report, which is restricted in its scope, as to whether there are any hidden problems with the ground upon which the property is built, nor are we able to comment on the possibility or otherwise of the property being affected by any other matters. Your solicitors should check this aspect.

1 2 3 NI

H1 Garage

N/A

H2 Other

The timber shed was locked and could not be opened due to the key not being provided. The condition of the shed is fair but further investigation is required. 3

Photo - 87

Photo - 88

H3 General

The property has an open garden area to the front with no boundary lines. To the rear is a slabbed and shingled area with timber fencing and gate to the rear and garden walling to the sides.

The fences appear ok however, maintenance is required to the wall. The gate external joinery is missing and damaged with the lock being up side down.

Patios and paved areas can provide external areas that can enhance the quality of a property's garden. However, they must be kept in satisfactory condition otherwise this could present a safety hazard to users. The height and type of the fence can be controlled by the local authority.

Property address

XXX

Grounds (including shared areas for flats) (continued)

Photo - 89

Photo - 90

Photo - 91

Photo - 92 Requires maintenance

Property address

XXX

H

Grounds (including shared areas for flats) (continued)

Photo - 93 Requires repair

Photo - 94

Photo - 95

Property address

XXX

I

Issues for your legal advisers

We do not act as 'the legal adviser' and will not comment on any legal documents. However, if during the inspection we identify issues that your legal advisers may need to investigate further, we may refer to these in the report (for example, check whether there is a warranty covering replacement windows).

I1 Regulation

Building regulations are statutory instruments that seek to ensure that the policies set out in the relevant legislation are carried out. Building regulations approval is required for most building work in the UK. Building regulations that apply across England and Wales are set out in the Building Act 1984 while those that apply across Scotland are set out in the Building (Scotland) Act 2003. The Act in England and Wales permits detailed regulations to be made by the Secretary of State. The regulations made under the Act have been periodically updated, rewritten or consolidated, with the latest and current version being the Building Regulations 2010.

It is suspected that the internal wall to separate the kitchen from the stair has been removed and an additional bedroom formed at ground floor level at some point in the past. Building Regulations sign off would have been required for these works.

The solicitor should check for certification issued by engineers/contractors registered under the governments competent person scheme website for the double glazing, electrical installation and gas installation.

In addition boundary wall ownership should be confirmed for maintenance and repair obligations.

A three storey building should have fire precautionary measures in place such as :-

a) Fire doors within the route of fire escape (corridors) protecting bedrooms.
This is no longer insisted for by building regulation provided the loft room has a satisfactory escape window
b) Interconnected smoke detectors on mains electric.
c) Fire protection to ceiling of the first floor.
d) Fire protection to the underside of the staircase.
e) Adequate floor joists.
f) Insulation and ventilation to the ceilings and roof cheeks if any.
g) Thickness of insulation should be around 150 mm of solid insulation to underside of the roof.
h) Staircase must comply with maximum pitch, handrails regulation, risers, goings and balustrades.
If the room is not used as a bedroom the ventilation and insulation regulation should still apply.
We have not checked for compliance to all of these requirements.

I2 Guarantees

N/A

I3 Other matters

N/A

Property address

XXX

J

Risks

This section summarises defects and issues that present a risk to the building or grounds, or a safety risk to people. These may have been reported and condition rated against more than one part of the property or may be of a more general nature, having existed for some time and which cannot be reasonably changed.

J1 Risks to the building

The rear pier should be repaired to prevent material loosening further and the structure breaking down and being a safety risk. The front pier has leak staining from the overflow. The overflow should be checked to ensure the issue has been addressed as long term leaking could lead to damp penetration.

External joinery should be mountaineers to prevent decay and damp penetration.

The guttering should be unblocked to prevent the risk of failure and damp penetration.

The back door key should be located.

The metal guarding to the Juliet balcony if not replaced should be treated to prevent further corrosion.

The roof covering and overall construction needs to be fully ascertained as there is no roof space ventilation which can lead to to condensation and decay.

The felt is damaged in the roof space which could lead to damp penetration. The felt could be asbestos based which could lead to health risks.

The cracking described for internal walls should be investigated further to confirm it is not structural.

The fire compartmentation and means of escape deviates from the current guidance as such a fire engineer should be employed to give specialist advice.

A lack of mechanical extraction as described in the report can lead to internal polluted air. Warm moist air from cooking, the bathroom and utility room could lead to it condensing on cooler surfaces leading to mould and finish damage from damp.

Self-clinging climbers such as Boston ivy and Virginia creeper (Parthenocissus sp.) do not usually cause damage to wall surfaces, but common or English ivy (Hedera helix sp.) supports itself by aerial roots and where these penetrate cracks or joints they may cause structural damage. Sound masonry is unaffected. Its dense cover can hide defects in the fabric of the building and hinder maintenance work. Ivy may also provide access for intruders and harbour pests such as mice. Where brickwork is sound, the main problem is to keep growth away from gutters and paint work. It has been suggested that vegetation attached to walls could lead to dampness resulting from slower drying conditions following rain. This may be plausible on a south-west facing wall where the rain is driven by prevailing winds. However, other sources suggest that such plants will have a slight drying effect on mortar and will also provide some degree of insulation in winter, particularly evergreen ivies covering exposed north and east-facing walls. Large climbers can pose a risk to buildings. Such problems are most likely with older property, those with shallow foundations and those built on clay soils.

J2 Risks to the grounds

The wall ownership should be confirmed for repair and maintenance obligations to prevent safety and stability issues overtime.

Property address

XXX

J

Risks (continued)

J3 Risks to people

The property may contain asbestos based material including the boiler flue. Which if damaged could cause health issues. Prior to any works at the property a suitably qualified asbestos surveyor should inspect and take samples for testing to ascertain the location of any asbestos.

Asbestos

Asbestos can be found in any building built or refurbished before the year 2000. Materials that contain asbestos are not dangerous unless they are disturbed or damaged and fibres are released into the air. It is when these fibres are inhaled they can cause serious diseases.

Asbestos can take many forms and was used in many areas including loose fill insulation, lagging, sprayed coatings, asbestos insulating boards (found in places such as partition walls, door panels, ceiling tiles, soffits, panels under windows, around baths, around boilers, floor tiles, textiles such as, fire blankets and composites such as, flash guards in fuse boxes and in toilet seats and cisterns, textured coating on walls and ceilings (artex), asbestos cement was used in places such as; roofs, wall panels/cladding, downpipes and gutters, flues, water tanks, fire surrounds and pipes.

It is recommended that before any removal, demolition or repair works are undertaken a full asbestos survey is carried out by a suitably qualified surveyor. Some works need to be undertaken by a licensed contractor and some works are notifiable to the HSE. All works should be undertaken in accordance with health and safety guidance and legislation and any waste containing asbestos correctly disposed of.

As noted previously the fire compartmentation and means of escape is unsatisfactory as such poses an escape risk in the event of a fire.

Glazing in critical locations should be safety glazing this includes verifying the side middle pane to the rear door.

The rusting inspection chamber lid could give way under weight.

J4 Other

Please see section I for suspected works requiring Building Regulations sign off.

Property address

XXX

Valuation

In my opinion the market value on 29TH AUGUST 2017 **as inspected was:**

£ 245,000	TWO HUNDRED AND FORTY-FIVE THOUSAND POINT ZERO ZERO POUNDS
	(amount in words)

Tenure: Freehold

Area of property (sq m): 3(9.025x4.407)=119.31 say 119m2

In my opinion the current reinstatement cost of the property (see note below) is:

£ 207,000	TWO HUNDRED AND SEVEN THOUSAND POINT ZERO ZERO POUNDS
	(amount in words)

In arriving at my valuation, I made the following assumptions.

With regard to the materials, construction, services, fixtures and fittings, and so on I have assumed that:

- an inspection of those parts that I could not inspect would not identify significant defects or a cause to alter the valuation
- no dangerous or damaging materials or building techniques have been used in the property
- there is no contamination in or from the ground, and the ground has not been used as landfill
- the property is connected to, and has the right to use, the mains services mentioned in the report; and
- the valuation does not take account of any furnishings, removable fittings or sales incentives. With regard to legal matters I have assumed that:
- the property is sold with 'vacant possession' (your legal advisers can give you more information on this term)
- the condition of the property, or the purpose the property is or will be used for, does not break any laws
- no particularly troublesome or unusual restrictions apply to the property, that the property is not affected by problems which would be revealed by the usual legal inquiries and that all necessary planning permissions and Building Regulations consents (including consents for alterations) have been obtained and complied with; and
- the property has the right to use the mains services on normal terms, and that the sewers, mains services and roads giving access to the property have been 'adopted' (that is, they are under local-authority, not private, control).

Any additional assumptions relating to the valuation

RATIONALE WOULD USUALLY BE INCLUDED HERE

Your legal advisers, and other people who carry out property conveyancing, should be familiar with these assumptions and are responsible for checking those concerning legal matters.
My opinion of the Market Value shown here could be affected by the outcome of the enquiries by your legal advisers (section I) and/or any further investigations and quotations for repairs or replacements. The valuation assumes that your legal advisers will receive satisfactory replies to their enquiries about any assumptions in the report.

Other considerations affecting value

See above
Valuation is based upon vacant possession.

Property address

XXX

K

Valuation (continued)

Note: You can find information about the assumptions I have made in calculating this reinstatement cost in the 'Description of the RICS HomeBuyer Service (Survey & Valuation)' provided. The reinstatement cost is the cost of rebuilding an average home of the type and style inspected to its existing standard using modern materials and techniques, and by acting in line with current Building Regulations and other legal requirements. This will help you decide on the amount of buildings insurance cover you will need for the property.

Property address

XXX

Surveyor's declaration

"I confirm that I have inspected the property and prepared this report"

Signature

Surveyor's RICS number

Qualifications

Company

Cosey Homes Chartered Surveyors

Address

36-40 Duke Street

Town

St Helens

County

Postcode

WA10 2JP

Phone number

03300535823

Website

www.coseyhomes.co.uk

Fax number

Email

Property address

XXX

Clients name

XXX

Date this report was produced

Thu 28th Sep 2017

RICS Disclaimer

1. This report has been prepared by a surveyor ('the Individual Surveyor') merely in his or her capacity as an employee or agent of a firm or company or other business entity ('the Company'). The report is the product of the Company, not of the Individual Surveyor. All of the statements and opinions contained in this report are expressed entirely on behalf of the Company, which accepts sole responsibility for these. For his or her part, the Individual Surveyor assumes no personal financial responsibility or liability in respect of the report and no reliance or inference to the contrary should be drawn.

   In the case of sole practitioners, the surveyor may sign the report in his or her own name unless the surveyor operates as a sole trader limited liability company.

   Nothing in this report excludes or limits liability for death or personal injury (including disease and impairment of mental condition) resulting from negligence.

2. This document is issued in blank form by the Royal Institution of Chartered Surveyors (RICS) and is available only to parties who have signed a licence agreement with RICS.

   RICS gives no representations or warranties, express or implied, and no responsibility or liability is accepted for the accuracy or completeness of the information inserted in the document or any other written or oral information given to any interested party or its advisers. Any such liability is expressly disclaimed.

! Please read the 'Description of the RICS HomeBuyer Report Service' (at the back of this report) for details of what is, and is not, inspected.

Property address

XXX

What to do now

Getting quotations

The cost of repairs may influence the amount you are prepared to pay for the property. Before you make a legal commitment to buy the property, you should get reports and quotations for all the repairs and further investigations the surveyor may have identified.

You should get at least two quotations from experienced contractors who are properly insured. You should also:

- ask them for references from people they have worked for;
- describe in writing exactly what you will want them to do; and
- get the contractors to put the quotations in writing.

Some repairs will need contractors with specialist skills and who are members of regulated organisations (for example, electricians, gas engineers, plumbers and so on). Some work may also need you to get Building Regulations permission or planning permission from your local authority.

Further investigations

If the surveyor is concerned about the condition of a hidden part of the building, could only see part of a defect or does not have the specialist knowledge to assess part of the property fully, the surveyor may have recommended that further investigations should be carried out to discover the true extent of the problem.

Who you should use for these further investigations

You should ask an appropriately qualified person, though it is not possible to tell you which one. Specialists belonging to different types of organisations will be able to do this. For example, qualified electricians can belong to five different government-approved schemes. If you want further advice, please contact the surveyor.

What the further investigations will involve

This will depend on the type of problem, but to do this properly, parts of the home may have to be disturbed and so you should discuss this matter with the current owner. In some cases, the cost of investigation may be high.

When to do the work

The condition ratings help describe the urgency of the repair and replacement work. The following summary may help you decide when to do the work.

- **Condition rating 2** – repairs should be done soon. Exactly when will depend on the type of problem, but it usually does not have to be done right away. Many repairs could wait weeks or months, giving you time to organise suitable reports and quotations.
- **Condition rating 3** – repairs should be done as soon as possible. The speed of your response will depend on the nature of the problem. For example, repairs to a badly leaking roof or a dangerous gas boiler need to be carried out within a matter of hours, while other less important critical repairs could wait for a few days.

Warning

Although repairs of elements with a condition rating 2 are not considered urgent, if they are not addressed they may develop into defects needing more serious repairs. Flat roofs and gutters are typical examples. These can quickly get worse without warning and result in serious leaks.

As a result, you should regularly check elements with a condition rating 2 to make sure they are not getting worse.

Description of the RICS HomeBuyer (Survey & Valuation) Service

The service

The RICS HomeBuyer (Survey & Valuation) Service includes:

- an inspection of the property (see 'The inspection')
- a report based on the inspection (see 'The report'); and
- a valuation, which is part of the report (see 'The Valuation')

The surveyor who provides the RICS HomeBuyer (Survey & Valuation) Service aims to give you professional advice to help you to:

- make an informed decision on whether to go ahead with buying the property
- make an informed decision on what is a reasonable price to pay for the property
- take account of any repairs or replacements the property needs; and
- consider what further advice you should take before committing to purchase the property.

The inspection

The surveyor inspects the inside and outside of the main building and all permanent outbuildings, but does not force or open up the fabric. This means that the surveyor does not take up carpets, floor coverings or floorboards, move furniture, remove the contents of cupboards, roof spaces, etc., remove secured panels and/or hatches or undo electrical fittings.

If necessary, the surveyor carries out parts of the inspection when standing at ground level from public property next door where accessible. The surveyor may use equipment such as a damp-meter, binoculars and torch, and may use a ladder for flat roofs and for hatches no more than 3 metres above level ground (outside) or floor surfaces (inside) if it is safe to do so.

Services to the property

Services are generally hidden within the construction of the property. This means that only the visible parts of the available services can be inspected, and the surveyor does not carry out specialist tests. The visual inspection cannot assess the efficiency or safety of electrical, gas or other energy sources; plumbing, heating or drainage installations (or whether they meet current regulations); or the inside condition of any chimney, boiler or other flue.

Outside the property

The surveyor inspects the condition of boundary walls, fences, permanent outbuildings and areas in common (shared) use. To inspect these areas, the surveyor walks around the grounds and any neighbouring public property where access can be obtained.

Buildings with swimming pools and sports facilities are also treated as permanent outbuildings, but the surveyor does not report on the leisure facilities, such as the pool itself and its equipment, landscaping and other facilities (for example, tennis courts and temporary outbuildings).

Flats

When inspecting flats, the surveyor assesses the general condition of outside surfaces of the building, as well as its access areas (for example, shared hallways and staircases). The surveyor inspects roof spaces only if they are accessible from within the property. The surveyor does not inspect drains, lifts, fire alarms and security systems.

Dangerous materials, contamination and environmental issues

The surveyor does not make any enquiries about contamination or other environmental dangers. However, if the surveyor suspects a problem, he or she should recommend a further investigation.

The surveyor may assume that no harmful or dangerous materials have been used in the construction, and does not have a duty to justify making this assumption. However, if the inspection shows that these materials have been used, the surveyor must report this and ask for further instructions.

The surveyor does not carry out an asbestos inspection and does not act as an asbestos inspector when inspecting properties that may fall within the Control of Asbestos Regulations 2012. With flats, the surveyor assumes that there is a 'dutyholder' (as defined in the regulations), and that in place are an asbestos register and an effective management plan which does not present a significant risk to health or need any immediate payment. The surveyor does not consult the dutyholder.

The report

The surveyor produces a report of the inspection for you to use, but cannot accept any liability if it is used by anyone else. If you decide not to act on the advice in the report, you do this at your own risk. The report focuses on matters that, in the surveyor's opinion, may affect the value of the property if they are not addressed.

The report is in a standard format and includes the following sections.

A Introduction to the report
B About the inspection
C Overall opinion and summary of the condition ratings
D About the property
E Outside the property
F Inside the property
G Services
H Grounds (including shared areas for flats)
I Issues for your legal advisers
J Risks
K Valuation
L Surveyor's declaration
What to do now
Description of the RICS HomeBuyer (Survey & Valuation) Service
Typical house diagram

Condition ratings

The surveyor gives condition ratings to the main parts (the 'elements') of the main building, garage and some outside elements. The condition ratings are described as follows.

Condition rating 3 – defects that are serious and/or need to be repaired, replaced or investigated urgently

Condition rating 2 – defects that need repairing or replacing but are not considered to be either serious or urgent. The property must be maintained in the normal way.

Condition rating 1 – no repair is currently needed. The property must be maintained in the normal way.

NI – not inspected.

The surveyor notes in the report if it was not possible to check any parts of the property that the inspection would normally cover. If the surveyor is concerned about these parts, the report tells you about any further investigations that are needed.

The surveyor does not report on the cost of any work to put right defects or make recommendations on how these repairs should be carried out. However, there is general advice in the 'What to do now' section at the end of the report.

Energy

The surveyor has not prepared the Energy Performance Certificate (EPC) as part of the RICS HomeBuyer Service for the property. If the surveyor has seen the current EPC, he or she will present the energy-efficiency and environmental impact ratings in this report. The surveyor does not check the ratings and cannot comment on their accuracy.

Issues for legal advisers

The surveyor does not act as 'the legal adviser' and does not comment on any legal documents. If, during the inspection, the surveyor identifies issues that your legal advisers may need to investigate further, the surveyor may refer to these in the report (for example, check whether there is a warranty covering replacement windows).

This report has been prepared by a surveyor ('the Individual Surveyor') merely in his or her capacity as an employee or agent of a firm or company or other business entity ('the Company'). The report is the product of the Company, not of the Individual Surveyor. All of the statements and opinions contained in this report are expressed entirely on behalf of the Company, which accepts sole responsibility for these. For his or her part, the Individual Surveyor assumes no personal financial responsibility or liability in respect of the report and no reliance or inference to the contrary should be drawn.

Description (continued)

In the case of sole practitioners, the surveyor may sign the report in his or her own name unless the surveyor operates as a sole trader limited liability company.

Nothing in this report excludes or limits liability for death or personal injury (including disease and impairment of mental condition) resulting from negligence.

Risks

This section summarises defects and issues that present a risk to the building or grounds, or a safety risk to people. These may have been reported and condition rated against more than one part of the property or may be of a more general nature, having existed for some time and which cannot reasonably be changed.

If the property is leasehold, the surveyor gives you general advice and details of questions you should ask your legal advisers.

The valuation

The surveyor gives an opinion on both the market value of the property and the reinstatement cost at the time of the inspection (see the 'Reinstatement cost' section).

Market value

'Market value' is the estimated amount for which an asset or liability should exchange on the valuation date between a willing buyer and a willing seller in an arm's length transaction, after proper marketing wherein the parties had each acted knowledgeably, prudently and without compulsion'

When deciding on the Market Value, the surveyor also makes the following assumptions.

The materials, construction, services, fixtures and fittings, and so on

The surveyor assumes that:

- an inspection of those parts that have not yet been inspected would not identify significant defects
- no dangerous or damaging materials or building techniques have been used in the property
- there is no contamination in or from the ground, and the ground has not been used as landfill
- the property is connected to, and has the right to use, the mains services mentioned in the report and
- the valuation does not take account of any furnishings, removable fittings and sales incentives of any description.

Legal matters

The surveyor assumes that:

- the property is sold with 'vacant possession' (your legal advisers can give you more information on this term)
- the condition of the property, or the purpose that the property is or will be used for, does not break any laws
- no particularly troublesome or unusual restrictions apply to the property, that the property is not affected by problems which would be revealed by the usual legal enquiries and that all necessary planning and Building Regulations permissions (including permission to make alterations) have been obtained and any works undertaken comply with such permissions; and
- the property has the right to use the mains services on normal terms, and that the sewers, mains services and roads giving access to the property have been 'adopted' (that is, they are under local-authority, not private, control).

Standard terms of engagement

1 **The service** – the surveyor provides the standard RICS HomeBuyer (Survey & Valuation) Service ('the service') described in the 'Description of the RICS HomeBuyer (Survey & Valuation) Service', unless you and the surveyor agree in writing before the inspection that the surveyor will provide extra services. Any extra service will require separate terms of engagement to be entered into with the surveyor. Examples of extra services include:
   - costing of repairs;
   - schedules of works;
   - supervision of works;
   - re-inspection;
   - detailed specific issue reports; and

2 **The surveyor** – the service is to be provided by an AssocRICS, MRICS or FRICS member of the Royal Institution of Chartered Surveyors, who has the skills, knowledge and experience to survey, value and report on the property and is a member of the RICS Valuer Registration Scheme.

3 **Before the inspection** – you tell the surveyor if there is already an agreed or proposed price for the property, and if you have any particular concerns (such as plans for extension) about the property.

4 **Terms of payment** – you agee to pay the surveyor's fees and any other charges agreed in writing.

5 **Before the inspection** – nothing in this clause 5 shall operate to exclude, limit or otherwise affect your rights to cancel under the Consumer Contracts (Information, Cancellation and Additional Charges) Regulations 2013 or the Consumer Rights Act 2015, or under any such other legislation as may from time to time be applicable. Entirely without prejudice to any other rights that you may have under any applicable legislation, you are entitled to cancel this contract in writing by giving notice to the surveyor's office at any time before the day of the inspection, and in any event within fourteen days of entering into this contract. Please note that where you have specifically requested that the surveyor provides services to you within fourteen days of entering into the contract, you will be responsible for fees and charges incurred by the surveyor up until the date of cancellation.

6 **Liability** – the report is provided for your use, and the surveyor cannot accept responsibility if it is used, or relied upon, by anyone else.

Complaints handling procedure

The surveyor will have a complaints handling procedure and will give you a copy if you ask. The surveyor is required to provide you with contact details, in writing, for their complaints department or the person responsible for dealing with client complaints. Where the surveyor is party to a redress scheme, those details should also be provided. If any of this information is not provided, please notify the surveyor and ask that it be supplied.

Note: These terms form part of the contract between you and the surveyor.

This report is for use in England, Wales, Northern Ireland, Channel Islands and Isle of Man.

Typical house diagram

This diagram illustrates where you may find some of the building elements referred to in the report.

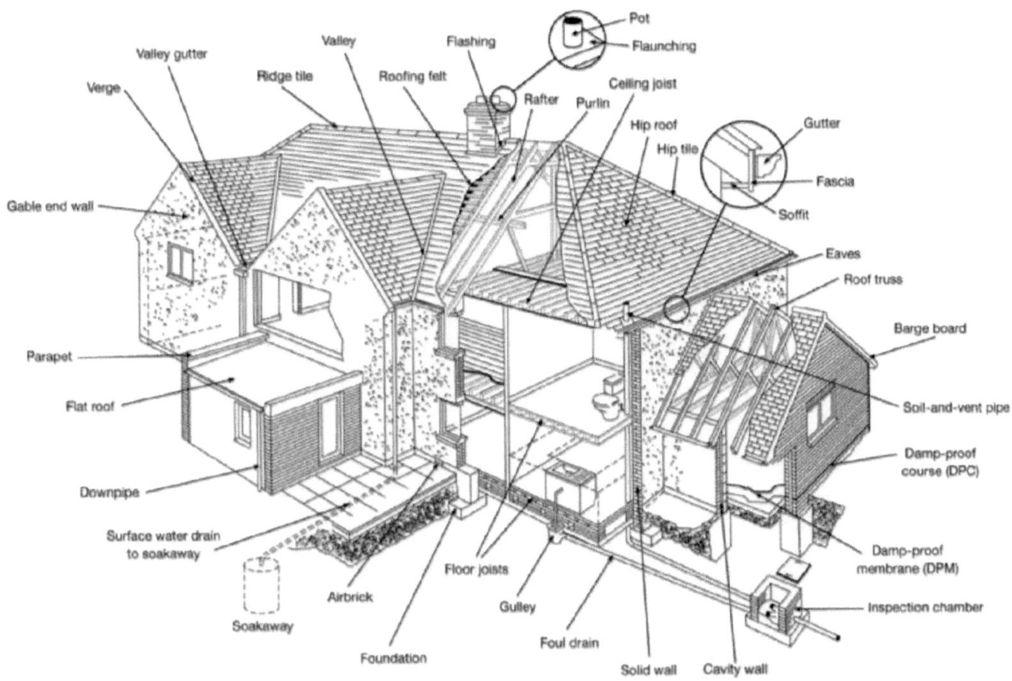

Appendix 3

Photographic schedule of condition

Reception

DESCRIPTION	CONDITION	ACTION REQUIRED
Ceilings: Painted plaster Old style light fitting	Marked	Prepare, repair and redecorate Replace lighting with modern lighting
Walis: Painted plaster	Marks, pattens, staining and old fixing points 	Make good. Prepare, repair and redecorate
Floors: Carpet	Worn and marked	Repair replace
Detailing: Window Two Plastic double glazed Sliding sash		Clean

Appendix 4

Project option ITT proposal form

Name: Jason Challender

Specialist Area: Construction Studies/Building Surveying

Topic

Improving the teaching of Construction Studies students as far as undertaking building surveys.

Summary:

Personal Justification:

I have reflected on my own education whilst at College and University to examine whether students on leaving education are being fully prepared for employment and thereby competent to undertake their chosen vocations.

My own personal experience, as a graduate in Building Surveying in 1989, was that on leaving University I was not equipped with the basic skills required to undertake a basic building survey, which was something that was expected of me in professional practice. The expertise and training that enabled me to undertake building surveys was gained whilst employed in industry through training from experienced surveying colleagues who had experienced similar dilemmas at a similar stage of their careers.

This event has led me to ask the question as to whether there is a gap in provision as far as teaching and learning of students on construction related courses are concerned with particular emphasis on building surveys. Simple reflective practice from my own experiences, those of my past and present colleagues in industry and my recent interface with Construction Studies lecturers and students would appear to suggest that there is currently a large divide between classroom teaching and academic qualifications and the basic skills required in professional practise.

For this assignment I have researched whether there is any data, case studies, reports or any other evidence that have been produced to support and bring meaning to my own opinions and findings on this matter. Several research documents have presented themselves to me in this respect and these will hopefully bring creditability to my argument and justify my views accordingly.

As a qualified practitioner with over 20 years experience in industry I will then seek to explore why teaching and learning has not been adapted to meet the needs of the Building Surveying profession over the years. Furthermore, I will suggest and develop an innovative project that can be introduced and piloted on to the National Diploma Construction Studies Course at Stockport College which will seek to enhance teaching and learning for the benefit of the students and other tutors. Following its integration and implementation I will then invite feedback and evaluate whether it has in fact made a difference and enhanced vocational related training for construction students.

Project:
To develop a simple concise handbook as a guide and template to assist students on construction related courses in undertaking building surveys and preparing survey reports thereafter. It will set down clear and concise elemental headings, following the standard format of the RICS Homebuyers Survey Report introduced in 2001, to give structure and relevance to all student reports. It will attempt to steer students through a logical sequence as far as the various stages and the methodology of the survey are concerned from start to finish. Furthermore, as students are generally not familiar with undertaking surveys of this kind, it will prompt responses from the learners by way of a potential list of relevant statements related to description, condition, defect and remedial works which students can use as checklist for their report.

I have agreed with the course leader of the Building Surveying Module (as part of the ND Construction Course) at Stockport College that the project can be piloted with students using the handbook as a template to assist them in undertaking building surveys of their own homes. This will then form part of their course assignment. I intend to develop a questionnaire that will be completed by participating students and lecturers to obtain feedback as to the usefulness of the survey handbook. This will assist me in my evaluation of the project which is an integral part of this module.

The basis of this handbook, as a reference tool in carrying out the survey on site, will hopefully attempt to give all the necessary information that the student requires to enable them to write up the report at a later stage. It will also provide learners with a basic familiarity and understanding as far as undertaking surveys of this type are concerned. This will then hopefully benefit them in their studies and better prepare and equip them for their chosen vocations in life.

Approved Date

Smart Plan

Specific Objectives:

Create a student Building Surveyors Handbook to assist learners through the survey process and prepare a Home Buyers Survey Report.

Students will become familiar with the RICS Home Buyers Report format to better prepare them for professional practice.

Close current gap between education/training gained by students at College and those skills required by industry.

Measure: completion Date

A group of individuals including students, College course leaders and professional practitioners will be asked to pilot the project by using the handbook to conduct an RICS Home Buyers Survey of their own homes. Following this implementation process these individuals will then complete a questionnaire that I have specifically designed for the purpose of obtaining their feedback on the handbook. This will then enable me to assess and measure the usefulness of the handbook against a list of predetermined criteria that I have devised. I am aiming to have the assignment submission completed on 25th May 2010.

Agreed as achievable

Name . . . . . . Jason Challender . . . . Tutor . . . . . .
Nigel Starr . . . . Date . . . . . . 14th April 2010 . . . .

Realistic

What resources are required:

I will research via the internet, construction journals, the Construction Skills Council and the Royal Institution of Chartered Surveyors to find material which confirms that there is indeed a gap in provision as far as training of Building Surveyors is concerned.

I will liaise and consult with various individuals who I have access to in my current position as Project Director of Stockport College including professional practitioners, course leaders and students to initially assess the worthiness of the project. I will then prepare the Student Building Surveyors Handbook and implement the project, as a pilot study, with these same individuals and obtain the necessary feedback from them.

I believe all these resources will be accessible and thereby realistic within the required timescale.

What knowledge will you need:

I will require my own knowledge in how building surveys are required to be undertaken. I will use this knowledge to create a multiple choice of potential defects for any particular element of the building. This will hopefully assist students when inspecting such elements as roofs, walls and floors and deciding which of the listed defects is most appropriate to represent the condition of that particular element.

Time – how much time do you estimate it will take?

I have estimated that this assignment will involve 73 hours of my time between 10th March 2010 and 25th May 2010 when I am aiming to complete my submission.

Name: *PLAN*

Date	*Time Allocated*	*Intended Action (adjust and comment on actual performance continue on another page if necessary)*
10–03–10	4 hours	Apply reflective practise to describe and analyse the events that
17–03–10	5 hours	have led me to conclude that my project is valid and credible. I will
24–03–10	5 hours	list reasons why I have embarked on my chosen project and what
06–04–10	8 hours	I am looking to achieve from it.
07–04–10	4 hours	I will be researching from various sources whether there are
08–04–10	8 hours	organisations, professional practitioners, teachers or educational
09–04–10	4 hours	establishments who have concluded similar findings as myself
12–04–10	4 hours	in terms of current gaps in educational provision. This will be
26–04–10	3 hours	intended to support and validate my arguments and give credibility
27–04–10	4 hours	to my chosen project.
29–04–10	4 hours	Further research
05–04–10	4 hours	Prepare draft Student Building Surveyors Handbook
12–05–10	4 hours	Consult with staff, students and professional practitioners on draft
16–05–10	4 hours	handbook proposal and obtain initial feedback
22–05–10	4 hours	Prepare final version of handbook
25–05–10	4 hours	Prepare questionnaire to obtain feedback
		Make all necessary arrangements for implementation of handbook, using pilot study.
		Collect feedback from pilot study, via questionnaires and analyse data.
		Write up assignment
		Write up assignment
		Write up assignment
		Write up assignment
		Write up assignment
		Write up assignment
		Complete Assignment

Appendix 5

Relevant standard by LLUK for professional knowledge and practice

Project Standards and Professional Practice **Name: Jason Challender**

Ref.	*Prof. Knowledge/understanding*	*Professional Practice*
AK4.1	Principles, frameworks and theories which underpin good practice in learning and teaching.	**AP4.1** Use relevant theories of learning to support the development of practice in learning and teaching.

How met and where
Follow LLUK standards and update knowledge through training and CPD. Motivation theory covered in relation to learning.
Reference to Driscoll's model in assignment report

AK4.3	Ways to reflect, evaluate and use research to develop own practice, and to share good practice with others.	**AP4.3** Share good practice with others and engage in continuing professional development through reflection, evaluation and the appropriate use of research.

How met and where
Use of reflective practice, research and evaluation in developing surveyors handbook. I shared good practice by developing the handbook and at the implementation stage. Assignment report and bibliography

CK1.1	Own specialist area including current development.	**CP1.1** Ensure that knowledge of own specialist area is current and appropriate to the teaching context

How met and where.
I undertake internal and external training courses as part of my CPD. This updates me on the latest issues that relate to my specialist area of construction, such as recent legislation and building regulations. I also undertake reading to ensure that my specialist area is appropriate to teaching. See bibliography.

DK3.1	Ways to evaluate own role and performance in planning and learning.	**DP3.1** Evaluate the success of planned learning activities.

How met and where
Evaluation based on reflection as well as learner and colleague feedback. The evaluation and conclusion discusses and assesses how effective the handbook was and what has been learnt from this. See report and appendix

Project Values and Comments **Name: Jason Challender**

Ref.	*Description*	*How met*	*Where*
AS4	Reflection and evaluation of their own practice and their continuing professional development as teachers.	Reflection of my own experience, research carried out and the introduction of an innovative teaching and learning aid.	Project proposal and report.
AS7	Improving the quality of their practice	Identifying a need amongst BS students. Design and implementation of the handbook and critical analysis.	Report
BS2	Applying and developing own professional skills to enable learners to achieve their goals.	Using questioning and feedback to inform practice	Project proposal
CS1	Understanding and keeping up to date with current knowledge in respect of own specialist area.	I regularly attend CPD in my own profession as a Project Manager. This keeps me abreast of all latest development in regulations. A recent course attended updated me on current DDA legislation, which has supported me in promoting inclusive learning and safeguarding against discrimination in the working place.	CPD Bibliography
CS4	Developing good practice in teaching own specialist area.	Reflective practice using LLUK overarching standards to strive for continuous improvement	Report
FS3	Maintaining own professional knowledge in order to provide information on opportunities for progression in own specialist area.	Informed learners of possibilities for progression within Building Surveying profession	Report

Appendix 6

Session plan linked with building surveying handbook

Venue & Room no.: 2nd floor GW | Tutor Jason Challender | No. students on register: 9 | Date:08/04/2010 | Time: 10.30am-12.00am

Lesson 36

Focus for this session: National Diploma Construction Students Yr 2

Link(s) to previous session(s): None

Group Profile: *(to include, for example: age, gender, ethnicity, individual learning needs indicated in diagnostic initial assessments, etc.)*

The group consists of a general mix of male and female students with a wide range of ages and ethnic backgrounds. Initial assessments indicated that only most member of the group have previous knowledge of the subject matter through research on the Colleges website and have a basic understanding of construction issues.

Health & Safety:

Students are aware of H and S emergency procedures including fire evacuation

General Objectives: *At the end of the session learners should –*
Be familiar with the Building Surveyors Student Handbook for undertaking a building survey

Specific Objectives/Learning Outcomes: *At the end of the session, learners should be able to –*
1. Name each section of the Homebuyers Survey
2. Explain how the handbook will assist students in compiling building information
3. Complete the handbook information requirements as part of the student assignment

Every Child/Learner Matters: *(identify how this session plan meets the needs of ECM policy)*

Be healthy	Room temperature monitored, and room layout planned to best meet students space requirements
Stay safe	H and S information/fire evacuation procedures given
Enjoy and achieve	Students will participate in activities
Make a positive contribution	Students will participate in activities
Achieve economic well-being	Lessons will have direct relationship with vocational positions

Resources: handouts

Time	*Sp. Objective/ Learning Outcome*	*Teacher Activity*	*Learner Activities*	*Functional Skills*	*Assessment*	*Resources*
10.30		Take register and recap on previous session. Introduction to session and explain the purpose of the Student Handbook for Building Surveys that I have devised as a teaching and learning aid	Ideally learners will take part in recap	Speaking/Listening	Q/A	Register Handouts
10.45	1.	Explain each section of the handbook and how it relates to an RICS Homebuyers Building Survey in its content	Students will observe and then compare the handbook with a re-alia survey document to compare and contrast the format.	Speaking/Listening Reading for understanding Note taking	Q/A Oral	Handouts
11.15	2.	Explain to students the advantages of the handbook and the reasons I have devised it to assist them in carrying out their own survey reports	Students will suggest their own advantages and benefits that the handbook can create for them on a personal level.	Speaking/Listening Reading for understanding Note taking	Q/A Oral	Handouts
11.30	3.	Explain to students how to complete the various sections of the handbook and the level of information that is requested from them under each section/ heading. Student will be asked to complete the handbook and a questionnaire for feedback on it for the next session.	Students will read and analyse each section of the handbook to become familiar with it, in order to complete the contents at a later date.	Speaking/Listening Reading for understanding Note taking	Q/A Oral	Handouts

Key Skills Please	*Communications*	*App. Of Number*	*ICT*	*Working with Others*	*Improving Own Learning & Performance*	*Problem Solving*

Appendix 7

Striving for a sustainable Salford

Striving for a Sustainable Salford

Environmental Sustainability Plan 2018-2030

August 2022 version 1.5

Environmental Sustainability Team

University of Salford MANCHESTER

Contents

Document Control Information			
Author	**Summary of changes**	**Version**	**Authorised & Date**
Marta Strzelecka	Review of document, no changes required.	V1.5 18/08/2022	N/A
Bec Bennett	Added suggestions from FRC and updated performance Appendix A	V1.5 14/01/2021	University Council 22th January 2021
Bec Bennett	Added Foreword	V1.4 22/10/2020	Huw Williams Chief Operating Officer 22nd October 2020
Bec Bennett	Updated 2019/20 data and targets	V1/3 08/10/2020	Vice Chancellor's Executive Team 13th October 2020
Bec Bennett	Updated waste and water targets and performance section	V1.2 02/09/2020	
Bec Bennett	Updated from EST comments, inclusion of SDGs	V1.1 14/04/2020	Environmental Projects Board
Bec Bennett	Creation of document draft	V1.0	

Our Mission

By pioneering exceptional industry partnerships we will lead the way in real world experiences preparing students for life.

As a civic institution, our mission is to:

- Educate the next generation of modern industrialists, innovators, creators, entrepreneurs and leaders
- Develop the skills and knowledge needed to capitalise on the next industrial revolution
- Work in collaboration with public and private sector partners to address local and global economic and societal challenges

Our Vision for Sustainability

By embedding sustainability in all aspects of University life, we will enable our University community to have maximum beneficial impacts for society and environment.

Striving for a Sustainable Salford

At the University of Salford we recognise our responsibility, and our opportunity, to have a positive impact on society. We've always played a major role in improving the lives of our local communities, and today we take this challenge more seriously than ever. Our research is focused on meeting the four grand challenges of the Industrial Strategy: Data and AI; Clean Growth; Future of Mobility; and Ageing Society.

In addition, we have a responsibility to enable our students and staff to contribute to a sustainable world and fully recognise our responsibility to address the direct negative impacts of our own operations. We are proud to be supporting the Greater Manchester carbon revolution which will transform Greater Manchester into a world-leading greener, cleaner, climate resilient city region, improving the health and quality of life for millions of people and protecting our green spaces and environment for future generations. Sustainable development is a main focus of our Campus Masterplan, which is being delivered alongside a broader vision for the Salford Crescent Development in partnership with Salford City Council.

This plan consolidates our existing action towards sustainability. Through this we will seek to build on our current progress and challenge ourselves more. As Chief Operating Officer I fully support this plan, will ensure sustainability is considered in our operational decision-making and call on the whole University to play their part in striving for a Sustainable Salford.

Huw Williams

Chief Operating Officer

UN Sustainable Development Goals

The Sustainable Development Goals, or Global Goals, are a call for action by all countries to promote prosperity while protecting the planet. They recognise that ending poverty must go hand-in-hand with strategies that build economic growth and address a range of social needs including education, health, social protection, and job opportunities, while tackling climate change and environmental protection.

We are aligning our Environmental Sustainability Plan to the Global Goals by mapping each area to the relevant goals.

About this Plan

This Plan provides narrative to the University of Salford policies and objectives to improve environmental sustainability performance. At the University of Salford we are proud of our place in the world. Situated in the heart of the vibrant Greater Manchester conurbation we play a significant role in the community. While our heart is anchored in the North West, we have a powerful global reach. We focus on equipping our students, our future leaders, for the world of work.

We recognise the global climate crisis and that climate change is one of the biggest challenges facing our society across the globe. As a higher education provider we have a major role to play in enabling our students and staff to respond positively to global challenges.

We also recognise our responsibilities for the direct impact on the environment; as one of the largest organisations in the Greater Manchester region offering a place of study and work to over 18,500 students and 2,500 staff across 27 buildings and around 70 hectares. In the 2016/17 Academic Year (the baseline year selected for the majority of this Plan), energy and water costs were around £2.1 million per year. Total carbon emissions from energy use for the same period were 8,901 tCO_2e and waste from University operations was 900 tonnes. We have an obligation to reduce our negative impact on our environment but also an opportunity to demonstrate environmental sustainability through our operations and using our campus environment as a living laboratory.

Figure 1 Our 2016/17 environmental impact

This plan sets out how we will demonstrate management of our environmental sustainability including how we will mitigate climate change, be resource efficient and contribute towards sustainable development through the operation of our campus and how this will contribute to education and research for sustainability.

Our Environmental Sustainability Policy sets the framework for our environmental objectives. This plan sets the key actions we will take, how we will monitor and review our progress. This plan is reviewed at least every two years and any amendments to objectives, targets and KPIs will be highlighted.

Our Story so Far

We have built on our first Carbon Management Plan launched in 2011 and our Environmental Sustainability Policy, adopted in 2015, to ensure that our commitments around environmental sustainability remain current, challenging and reflect our Campus Masterplan. Most notably through the adoption of significant objectives to support the Greater Manchester science-based carbon target and ambition to become the greenest city region in the UK. Our new Energy, Water and Carbon Management Plan builds on our significant progress in carbon reduction (65% reduction since 2005/6) to ensure we can achieve our objective of an 81% reduction by 2030 and work towards net zero carbon by 2038. We have also committed to eliminate avoidable single use plastics from our campus.

CO_2 65%

We recognise the importance of our green space on campus and have committed to maintaining these to a high standard including encouraging biodiversity. In recognition of this we received a prestigious Green Flag Award for our Peel Park and Frederick Road Campuses, well ahead of our target date. Our Campus Masterplan will also look for opportunities for new green spaces in our urban environment to help promote learning, wellbeing and healthy lifestyles.

In 2018 we completed development and implementation of our environmental and energy management into a system through the EcoCampus framework and are proud to have subsequently been awarded EcoCampus Platinum, ISO 14001 and ISO 50001 certifications.

We are proud of our achievements so far but we are committed to doing more. We are currently positioned 64th, a 2:2 classification, (2019) in the People and Planet University League, our aim is to increase to the 1st class banding.

Environmental Management

An environmental management system is a way for an organisation to demonstrate that it addresses and minimises its environmental impacts, manages its legal compliance and continuously improves its environmental performance. An Environmental Management System is a systematic approach to managing an organisation's impacts on the environment. An effective EMS will involve the following:

- An initial assessment of how the organisation's activities, products and services might affect the environment
- Making sure they comply with all relevant environmental regulations
- Gathering data e.g. energy use, waste, water, raw materials
- Developing and communicating an environmental policy
- Procedures for controlling activities with significant environmental impacts
- Identifying risks and opportunities associated with environmental issues
- Setting targets and measuring progress
- Defining roles and responsibilities for all employees
- Training and awareness
- Periodic internal auditing
- Management Review and commitment from Leadership

The University Estates and Facilities Division leads on the Environmental and Energy Management System (EEMS) which is certified to both BS EN ISO 14001: 2015 and BS EN ISO 50001:2018 standards covering all of our physical estates and operations. The EEMS was developed over a number of years using the EcoCampus approach with awards from bronze to platinum.

We will meet our environmental compliance obligations, protect the environment in which we operate our estates and facilities, and prevent pollution by reducing and eliminating pollution sources

We will achieve this by

- developing and maintaining an Environmental and Energy Management System externally accredited (ISO 14001:2015 and ISO 50001:2018 standards) for Estates and Facilities Division
- Maintaining a register of environmental compliance obligations and regularly evaluating our compliance
- Reducing pollution risk (emissions & discharges) to land, water and groundwater

We will monitor this by

- ISO 14001:2015 and ISO 50001:2018 certification
- Number of major non-conformities related to a breach in compliance obligations
- Number of pollution incidents

Relevant Documents

- Environmental Sustainability Policy

Energy, Water & Carbon

Carbon emissions are broken down into three categories by the Greenhouse Gas Protocol in order to better understand the source.

Scope 1 – All Direct Emissions from the activities of an organisation or under their control. Including fuel combustion on site such as gas boilers and fleet vehicles.

Scope 2 – Indirect Emissions from electricity purchased and used by the organisation. Emissions are created during the production of the energy and eventually used by the organisation.

Scope 3 – All Other Indirect Emissions from activities of the organisation, occurring from sources that they do not own or control. These are usually the greatest share of the carbon footprint, covering emissions associated with business travel, procurement, waste and water.

In 2011, the University of Salford launched their first Carbon Management Plan with targets to reduce scope 1 and 2 carbon emissions by 43% by September 2020 compared to a 2005/06 baseline. As of end of academic year 2017/18 the University had already reduced its scope 1 and 2 carbon emissions by 64%. This has been achieved partly through reductions in energy use at the University through investment, partly through reductions in the grid electricity emissions factor, and also partly through the University's divestment of its student accommodation in 2007/08.

Our Energy, Water and Carbon Management Plan builds on our significant progress in carbon reduction to ensure we can achieve our objective of a scope 1 and 2 81% reduction by 2030 and work towards net zero carbon by 2038 in line with the Greater Manchester carbon reduction strategy.

Scope 3 carbon emissions account for just over 80% of the University's total emissions and we recognise the impact of these emissions so have committed to improving our monitoring and measuring of these and setting reduction targets where possible.

We will reduce the energy and water consumption of our buildings to meet the targets as established in the Energy, Water and Carbon Management Plan

We will achieve this by

- Improved monitoring of energy, carbon and water data
- Eliminating or minimising energy and water waste
- Improving energy and water efficiency
- Monitoring our scope 3 carbon emissions

We will monitor this by

- Percentage reduction of scope 1 and 2 carbon emissions from baseline
- Energy consumed by gross internal floor area
- Percentage reduction in water use (per staff and student FTE) from baseline
- Publication of an annual scope 3 carbon emissions report

Relevant Documents

- Energy and Water Policy
- Energy, Water and Carbon Management Plan
- Scope 3 Carbon Emissions Report

6 CLEAN WATER AND SANITATION

7 AFFORDABLE AND CLEAN ENERGY

13 CLIMATE ACTION

Waste Management

Waste and recycling is one of the most visible areas in the sustainability agenda and can demonstrate our commitment to staff, students and visitors as well as offering an opportunity for people to easily get involved in environmental improvement. The "waste hierarchy" ranks waste management options according to what is best for the environment. It gives top priority to preventing waste in the first place. When waste is created, it gives priority to preparing it for re-use, then recycling, then recovery, and last of all disposal (e.g. landfill).

Waste management and recycling are co-ordinated by the University Estates and Facilities Division. In the academic year 2016/7 we produced approximately 900 tonnes of waste across the University (excluding construction waste). We have worked with our waste contractors to divert as much waste as possible from landfill – the majority of our waste is sent for energy recovery. We operate a reuse scheme through Warp-it which facilitates reuse within the University and wider. Our recycling rate in 2016/17 was 38%, we are working to increase this as well as reducing our total waste output.

Along with other universities and colleges in Greater Manchester, The University of Salford will work to eradicate avoidable single-use plastics from catering, labs and stationery.

On plastics, we have already worked with our caterers, Chartwells (Salfood) and the Students' Union on a number of actions, including incentivised use of re-usable cups and non-plastic cutlery and takeaway containers. Chartwells and the Students' Union have also added a zero-waste element to one of their shops which sells dried goods. Instead of plastic packaging, customers bring their own containers or bags to fill up products and pay by weight or item.

We will apply the principles of the waste hierarchy to prioritise reduction, reuse and recycling to reduce the impacts of waste management

We will achieve this by

- Reducing total waste sent for recycling and disposal
- Increasing waste recycled
- Eliminating avoidable single use plastics on campus
- Encouraging contractors to apply the principles of the waste hierarchy on capital projects

We will monitor this by

- Percentage reduction in waste from baseline
- Percentage waste recycled
- Number of avoidable single use plastics on campus

Relevant Documents

- Waste Management Plan

Sustainable Procurement

Sustainable Procurement is the process of recognising the negative and positive environmental, social and economic impacts of purchased goods, works and services. It is a cradle to grave approach and should consider the entire supply chain rather than just the individual product or service. The UN Marrakech Task Force define Sustainable Procurement as "Sustainable Procurement is a process whereby organisations meet their needs for goods, services, works and utilities in a way that achieves value for money on a whole life basis in terms of generating benefits not only to the organisation, but also to society and the economy, whilst minimising damage to the environment".

The Estates and Facilities Division at University of Salford has significant spend covering everything from facilities management, minor works and regulatory maintenance and testing to major building projects and infrastructure development. With this amount of spend and potential environmental impact, the need for staff involved in purchasing to ensure that spend provides value is evident. Environmental and social responsibility is considered as part of our value for money process and appropriately considered in the procurement of supplies, services or works. For example we refer to the Energy and Water Technology Lists in relevant purchases.

Through our membership of North West Universities Purchasing Consortium, we are members of Electronics Watch, an independent monitoring organisation set up with the purpose of protecting the rights of workers within electronics supply chains. Through this membership we can collaborate to have a positive impact on workers through the electronics we purchase. We are also exploring ways to ensure transparency in supply chains of other goods.

We will develop and apply an ethical and sustainable procurement system to ensure social and environmental impacts of purchases are appropriately considered

We will achieve this by

- Implementing a Sustainable Purchasing Policy and Plan for Estates & Facilities
- Ensuring all Estates & Facilities tender evaluations include sustainability
- Meeting Level 4 in all areas of the Flexible Framework
- Developing and implementing a Sustainable Food Policy

We will monitor this by

- Flexible Framework Self-Assessment
- Reviewing Estates & Facilities Invitation to Tenders for sustainability

Relevant Documents

- Sustainable Procurement Plan
- Sustainable Food Policy

Sustainable Construction

As a Future Enabler, one of the key pan-university projects that will enable us to deliver the university's vision, is our Campus Masterplan. Central to the plans is an energy strategy that takes a major step towards a zero carbon future, while providing high quality spaces for residential, teaching, research and commercial uses that are cost-effective to run. University of Salford buildings will undergo refurbishment to make them highly energy efficient, with maximum potential for renewable energy generation. We recognise there may be competing priorities and financial pressures, therefore we have developed a policy and design standard to guide our development teams on our sustainability principles.

We are committed to managing design, construction, refurbishment and post completion occupancy of its buildings in order to reduce environmental impact, enhance the wellbeing of staff and student users of the building, minimise operating costs and comply with all relevant sustainable building legislation.

We will ensure sustainability is a core principle of the Estate Masterplan by embedding sustainable construction practices and performance specifications into plans for new buildings and refurbishments

We will achieve this by

- Implementing our Sustainable Construction Policy
- Aiming for BREEAM Excellent for high value new builds where possible
- Considering the use of SKA methodology targeting at least Silver for lower value new builds and refurbishments
- Implementing our Energy Design Standard on all projects regardless of value

We will monitor this by

- Percentage of construction projects meeting sustainable construction policy requirements

Relevant Documents

- Sustainable Construction Policy
- Energy Design Standard

Travel & Transport

Our latest Travel Plan covered the period 2012-2017. We have recently reviewed this Plan and are aiming to relaunch an updated version as soon as possible.

Our travel surveys show that the majority of students travel by non-car modes (75%) and 49% of staff travel by non-car modes. Single occupancy car use has increased for both staff and students slightly overtime and is likely to be a consequence of reduced car sharing promotion and allowance of unrestricted parking for all users. The surveys also show that a high number of staff and students are unaware of the existing tools and infrastructure measures.

Our ambition is to consolidate and rationalise car-parking across campus and to reduce internal car circulation to promote a healthier, more pedestrian-friendly campus through our Campus Masterplan. The vision also promotes greater levels of cycling in and around the site. Improved pedestrian connections will be key to improving permeability across the campus – think tree-lined boulevards, public squares and a generally more well-defined network of routes and open spaces.

We will develop a travel plan which encourages alternative methods to single-occupancy car journeys and minimises the environmental impact of the University fleet and business travel

We will achieve this by

- Developing an updated Sustainable Travel Plan for 2021-2030
- Improving facilities for cyclists on campus
- Supporting the use of electric vehicles by staff and students
- Increasing the use of electric vehicles on campus by Estates & Facilities

We will monitor this by

- Published Sustainable Travel Plan
- Number of cycle parking spaces
- Number of electric vehicle charge points on campus
- Percentage of electric vehicles in Estates & Facilities fleet

Relevant Documents

- Sustainable Travel Plan

3 GOOD HEALTH AND WELL-BEING

11 SUSTAINABLE CITIES AND COMMUNITIES

13 CLIMATE ACTION

Biodiversity

Biodiversity, or biological diversity, is the variety of life! This includes variety in habitats (e.g. grassland and woodland) as well as diversity in species and the ecosystems in which they occur. Biodiversity is important because of the essential contribution that it makes to the functioning of our planet and because of all the benefits that it provides, from foods and medicine to climate regulation. Contact with biodiversity and the natural world has also been linked to improvements in health and emotional well-being.

Despite being just a mile and a half from the City of Manchester, our main campus in Salford is a leafy, riverside site adjacent to Peel Park and the River Irwell. The habitats on the University campuses include lawned areas, memorial gardens, wildflower areas and woodlands. Many common British wildlife can be found on the campus and surrounding areas, as well as some not so common.

Our Campus Masterplan plans to fully utilise the green spaces in and around the campus, particularly Peel Park, David Lewis Playing Fields and The Meadows. Through our Masterplan, we will look for opportunities for new green spaces in our urban environment to help promote learning, wellbeing and healthy lifestyles.

We will protect and enhance the natural habitats and biodiversity on the University estate

We will achieve this by

- Implementing our Landscape Management Plan
- Maintaining our Green Flag Award

We will monitor this by

- Green Flag Award

Relevant Documents

- Landscape Management Plan

3 GOOD HEALTH AND WELL-BEING

11 SUSTAINABLE CITIES AND COMMUNITIES

15 LIFE ON LAND

17 PARTNERSHIPS FOR THE GOALS

Community Engagement

Communication and engagement are important elements of our Environmental and Energy Management System (EEMS) and our Environmental Sustainability Policy.

To achieve our EEMS objectives and reduce the impact of the University's activities on the environment, we need to raise awareness amongst staff, students and other interested parties to encourage and support them to take action and adopt positive behaviours as well as ensuring that any legal compliance requirements, our policies and environmental objectives are clearly communicated to staff, students and other interested parties.

Our colleagues' efforts in our Green Impact programme are a key component of our environmental sustainability aims. From energy-saving and waste reduction initiatives to greening offices with plants and Fairtrade coffee mornings our Green Impact teams lead a wide range of creative activities that reduce the environmental impact of our day to day work. Green Impact is an international scheme which brings staff and students together to increase sustainability across campus. Alongside the Green Impact teams, the scheme is facilitated by student auditors who assess project teams across our University to meet their green targets.

We also work with key partners such as our Student's Union, catering teams and external partners such as TfGM and Salford Council on Go Green Salford, our four-week long annual campaign to raise awareness of the importance of sustainability and encourage positive change among our students and staff. Usually, taking place in February/March, it incorporates national campaigns such as the People and Planet, Go Green Week, Student Volunteering Week and Fairtrade Fortnight.

We recognise the importance of networking with relevant sector and professional bodies to understand best practice. Raising the external profile of the University's activities and achievements in environmental sustainability and promoting any areas of good practice in environmental sustainability is also important. We have been shortlisted as finalists in the sector Green Gown Awards every year between 2011 and 2018, with a number of winning and highly commended entries.

We will embed care for the environment in the culture of our organisation as part of our goal to be a sustainable University and empower and motivate the whole University community to support our policy and strategy through appropriate education and communication

We will achieve this by

- Implementing our Communications and Engagement Strategy for sustainability
- Increasing the influence of the Green Impact initiative at the University
- Increasing engagement with students and staff at the University with sustainability

We will monitor this by

- Number of staff/students influenced in Green Impact
- Number of staff/students engaged with sustainability campaigns and events

Relevant Documents

- Environmental Sustainability Communications and Engagement Strategy

Teaching, Learning & Research

One of the principles of the University Industry Collaboration Strategy is to 'promote sustainability and social responsibility'. We take a broad definition of 'industry' and when we refer to industry partnerships and collaboration this includes work with private, public and third sector organisations of all sizes. From our inception, we have been committed to providing real world solutions to tomorrows challenges and the application of our research through partnership working and our current Research & Knowledge Exchange Strategy continues this tradition.

For example, our Applied Buildings and Energy Research Group (ABERG) is focused on establishing an evidence base to better understand and address the issues of energy consumption in buildings. The team covers a wide range of disciplines including building performance, electrical engineering, construction management, design and social sciences. ABERG is also the home of the unique Salford Energy House, a full sized two-bedroom terraced house built inside an environmental chamber that can replicate almost any weather conditions. With Energy House 2.0 coming soon we are looking to do even more in the area of Smart Living.

Whilst this strategy is focused on the environmental sustainability of our University operations we aim to support and encourage teaching, learning and research through this. We aim to 'practice what we preach' by using our campus as a Living Laboratory. We are already partners in a multi-million pound European contract for a project to develop 'green' resilience to extreme climate hazards. The IGNITION project backed by £4 million from the EU's Urban Innovation Actions initiative, will see the University host a 'living laboratory' for urban green infrastructure including roofs, walls and gardens. The project is led by Greater Manchester Combined Authority and supported by 11 partners, including the Environment Agency, and comes in response to increasing flooding, heat hazards and worsening air quality.

In our teaching and learning, one of our principles of a curriculum for Industry Collaboration is 'Education for ethical behaviour', which presents an approach to student education that seeks to balance human and economic well-being, knowledge, and competencies with cultural traditions, respect for other people, and for the earth and its natural resources. In practice this requires design of academic programmes with consideration for sustainability including all the Sustainable Development Goals.

We will support our University Strategy in providing real world experiences preparing students for life and supporting our research to provide real world solutions to tomorrow's challenges

We will achieve this by

- Continuing to share our environmental sustainability experience and challenges through guest lectures
- Offering our environmental sustainability challenges as live briefs for students
- Encourage and support the implementation of a living laboratory on the campus showcasing sustainability solutions

We will monitor this by

- Number of environmental sustainability guest lectures delivered
- Number of environmental sustainability live briefs/student projects supported
- Number of demonstrator environmental sustainability projects on campus

Relevant Documents

- University of Salford Strategy 2016-2021
- Programme Design, Approval, Amendment, Review and Withdrawal Policy

SUSTAINABLE DEVELOPMENT GOALS

We have the ability to contribute to all the Global Goals through our teaching, learning and research.

16 PEACE, JUSTICE AND STRONG INSTITUTIONS

17 PARTNERSHIPS FOR THE GOALS

Our Governance

We manage and control environmental sustainability on our campus and in our operations through our Environmental and Energy Management System (EEMS) which is certified to ISO 14001:2015 and ISO 50001:2018. This provides the framework for continual improvement in our environmental performance. We will deliver, review and monitor this plan through our EEMS.

Environmental sustainability is embedded within the University through high-level commitment from University leadership and our policies. A number of University groups provide strategic direction, support and oversight for implementation of the University policies.

The Environmental Sustainability Team lead the delivery of this plan, directly reporting to the Environmental Projects Board, chaired by the Director of Estates & Facilities, who report to University senior management groups via the Estate Programme Group and Sustainability Strategy Group.

The Environmental Sustainability Team manage budgets for energy management, waste management and sustainability communications and engagement. Business cases and requests for further funding are submitted via the Environmental Projects Board through the central University financial process.

We have created a number of policies and detailed plans to support our continual improvement aims and progress is reported quarterly to the Environmental Projects Board and publicly annually.

Environmental Sustainability Policy

Energy and Water Policy

Sustainable Construction Policy

Estates Sustainable Procurement Policy

Sustainable Food Policy

Energy, Water & Carbon Management Plan

Sustainable Procurement Plan

Waste Management Plan

Landscape Management Plan

Sustainable Travel Plan

Communications and Engagement Strategy

Sustainable Food Plan

Appendix A

Environmental Sustainability Team

Environmental Sustainability Performance

University of Salford MANCHESTER

Objective	Key Performance Indicator	Baseline year	2017-18	2018-19	2019-20	Target/date	Tracking 2019/20
Environmental Management							
Maintain ISO 14001 and ISO 50001 certification	ISO 14001 and ISO 50001 certification	N/A	N/A	Achieved certfication	Maintained certfication	Maintain certification	Achieved
Reduce pollution risk (emissions & discharges) to land, water and groundwater	Number of pollution incidents	N/A	0	0	0	0 annually	Achieved
Maintain compliance with environmental legislation and other requirements	Number of major non-conformities related to a breach in compliance obligations	N/A	0	0	0	0 annually	Achieved
Energy, Water & Carbon							
Reduce Scope 1 and 2 carbon emissions	% reduction of scope 1 and 2 carbon emissions	2005/06	64%	66%	70%(TBC)	81% reduction by 2030	On Track
Improve energy efficiency	kWh energy consumed per m2 gross internal floor area	2016/17	193 kWh/m2	183 kWh/m2	165kWh/m2	155kWh/m2 by 2030	On Track
Reduce water consumption	% reduction in water use (per staff & student FTE)	2016/17	24%	37%	Data not available	20% reduction by 2023 NEW target 20% reduction by 2025 from 2018/19	On Track
Waste Management							
Reduce waste disposal	% reduction in waste from baseline	2016/17	9%	29%	43% (TBC)	20% reduction by 09/2020 NEW target: 12% reduction by 2025 from 2018/19	Achieved
Increase waste recycled	% waste recycled	2016/17	38%	40%	38% (TBC)	50% by 09/2020 New target: 65% by 2025	Not Achieved
Eliminate avoidable single use plastics	Number of avoidable single use plastics	2019/20	N/A	N/A	Data not available	Zero by end 2022	On Track

Environmental Sustainability Team

Environmental Sustainability Performance

University of Salford MANCHESTER

Objective	Key Performance Indicator	Baseline year	2017-18	2018-19	2019-20	Target/date	Tracking 2019/20
Sustainable Procurement							
Develop and embed sustainable purchasing policy in Estates	Flexible Framework Self Assessment All Estates tender evaluations to include sustainability	2016/17	N/A	N/A	TBC	Meet level 4 in all areas of the Flexible Framework All Estates tender evaluations to include sustainability assessment	On Track
Sustainable Construction							
Ensure sustainability considered within construction through implementation of the sustainable construction policy	% of construction projects meeting sustainable construction policy requirements	2018/19	N/A	100%	100%	100% of construction projects meeting sustainable construction policy requirements	Achieved
Travel & Transport							
Improve facilities for cyclists on campus	Number of cycle parking spaces	2016/17	410	362	362	500 cycle parking spaces on campus by end 2023	On Track
Support use of Electric Vehicles by staff and students	Number of Electric Vehicle charge points on campus	2016/17	6	6	6	Increase number of EV charge points on campus [TBD]	On Track
Increase use of Electric Vehicles on campus by Estates	% Electric Vehicles in estates fleet	2016/17	0%	0%	0%	20% of Estates fleet by end 2025	On Track
Community Engagement							
Increase influence of Green Impact at the University	Number of staff/students influenced in Green Impact	2016/17	21%	156%	Data Not Available	Increase in number of staff/students influenced	On Track
Increase engagement with staff and students at the University with sustainability	Number of staff/students engaged with sustainability campaigns & events	2019/20	N/A	N/A	1,378	Increase in number of staff/students engaged	On Track
Biodiversity							
Maintain Green Flag Award	Green Flag Award	2018/19	N/A	Green Flag Achieved	Green Flag Achieved	Maintain Green Flag Award annually	Achieved

Appendix 8

University of Salford energy, water and carbon management plan

1. Executive summary

In 2011, we launched our first Carbon Management Plan with targets to reduce scope 1 and 2 carbon emissions by 43% by September 2020 compared to a 2005/06 baseline. As of end of academic year 2017/18 we had already reduced our scope 1 and 2 carbon emissions by 64%. This has been achieved partly through reductions in energy use on campus through investment, partly through reductions in grid electricity emissions factor, and partly through the divestment from our student accommodation in 2007/08.

Our new Energy, Water and Carbon Management Plan builds on our significant progress in carbon reduction to ensure we can achieve our objective of a scope 1 and 2 81% reduction by 2030 and work towards net-zero carbon by 2038 in line with the Greater Manchester carbon reduction strategy. Environmental sustainability is a core element of our ambitious University Campus Masterplan, launched in the last year. Central to the plans is an energy strategy that takes a major step towards a zero-carbon future, while providing high quality spaces for residential, teaching, research and commercial uses that are cost-effective to run. University of Salford buildings will undergo refurbishment to make them highly energy efficient, with maximum potential for renewable energy generation. In the meanwhile, we will carry out extensive audits on our existing buildings to identify opportunities for improvements, implement where technology, scale and finances allow and develop and submit businesses cases for funding for larger projects.

The 2005/06 baseline is to be used for absolute carbon emissions, and this has therefore been adopted under the Energy, Water and Carbon Management Plan (EWCMP). Other relevant baselines are included in the Objectives, Targets and Action Plan include the academic year 2016/17 as a baseline for energy, gas and water consumption, to be used in the University's ISO50001 Energy Management System.

Carbon emissions are broken down into three categories by the Greenhouse Gas Protocol in order to better understand the source.

Scope 1 – All Direct Emissions from the activities of an organisation or under their control. Including fuel combustion on site such as gas boilers and fleet vehicles.

Scope 2 – Indirect Emissions from electricity purchased and used by the organisation. Emissions are created during the production of the energy and eventually used by the organisation.

Scope 3 – All Other Indirect Emissions from activities of the organisation, occurring from sources that they do not own or control.

2. Background and context

2.1 *Background*

Situated in Salford, to the North West of Manchester City Centre, we have three campuses with 27 principal buildings providing teaching to around 16,000 students. In the 2016/17 Academic Year (the "audit" year selected for this Plan), we spent around £2.1 million on energy and water, of which electricity is the dominant cost at around 73%, with gas making up around 19%, and water and district heating and cooling the remainder at around 8%. Total carbon emissions from energy use for the same period were 8,901 tCO2.

Energy costs are expected to rise significantly over the next three years (and beyond), with electricity in particular expected to rise as a result of measures announced under the Electricity Market Reform, and gas as a result of the closure of the CRC Energy Efficiency Scheme following the 2018/19 compliance year, and the recovery of the lost revenue by HM Treasury by the loading of the Climate Change Levy, particularly onto gas. We calculate that CCL changes alone will increase costs by around 11% in the next 3 years, and by at least 15% up to 2025. Upwards pressure on commodity prices because of investment decisions in new electricity generation and distribution systems and gas resource depletion and geopolitical factors make it likely that energy prices will rise still further over the period.

We launched our first Carbon Management Plan in 2011, which included a commitment to reduce Scope 1 and 2 carbon emissions by 43% by 2020 compared to a 2005/06 baseline. As of 2016/17 we had already reduced our carbon emissions by 55% over the period and hence initiated the development of a new Energy Water and Carbon Management Plan (EWCMP).

We have recently developed and published a Masterplan in conjunction with Salford City Council. Our reason for developing the new Masterplan are to improve the student experience and placemaking. Research shows that today's students are environmentally engaged and seek learning environments which are sustainable and efficient. This has been incorporated into the Campus Master Plan, particularly through an Energy Strategy and recognition of a physically green campus. The Campus Master Plan Energy Strategy recommends an all-electric approach for the campus by 2030 in order to achieve the targeted carbon reduction emissions. This will require an increase in efficiency in new and existing buildings, use of low carbon and renewable energy technologies on campus as well as taking advantage of decarbonisation of the grid.

2.2 *Methodology*

In developing this Plan, we have partnered with an external consultancy. First, we held an Energy Risk and Opportunity Workshop in July 2018 together involving colleagues from across the University. We also conducted a PESTLE (Political, Economic, Social, Technological, Legal and Environmental) analysis to determine the internal and external issues that can affect our ability to achieve the intended outcomes of our Energy Management System, which are:

- To enhance the energy performance of all activities of the University of Salford Estates and Facilities Division.
- To fulfil compliance obligations.
- To comply with the policy and achieve environmental objectives.

These activities have helped us develop a list of energy-related issues and associated risks and opportunities, together with actions to address them. In total over 60 issues were identified, of these 20 were prioritised as potentially having a significant impact (see the following table).

Internal Issues	*External Issues*
Systems and Procedures: Budget/Funding Process for Energy Saving Initiatives Setting of Utilities Budget Procurement Systems and Frameworks Consultation on Technological Issues	Compliance Issues: Minimum Energy Efficiency Standards (MEES) Requirement for EPCs/DECs Changing Legislation and Awareness of Legislation (Existing, New and Amended)
Development Activities and New Projects: Implementation of Master Plan Communication and Visibility of Projects Consultation Process During a Building Design Ability of New Buildings to Achieve Expected Energy Performance Buildings and Their Use: Managing Building User's Expectations Building Manager's Engagement in Energy Performance Old Systems Within Buildings, e.g., Heating, Lighting, IT Heating and Cooling Expectations/Policy Old Inefficient 1960s Buildings Building Management System Issues: Access to BMS for Relevant Parties BMS Performance BMS Sensors in Wrong Place, e.g., on West- or East-Facing Walls or Near Heat Sources Training and Competency of Staff on Installed Building Systems e.g. BMS	

We also developed an Energy and Water Review tool to review all of our current and historic energy use, costs and associated carbon emissions.

The purpose of the energy review was:

- To establish an appropriate baseline for measurement of energy use and carbon emissions.
- To derive suitable energy, water and carbon benchmarks and key performance indicators (KPIs) which can be used to set targets and measure performance going forward.
- To accurately quantify energy use, water use and carbon emissions for the university's main campuses, buildings and end users, and to use this as the basis of derivation of areas of significant energy use and opportunities for energy performance improvement in line with the requirements of ISO 50001:2018.
- To establish relationships between energy and water use and key driving variables, in order to develop a forecasting methodology. This in turn will be used to calculate future energy and water use and carbon emissions, and then used to establish what level of reduction is required to meet any future targets.

3. Energy and water review

3.1 Baseline and key targets

Through the Higher Education Funding Council for England (HEFCE) carbon management programme we initially adopted the sector target published in January 2010. HEFCE advocated targets for the higher education sector for reduction in Scope 1 and 2 carbon emissions of 43% by 2020 when compared to a 2005 baseline.

We adopted the 2005/06 academic year as our baseline for carbon emissions accordingly. To the 2018/19 academic year, we had achieved an actual reduction of around 64%. This has been achieved partly through reductions in energy use on campus through investment, partly through reductions in grid electricity emissions factor and partly through our divestment of our student accommodation in 2007/08.

The 2005/06 baseline is to be used for absolute carbon emissions, and this has therefore been adopted under the Energy, Water and Carbon Management Plan. Other relevant baselines are included in the Objectives, Targets and Action Plan (see Section 4) and include the academic year 2016/17 as a baseline for energy, gas and water consumption, to be used in our ISO50001 Energy Management System.

Our Campus Masterplan Energy Strategy suggested a target to reduce carbon emissions by 81% by 2030, through adoption of an energy strategy which would see the campus move to an all-electric model, with heat pumps to provide heating and cooling, and with grid electricity supplemented by the limited implementation of building-mounted solar PV. This target and an aspiration to move towards net-zero carbon by 2038 was adopted and published via the mayor of Manchester's Green Summit in March 2018.

Our Carbon Reduction Objective:

To reduce scope 1 and 2 carbon emissions from 2005/6 baseline by 81% by 2030 and work towards net-zero carbon by 2038

3.2 Benchmarks and key performance indicators

Potential benchmarks and key performance indicators (KPIs) have been explored. Obvious KPIs are those that relate energy and water use and carbon emissions to a driving variable, such as floor area or student numbers. This subject is more complex than perceived as there is more than one variable affecting energy and water use. This is explored further in Section 3.5, 'Energy and water forecast'.

Data has been obtained from the Higher Education Statistics Authority (HESA) and used to compare our 2016/17 performance against that of our peers using suitable normalised KPIs i.e:

- Energy use/floor area/year (kWh/m^2/year)
- Water use/full time equivalent (staff and students)/year (m^3/FTE/year)
- Carbon emissions/floor area/year (kgCO_2e/m^2/year).

As these are the accepted KPIs used within the sector, they have been adopted as the overarching KPIs within our Environmental and Energy Management System.

For energy use, on a total kWh/m^2 (gross internal floor area) basis the University of Salford performs at around 193 kWh/m^2/year (119th out of 161) as shown in Figure 1.

For water, the university's performance is 5.7 m^3/FTE/year (123rd out of 161) as shown in Figure 2.

Document Ref: Author: Version No. And Date:

EWCMP Bec Bennett/Nia Prys-Williams V4.0 03/03/2020 8

For Scope 1 and 2 carbon emissions, the university's performance is 54.8 kgCO_2e/m^2/year (97th out of 161) as shown in Figure 3.

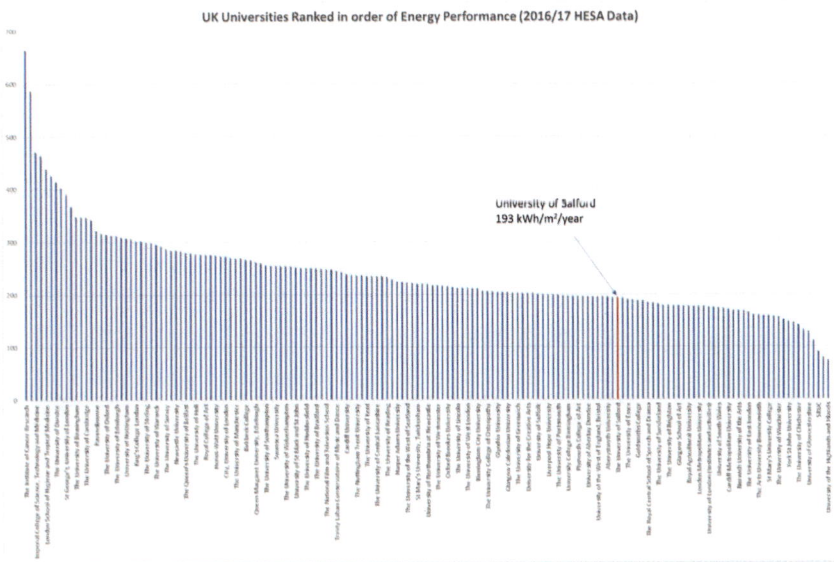

Figure 1 Comparative Energy Performance of UK Universities (HESA data)

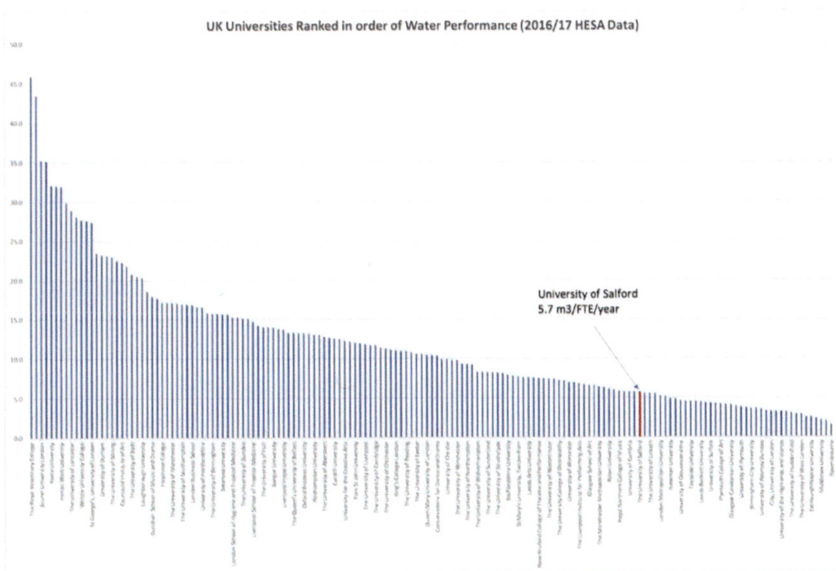

Figure 2 Comparative Water Performance of UK Universities (HESA data)

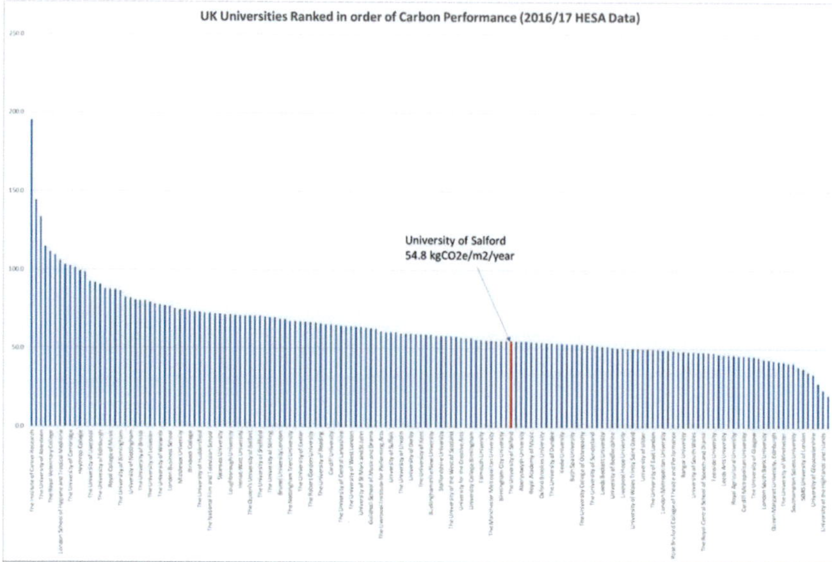

Figure 3 Comparative Carbon Performance of UK Universities (HESA data)

3.3 *Energy and water review results*

In line with the requirements of ISO 50001:2018, an Energy and Water Review has been carried out. The review used data from the 2016/17 academic year (year ending 31st July 2017), which was the last year for which full data was available at the time the review as carried out (May/June 2018). A review of available historical data was also undertaken as part of the Energy Review dating back to the baseline year 2005/06, obtained from the university's HESA returns and is included in section 3.5, Figure 14.

The purpose of the review is to accurately quantify the energy and water use and carbon emissions, and then to determine the areas of significant energy use (SEU). SEUs have determined on the basis of both individual buildings (as was already standard practice at the university, in accordance with existing metering arrangements) and also by end user, i.e., heating, cooling, lighting, etc. The intention of this approach is to allow insights into where energy is used across the campus and within individual buildings, and therefore where practical measures to reduce energy and water use might be applied in future. In most cases, there is no metering available to derive the breakdown of energy use by end user at building level, so an approach of using available benchmark data adjusted for individual buildings (based on knowledge gained through a campus walkover) was adopted, backed by submetre data where it was available.

Data has been drawn from the university's SystemsLink and Active Energy Manager data platforms. Data includes both main (fiscal) meters for gas, electricity and water and submetres.

The overview of the university's energy and water use, cost and emissions is shown in Table 1.

Table 1 Overview of Current Utilities Use, Costs and Carbon Emissions (2016/17 data)

Utility	Total Usage	Units	% Use	Total Cost (£)	%	Total Emissions tCO2/year	%

Gas 16,223,546 kWh 49.6% Electricity 14,858,829 kWh 48.6%
18.7% 73.2% 0.7% 0.7% 93.3%
6.7%
100.0% Utilities Cost Breakdown
Gas Electricity Water District Heating District Cooling

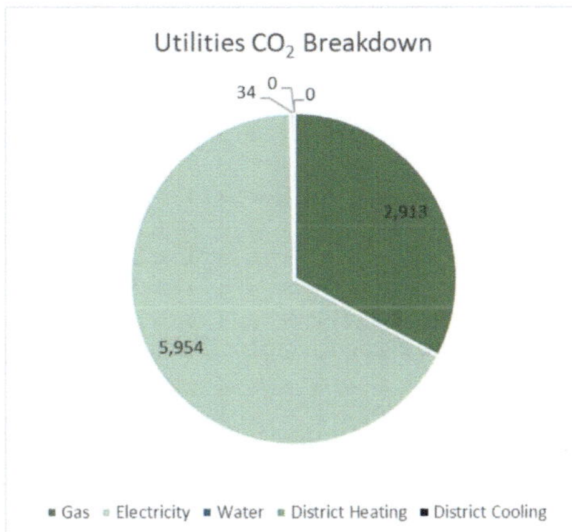

Figure 4 Breakdown of Utilities by Cost

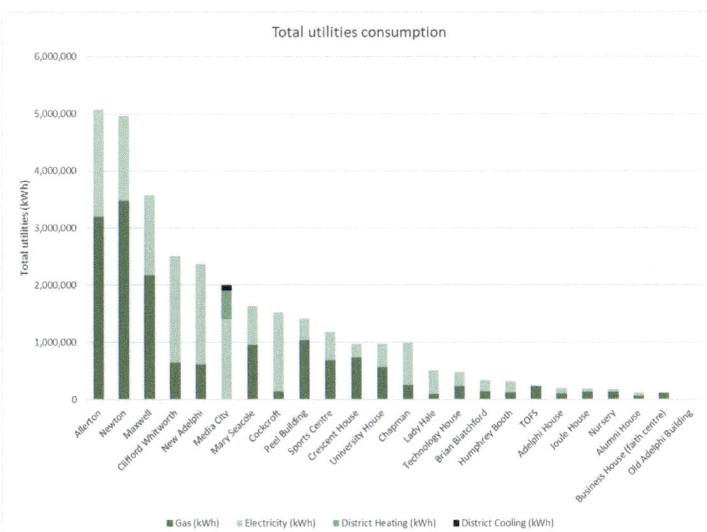

Figure 5 Building Energy Consumption, ranked largest to smallest

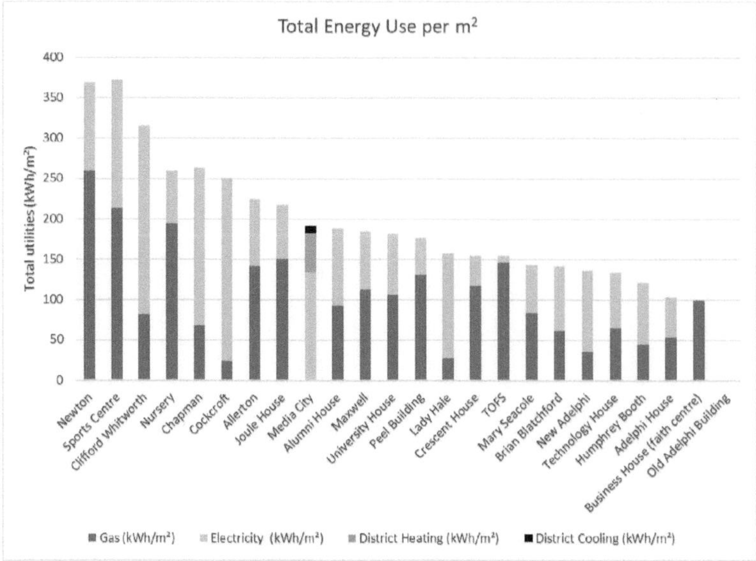

Figure 6 Building Energy Consumption per m² floor area, ranked largest to smallest

Document Ref: Author: Version No. And Date:
EWCMP Bec Bennett/Nia Prys-Williams V4.0 03/03/2020 11

Top 5 buildings =
over 50% of all energy use
Top 13 buildings = 90% of all energy use

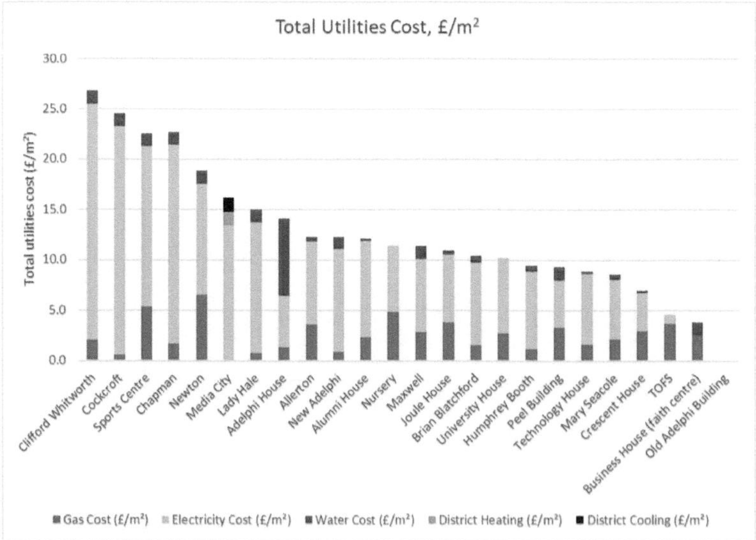

Figure 7 Building Energy Cost per m² floor area, ranked largest to smallest

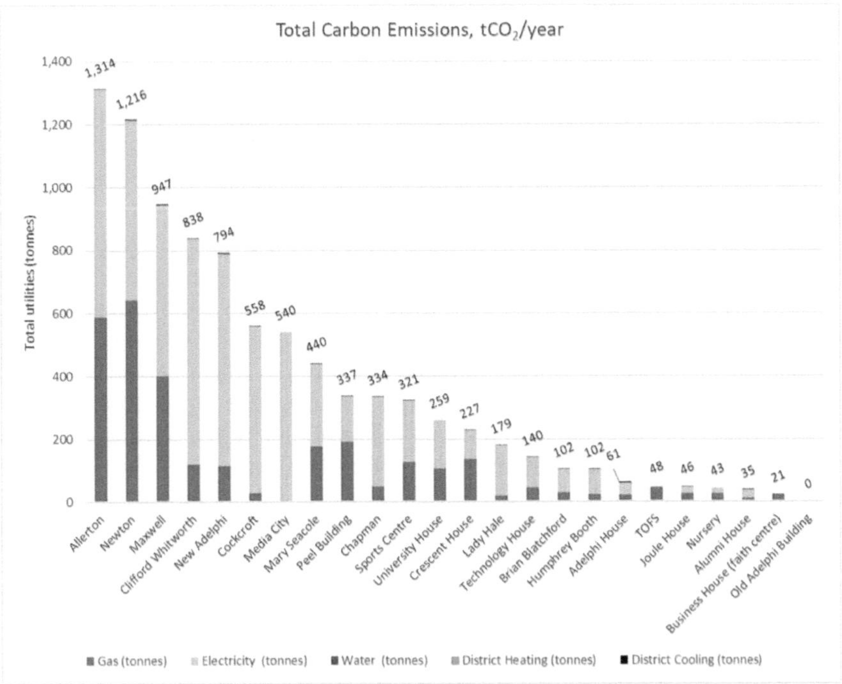

Figure 8 Building Energy-Related CO_2 Emissions, ranked largest to smallest

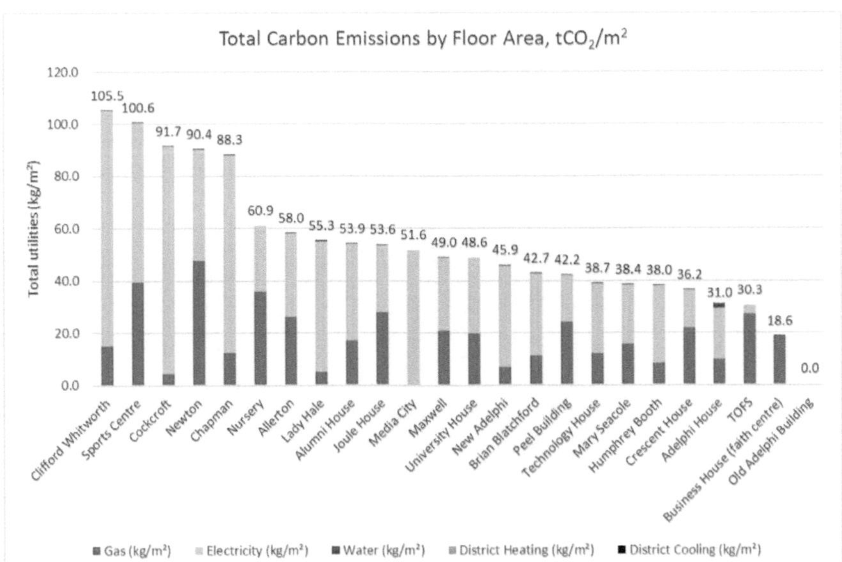

Figure 9 Building Energy-Related CO_2 Emissions, ranked largest to smallest

In terms of significant energy uses, as noted previously, analysis may be made by both building and by end use.

In building terms, it is obvious and notable that the largest buildings are the highest users, with the top five largest buildings using around 50% of our total annual energy use, cost and emissions, and the largest half of the estate being responsible for around 90% of our energy use, cost and resultant emissions. As such these are deemed our significant energy uses (SEUs) under the ISO50001 Energy Management System, since they are likely to offer considerable potential for energy performance improvements. These are the subject of detailed energy audits as detailed in the Action Plan presented in Section 4.

The Energy Review tool also allows the user to 'slice' the data at individual building levels in order to identify significant energy uses by end use category, for example heating, domestic hot water, cooling, fans and pumps, lighting, data centres, office equipment, laboratories, swimming pools and so on. Whilst not absolutely accurate (due to the absence of sufficient end user submetering), it does serve as a useful indicator of which services are the significant energy uses, and therefore where future energy and carbon emissions reduction efforts might best be targeted. It is not the intention that this approach be used as a formal SEU measure in the ISO50001 system at this stage, however, this could be adopted at a later date when the level of submetering allows for energy used by such end users to be more accurately measured. For the time being, the analysis presented in the following tables and charts is used to inform thinking on the targeting of energy.

End User	Gas (kWh/year)	Electricity (kWh/year)	Total Energy (kWh/year)

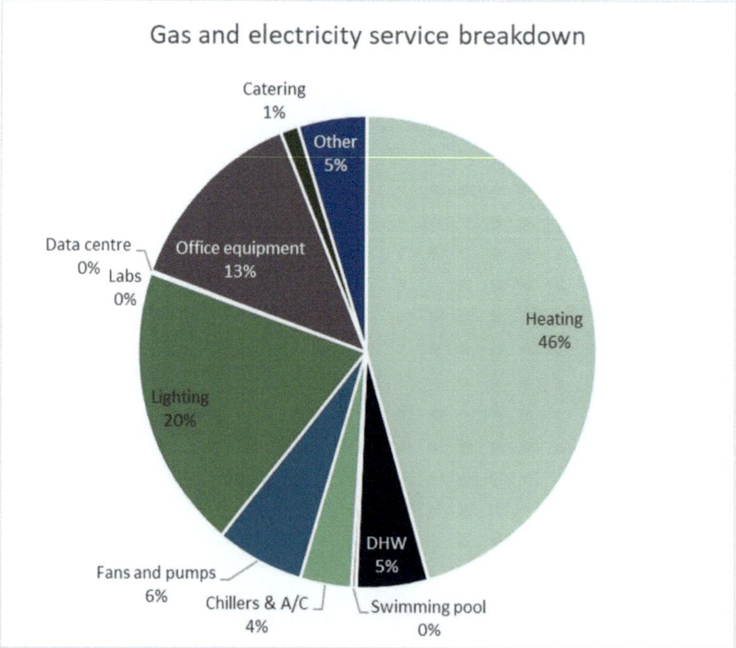

Figure 10 – Breakdown of Energy Use by End User across all buildings

As can be seen, in SEU terms, the largest energy users ranked by end use category are:

End User	Breakdown of Use %

Fans and pumps 6% Domestic Hot Water 5% Chillers & A/C 4% Catering 1% Swimming pool <1% Data centre <1% Labs <1% Other 6% Total 100%

The previous table indicates that the SEUs in terms of end users are space heating, lighting, IT and office equipment, fans and pumps, DHW and chillers and A/C, which together account for over 90% of overall consumption. Other users can be considered insignificant at this stage at a campus level as their percentage consumption is 1% or less of the total (unless they are found to have significant potential for energy performance improvement as identified through detailed energy audits).

3.4 Energy market context and risk assessment

The current market for electricity, gas and water has been examined in the context of fuel supplies and availability, anticipated market condition and future pricing, legislative impact and taxation, together with an examination of the risks and opportunities faced by the university. The output of the assessment is included in Appendix 1 for information.

One point worth noting is that the unit costs of electricity and gas are forecast to rise markedly in future. According to forecasts carried out by the Department for Business, Energy and Industrial Strategy (BEIS), the cost of electricity is forecast to rise from current (2018) levels by around 28% to 2028, before falling back to an overall rise of around 8% by 2035. Gas prices are forecast to rise by around 90% between now and 2030 before levelling off.

The forecast rises are shown in Figure 11.

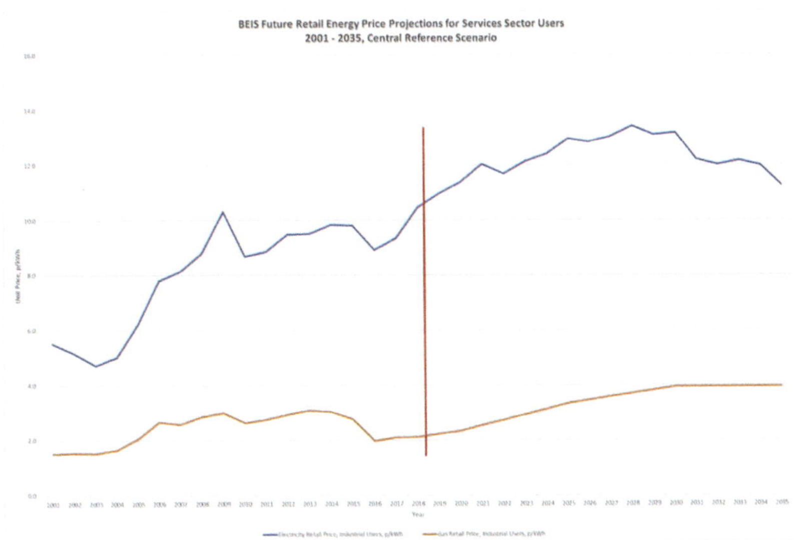

Figure 11 Forecast Rise in Electricity and Gas Unit Costs to 2035 (BEIS, Services Sector, Central Reference Scenario)

3.5 Energy and water forecast

Following the energy review, historic energy and water use and carbon emission data along with data on floor area, staff and student numbers and turnover has been obtained from the university's HESA (formerly HEFCE) returns. Together with heating and cooling degree day data, this has been used to explore the relationship between energy and water use and the various driving variables. It has been found that the strongest correlation for energy use (gas and electricity) is a combination of floor area, staff and student numbers (expressed as FTEs) and heating degree days (for water any combination tried gives a much weaker correlation though there are uncertainties in the data which make this analysis less reliable).

Our campus master plan has been considered for planned new build and refurbishment works in order to obtain a reasonably accurate projection of floor area going forward to 2030/31 and forecast student numbers from 2017/18 to 2022/23 (and an estimate has been used thereafter to 2030/31). Degree days are difficult to predict, so an average figure has been assumed.

Using the forecast tool an estimate of gas and electricity to 2030/31 has been generated. The resulting profile (showing total energy use) is shown in Figure 12:

This has then been overlaid with forecast gas and electricity price rises (derived from the BEIS Forecast referred to in section 3.4, Figure 11), forecast grid electricity emission factors (again derived from BEIS) and forecast rates of Climate Change Levy (currently set only as far ahead as 2019/20, so with an RPI-based inflation factor built in thereafter).

The result is a forecast of both energy costs and Scope 1 and 2 carbon emissions between now and 2030/31.

This is then used to model various energy use and emission reduction scenarios to give a 'top-down' view on what measures we will need to put in place to achieve our 81% carbon emissions reduction target by 2030/31, and also what would be a sensible interim target to be achieved by implementation of the EWCMP.

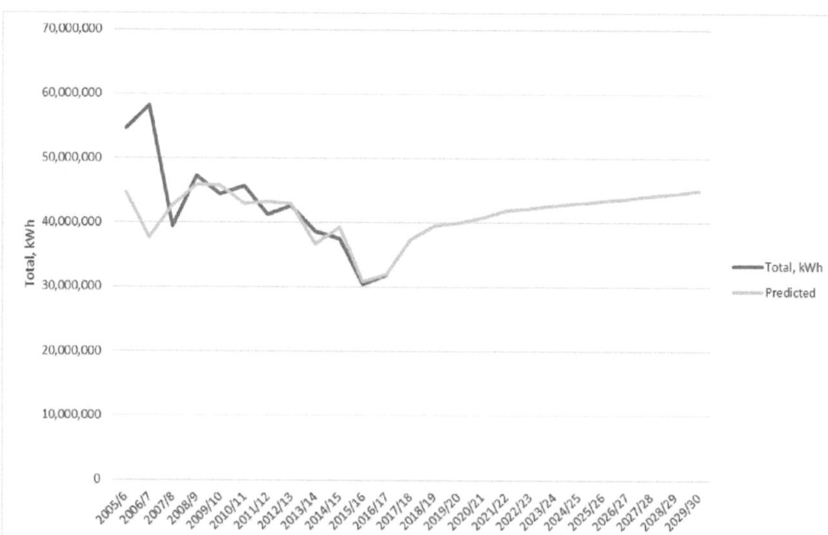

Figure 12 Predicted Energy Use to 2030/31

The model also provides a view of the energy and emissions 'value at stake', as discussed in section 3.6.

3.6 Value at stake

The value-at-stake analysis examines the consequences of 'doing nothing' in terms of energy and emissions management, versus the impact of 'doing something'. For the purposes of this analysis, we have taken the energy use as forecast by the methodology described in section 3.5, overlaid with forecast energy price rises, carbon emissions taxes and forecast grid electricity factors. The impact on overall utilities costs is shown in Figure 13.

It can be seen, that without action on energy, water and emissions management, costs are forecast to increase from current levels of £2.8 million per year, to £4.6 million per year by 2030/31, an increase of 64% at today's prices.

In the case of carbon emissions, the chart shows that, over the period to 2030/31, emissions will actually fall, driven largely by the successive decarbonisation of the electricity supply grid as forecast by BEIS, despite increasing consumption. Emissions from gas use, however, will continue to rise slightly in line with increasing consumption.

It can be seen, that without action on energy and emissions management, emissions overall are forecast to fall from current levels of around 10,000 tCO_2e per annum to around 6,600 tCO_2e per annum by 2030/31, a decrease of 35%. When compared to the 2005/06 baseline, the overall reduction is around 65%.

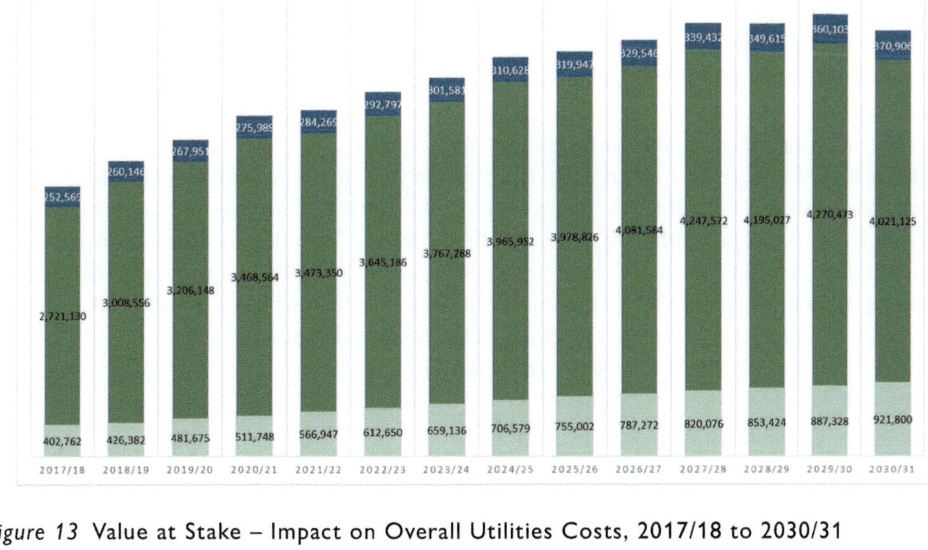

Figure 13 Value at Stake – Impact on Overall Utilities Costs, 2017/18 to 2030/31

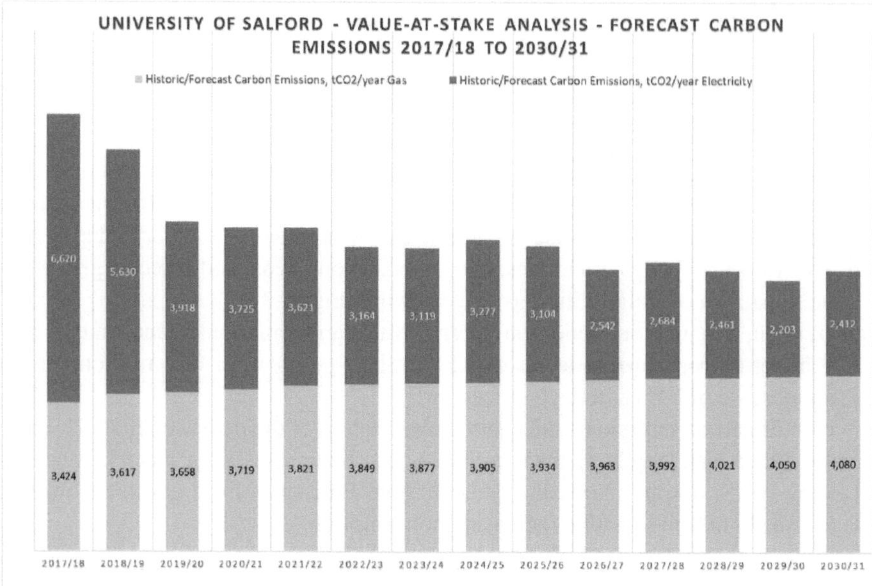

Figure 14 Value at Stake – Impact on Overall Carbon Emissions, 2017/18 to 2030/31

The question of what could be achieved through proactive measures to reduce energy and water use and consequent costs and carbon emissions is the true value at stake and is addressed in section 4.

4. Energy and water objectives, targets and action plan

4.1 Development of energy and water objectives and targets

The aim of our energy, water and carbon management plan is to review our current position and to develop appropriate objectives and targets for the five-year period 2017/18 to 2022/23. This will help set the direction to achieving our long-term targets of reducing carbon emissions by 81% by 2030 and net-zero carbon by 2038. Targets set under the EWCMP will be SMART (specific, measurable, actionable, relevant and time-bound).

The analysis undertaken during the development of the EWCMP has resulted in two approaches to setting potential new realistic and achievable targets over the coming five-year period to 2022/23 and beyond to 2030/31. The first is a 'top-down' view of the emissions 'glide-path' required to meet the 2030/31 target, taking into account the impact on energy use of increasing student numbers and expanding floor area, together with the anticipated impact of grid decarbonisation as used in the forecasting tool described in section 3.5.

The results of the top-down forecast are shown in Figure 15:

Gas

The UK operates a liberalised gas market, enabling consumers to change suppliers and tariffs at will (within normal contractual obligations), in order to take advantage of a competitive

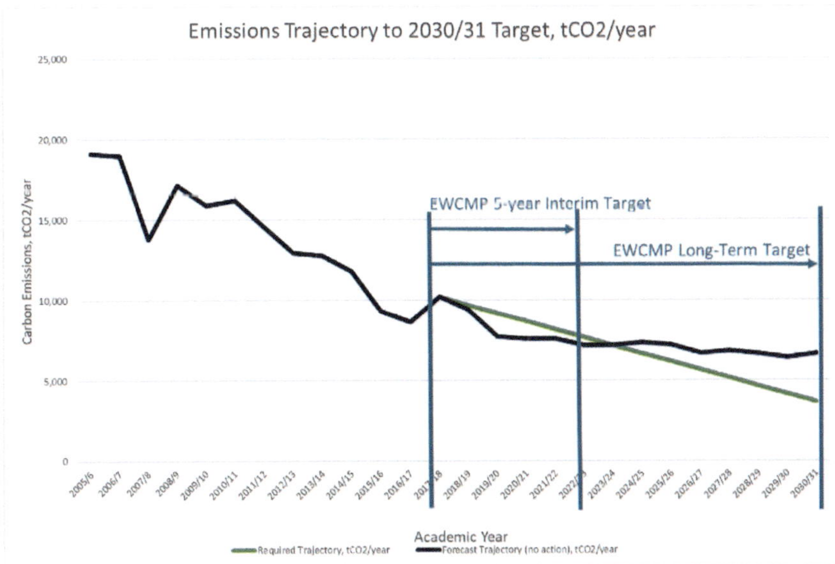

Figure 15 Historic Emissions and Future Trajectory to 2030/31

market. The market is dominated by a handful of larger players, with smaller operators at the margins. Competition and the availability of imported gas via transcontinental and subsea pipelines bringing gas from Russia and Central Asia, together with shipped imports of LNG from the Middle East and North Africa has resulted in historically low gas prices over the last 15 or 20 years; however, these are starting to increase. The main factor driving the increase is competition for gas as the developing world both industrialises and decarbonises. Overall, according to BEIS forecasts, gas prices are expected to double between 2018 and 2030.

Costs are made up of commodity costs (the cost of the gas supplied itself), plus non-commodity costs which cover the cost of gas storage and distribution, plus government taxes/levies to support the development of renewable energy and reduce carbon emissions. The main tax is the Climate Change Levy, which is expected to increase dramatically in the coming few years as government balances it up to meet that of electricity, in order to reflect the lower carbon impact of electricity now.

- In addition, VAT is charged on top of gas charges and levies.
- Rising prices as a result of increased demand.
- Potential increases in tariffs on imported gas (up to 50% of the UK's gas demand is imported through sub-sea interconnectors and shipping) as a result of Brexit. Geopolitical events causing market instability and rising prices Potential supply interruptions as a result of local, regional, national or international demand, geopolitical events or malicious disruption. Increases to CCL to offset closure of CRC scheme. Gas will increase at a greater rate than electricity to level up the disparity between the two.
- Purchase gas at lower prices.
- Move to all-electric campus to avoid gas costs and associated CCL. Reduce consumption through behavioural change and investment in energy efficient technologies.

Water

Water has traditionally been supplied on a regional basis, with consumers tied to regional water supply companies. The water market in England has been opened up to competition from April 2017, however, with businesses able to purchase water and waste water services from other providers on a national basis, or even elect to become a water supplier themselves ('self-supply', generally suited to organisations with large numbers of premises spread over a wide geographical area). The ultimate aim is to introduce competition between water service providers and reduce costs to consumers, although it is unclear how successful this has been to date.

- Rising prices caused by the water companies' need to invest to replace aging infrastructure. Potential supply interruption or restrictions as a result of water stress resulting from climate change.
- Reduce consumption through behavioural change and investment in water efficient technologies.
- Invest in rainwater harvesting or greywater recycling systems to reduce mains water demand. Investigate the viability of on-site abstraction from boreholes with on-site treatment.
- Use water metering to identify and eliminate water leaks.
- Reduce meter sizes to match actual demand to avoid meter size-related standing charges.
- Look at on-site primary wastewater treatment to reduce tariff charges. Check water tariffs to ensure that these are fair and reasonable, particularly with respect to meter standing charges, surface water drainage, water not otherwise returned to sewer (e.g. evaporation), etc.
- Investigate options for purchasing water and waste water services from other providers on the open market.

Appendix 9

Environmental and sustainability strategy document

Our mission

By pioneering exceptional industry partnerships we will lead the way in real-world experiences preparing students for life

As a civic institution, our mission is to:

- Educate the next generation of modern industrialists, innovators, creators, entrepreneurs and leaders
- Develop the skills and knowledge needed to capitalise on the next industrial revolution
- Work in collaboration with public and private sector partners to address local and global economic and societal challenges

Our vision for sustainability

By embedding sustainability in all aspects of organisation life, we will enable our organisation community to have maximum beneficial impacts for society and environment

Sustainable development goals

The Sustainable Development Goals, or Global Goals, are a call for action by all countries to promote prosperity while protecting the planet. They recognise that ending poverty must go hand in hand with strategies that build economic growth and address a range of social needs including education, health, social protection and job opportunities, while tackling climate change and environmental protection.

We are aligning our environmental sustainability plan to the Global Goals by mapping each area to the relevant goals.

About this plan

This plan provides narrative to the organisation policies and objectives to improve environmental sustainability performance. At the organisation we are proud of our place in the world. Situated in the heart of the vibrant Greater Manchester conurbation we play a significant role in the community. While our heart is anchored in the North West, we have a powerful global reach. We focus on equipping our students, our future leaders, for the world of work.

We recognise the global climate crisis and that climate change is one of the biggest challenges facing our society across the globe. As a higher education provider we have a major role to play in enabling our students and staff to respond positively to global challenges.

We also recognise our responsibilities for the direct impact on the environment, as one of the largest organisations in the Greater Manchester region offering a place of study and work to over 18,500 students and 2,500 staff across 27 buildings and around 70 hectares. In the 2016/17 academic year (the baseline year selected for the majority of this plan), energy and water costs were around £2.1 million per year. Total carbon emissions from energy use for the same period were 8,901 tCO_2e and waste from organisation operations was 900 tonnes. We have an obligation to reduce our negative impact on our environment but also an opportunity to demonstrate environmental sustainability through our operations and using our campus environment as a living laboratory.

This plan sets out how we will demonstrate management of our environmental sustainability including how we will mitigate climate change, be resource efficient and contribute towards sustainable development through the operation of our campus and how this will contribute to education and research for sustainability.

Our environmental sustainability policy sets the framework for our environmental objectives. This plan sets the key actions we will take, how we will monitor and review our progress. This plan is reviewed at least every two years and any amendments to objectives, targets and KPIs will be highlighted.

Figure 1 Our 2016/17 environmental impact

Our story so far

We have built on our first carbon management plan launched in 2011 and our environmental sustainability policy, adopted in 2015, to ensure that our commitments around environmental sustainability remain current, challenging and reflect our campus master plan. Most notably through the adoption of significant objectives to support the Greater Manchester science-based carbon target and ambition to become the greenest city region in the UK. Our new energy, water and carbon management plan builds on our significant progress in carbon reduction (65% reduction since 2005/6) to ensure we can achieve our objective of an 81% reduction by 2030 and work towards net-zero carbon by 2038. We have also committed to eliminate avoidable single use plastics from our campus by 2022.

We recognise the importance of our green space on campus and have committed to maintaining these to a high standard including encouraging biodiversity. In recognition of this we received a prestigious Green Flag Award for our Peel Park and Frederick Road Campuses, well ahead of our target date. Our campus master plan will also look for opportunities for new green spaces in our urban environment to help promote learning, well-being and healthy lifestyles.

In 2018 we completed development and implementation of our environmental and energy management into a system through the EcoCampus framework and are proud to have subsequently been awarded EcoCampus Platinum, ISO 14001 and ISO 50001 certifications.

We are proud of our achievements so far but we are committed to doing more. We are currently positioned 64th, a 2:2 classification, (2019) in the People and Planet Organisation League, our aim is to increase to the first-class band of the Environmental Management ISO 50001 league.

Environmental management

An environmental management system is a way for an organisation to demonstrate that it addresses and minimises its environmental impacts, manages its legal compliance and continuously improves its environmental performance. An environmental management system is a systematic approach to managing an organisation's impacts on the environment. An effective EMS will involve the following:

- An initial assessment of how the organisation's activities, products and services might affect the environment
- Making sure they comply with all relevant environmental regulations
- Gathering data e.g. energy use, waste, water, raw materials
- Developing and communicating an environmental policy

- Procedures for controlling activities with significant environmental impacts
- Identifying risks and opportunities associated with environmental issues
- Setting targets and measuring progress
- Defining roles and responsibilities for all employees
- Training and awareness
- Periodic internal auditing
- Management review and commitment from leadership

The Organisation Estates and Facilities Division leads on the Environmental and Energy Management System (EEMS) which is certified to both BS EN ISO 14001: 2015 and BS EN ISO 50001:2018 standards covering all of our physical estates and operations. The EEMS was developed over a number of years using the EcoCampus approach with awards from bronze to platinum.

We will meet our environmental compliance obligations, protect the environment in which we operate our estates and facilities, and prevent pollution by reducing and eliminating pollution sources.

We will achieve this by

- developing and maintaining an Environmental and Energy Management System externally accredited (ISO 14001:2015 and ISO 50001:2018 standards) for Estates and Facilities Division
- Maintaining a register of environmental compliance obligations and regularly evaluating our compliance
- Reducing pollution risk (emissions and discharges) to land, water and groundwater

We will monitor this by

- ISO 14001:2015 and ISO 50001:2018 certification
- Number of major nonconformities related to a breach in compliance obligations
- Number of pollution incidents

Relevant documents

- Environmental Sustainability Policy

- Energy, Water and Carbon

Carbon emissions are broken down into three categories by the Greenhouse Gas Protocol in order to better understand the source.

Scope 1 – All Direct Emissions from the activities of an organisation or under their control. Including fuel combustion on site such as gas boilers and fleet vehicles.
Scope 2 – Indirect Emissions from electricity purchased and used by the organisation. Emissions are created during the production of the energy and eventually used by the organisation.
Scope 3 – All Other Indirect Emissions from activities of the organisation, occurring from sources that they do not own or control. These are usually the greatest share of the carbon footprint, covering emissions associated with business travel, procurement, waste and water.

In 2011, the organisation launched their first carbon management plan with targets to reduce scope 1 and 2 carbon emissions by 43% by September 2020 compared to a 2005/06 baseline. As of end of academic year 2017/18 the organisation had already reduced its scope 1 and 2 carbon emissions by 64%. This has been achieved partly through reductions in energy use at the organisation through investment, partly through reductions in the grid electricity emissions factor, and also partly through the organisation's divestment of its student accommodation in 2007/08.

Our energy, water and carbon management plan builds on our significant progress in carbon reduction to ensure we can achieve our objective of a scope 1 and 2 81% reduction by 2030 and work towards net-zero carbon by 2038 in line with the Greater Manchester carbon reduction strategy.

Scope 3 carbon emissions account for just over 80% of the organisation's total emissions and we recognise the impact of these emissions so have committed to improving our monitoring and measuring of these and setting reduction targets where possible.

We will reduce the energy and water consumption of our buildings to meet the targets as established in the energy, water and carbon management plan.

We will achieve this by

- Improved monitoring of energy, carbon and water data
- Eliminating or minimising energy and water waste
- Improving energy and water efficiency
- Monitoring our scope 3 carbon emissions

We will monitor this by

- Percentage reduction of scope 1 and 2 carbon emissions from baseline
- Energy consumed by gross internal floor area
- Percentage reduction in water use (per staff and student FTE) from baseline
- Publication of an annual scope 3 carbon emissions report

Relevant documents

- Energy and Water Policy
- Energy, Water and Carbon Management Plan
- Scope 3 Carbon Emissions Report

Waste management

Waste and recycling is one of the most visible areas in the sustainability agenda and can demonstrate our commitment to staff, students and visitors as well as offering an opportunity for people to easily get involved in environmental improvement. The 'waste hierarchy' ranks waste management options according to what is best for the environment. It gives top priority to preventing waste in the first place. When waste is created, it gives priority to preparing it for reuse, then recycling, then recovery and last of all disposal (e.g. landfill).

Waste management and recycling are co-ordinated by the Organisation Estates and Facilities Division. In the academic year 2016/7 we produced approximately 900 tonnes of waste across the organisation (excluding construction waste). We have worked with our waste contractors to divert as much waste as possible from landfill – the majority of our waste is sent for energy recovery. We operate a reuse scheme through Warp-it which facilitates reuse within the organisation and wider. Our recycling rate in 2016/17 was 38%, we are working to increase this as well as reducing our total waste output.

Along with other universities and colleges in Greater Manchester, the organisation will work to eradicate avoidable single-use plastics from catering, labs and stationery by 2022.

On plastics, we have already worked with our caterers, Chartwells (Salfood) and the Students' Union on a number of actions, including incentivised use of reusable cups and non-plastic cutlery and takeaway containers. Chartwells and the Students' Union have also added a zero-waste element to one of their shops which sells dried goods. Instead of plastic packaging, customers bring their own containers or bags to fill up products and pay by weight or item.

We will apply the principles of the waste hierarchy to prioritise reduction, reuse and recycling to reduce the impacts of waste management.

We will achieve this by

- Reducing total waste sent for recycling and disposal
- Increasing waste recycled
- Eliminating avoidable single-use plastics on campus
- Encouraging contractors to apply the principles of the waste hierarchy on capital projects

We will monitor this by

- Percentage reduction in waste from baseline
- Percentage waste recycled
- Number of avoidable single-use plastics on campus

Relevant documents

- Waste Management Plan

- Sustainable Procurement

Sustainable procurement is the process of recognising the negative and positive environmental, social and economic impacts of purchased goods, works and services. It is a cradle to grave approach and should consider the entire supply chain rather than just the individual product or service. The UN Marrakech Task Force defines sustainable procurement as:

> a process whereby organisations meet their needs for goods, services, works and utilities in a way that achieves value for money on a whole life basis in terms of generating benefits not only to the organisation, but also to society and the economy, whilst minimising damage to the environment.

The Estates and Facilities Division at the organisation has significant spend covering everything from facilities management, minor works and regulatory maintenance and testing to major building projects and infrastructure development. With this amount of spend and potential environmental impact, the need for staff involved in purchasing to ensure that spend provides value is evident. Environmental and social responsibility is considered as part of our value for money process and appropriately considered in the procurement of supplies, services or works. For example we refer to the energy and water technology lists in relevant purchases.

Through our membership of North West Universities Purchasing Consortium, we are members of Electronics Watch, an independent monitoring organisation set up with the purpose of protecting the rights of workers within electronics supply chains. Through this membership we can collaborate to have a positive impact on workers through the electronics we purchase. We are also exploring ways to ensure transparency in supply chains of other goods.

We will develop and apply an ethical and sustainable procurement system to ensure social and environmental impacts of purchases are appropriately considered.

We will achieve this by

- Implementing a sustainable purchasing policy and plan for Estates and Facilities
- Ensuring all Estates and Facilities tender evaluations include sustainability
- Meeting Level 4 in all areas of the Flexible Framework
- Developing and implementing a sustainable food policy

We will monitor this by

- Flexible Framework Self-Assessment
- Reviewing Estates and Facilities Invitation to Tenders for sustainability

Relevant documents

- Sustainable Procurement Plan
- Sustainable Food Policy

Sustainable Construction

As a Future Enabler, one of the key pan-organisation projects that will enable us to deliver the organisation's vision is our campus master plan. Central to the plans is an energy strategy that takes a major step towards a zero-carbon future, while providing high-quality spaces for residential, teaching, research and commercial uses that are cost-effective to run. Organisation buildings will undergo refurbishment to make them highly energy efficient, with maximum potential for renewable energy generation. We recognise there may be competing priorities and financial pressures, therefore we have developed a policy and design standard to guide our development teams on our sustainability principles.

We are committed to managing design, construction, refurbishment and post completion occupancy of its buildings in order to reduce environmental impact, enhance the well-being of staff and student users of the building, minimise operating costs and comply with all relevant sustainable building legislation.

We will ensure sustainability is a core principle of the estate master plan by embedding sustainable construction practices and performance specifications into plans for new buildings and refurbishments.

We will achieve this by

- Implementing our sustainable construction policy
- Aiming for BREEAM Excellent for high-value new builds where possible
- Considering the use of SKA methodology targeting at least Silver for lower-value new builds and refurbishments
- Implementing our energy design standard on all projects regardless of value

We will monitor this by

- Percentage of construction projects meeting sustainable construction policy requirements

Relevant documents

- Sustainable Construction Policy
- Energy Design Standard

Travel and Transport

Our latest travel plan covered the period 2012–2017. We have recently reviewed this plan and are aiming to relaunch an updated version as soon as possible.

Our travel surveys show that the majority of students travel by non-car modes (75%) and 49% of staff travel by non-car modes. Single-occupancy car use has increased for both staff and students slightly overtime and is likely to be a consequence of reduced car sharing promotion and allowance of unrestricted parking for all users. The surveys also show that a high number of staff and students are unaware of the existing tools and infrastructure measures.

Our ambition is to consolidate and rationalise car parking across campus and to reduce internal car circulation to promote a healthier, more pedestrian-friendly campus through our campus master plan. The vision also promotes greater levels of cycling in and around the site. Improved pedestrian connections will be key to improving permeability across the campus – think tree-lined boulevards, public squares and a generally more well-defined network of routes and open spaces.

We will develop a travel plan which encourages alternative methods to single-occupancy car journeys and minimises the environmental impact of organisation fleet and business travel.

We will achieve this by

- Developing an updated sustainable travel plan for 2021–2030
- Improving facilities for cyclists on campus
- Supporting the use of electric vehicles by staff and students
- Increasing the use of electric vehicles on campus by Estates and Facilities

We will monitor this by

- Published sustainable travel plan
- Number of cycle parking spaces
- Number of electric vehicle charge points on campus
- Percentage of electric vehicles in Estates and Facilities fleet

Relevant documents

- Sustainable Travel Plan

Biodiversity

Biodiversity, or biological diversity, is the variety of life! This includes variety in habitats (e.g. grassland and woodland) as well as diversity in species and the ecosystems in which they occur. Biodiversity is important because of the essential contribution that it makes to the functioning of our planet and because of all the benefits that it provides, from foods and medicine to climate regulation. Contact with biodiversity and the natural world has also been linked to improvements in health and emotional well-being.

Despite being just a mile and a half from the City of Manchester, our main campus in Salford is a leafy, riverside site adjacent to Peel Park and the River Irwell. The habitats on the organisation campuses include lawned areas, memorial gardens, wildflower areas and woodlands. Many common British wildlife can be found on the campus and surrounding areas, as well as some not so common.

Our campus master plan plans to fully utilise the green spaces in and around the campus, particularly Peel Park, David Lewis Playing Fields and The Meadows. Through our master plan, we will look for opportunities for new green spaces in our urban environment to help promote learning, well-being and healthy lifestyles.

We will protect and enhance the natural habitats and biodiversity on the organisation estate.

We will achieve this by

- Implementing our landscape management plan
- Maintaining our Green Flag Award

We will monitor this by

- Green Flag Award

Relevant documents

- Landscape Management Plan

Community Engagement

Communication and engagement are important elements of our Environmental and Energy Management System (EEMS) and our environmental sustainability policy.

To achieve our EEMS objectives and reduce the impact of the organisation's activities on the environment, we need to raise awareness amongst staff, students and other interested parties to encourage and support them to take action and adopt positive behaviours as well as ensuring that any legal compliance requirements, our policies and environmental objectives are clearly communicated to staff, students and other interested parties.

Our colleagues' efforts in our Green Impact programme are a key component of our environmental sustainability aims. From energy-saving and waste reduction initiatives to greening offices with plants and Fairtrade coffee mornings our Green Impact teams lead a wide range of creative activities that reduce the environmental impact of our day to day work. Green Impact is an international scheme which brings staff and students together to increase sustainability across campus. Alongside the Green Impact teams, the scheme is facilitated by student auditors who assess project teams across our organisation to meet their green targets.

We also work with key partners such as our Student's Union, catering teams and external partners such as TfGM and Salford Council on Go Green Salford, our four-week long annual campaign to raise awareness of the importance of sustainability and encourage positive change among our students and staff. Usually, taking place in February/March, it incorporates national campaigns such as the People and Planet, Go Green Week, Student Volunteering Week and Fairtrade Fortnight.

We recognise the importance of networking with relevant sector and professional bodies to understand best practice. Raising the external profile of the organisation's activities and achievements in environmental sustainability and promoting any areas of good practice in environmental sustainability is also important. We have been shortlisted as finalists in the sector Green Gown Awards every year between 2011 and 2018, with a number of winning and highly commended entries.

We will embed care for the environment in the culture of our organisation as part of our goal to be a sustainable organisation and empower and motivate the whole organisation community to support our policy and strategy through appropriate education and communication.

We will achieve this by

- Implementing our communications and engagement strategy for sustainability
- Increasing the influence of the Green Impact initiative at the organisation
- Increasing engagement with students and staff at the organisation with sustainability

We will monitor this by

- Number of staff/students influenced in Green Impact
- Number of staff/students engaged with sustainability campaigns and events

Relevant documents

- Environmental Sustainability Communications and Engagement Strategy

Teaching, Learning and Research

One of the principles of the Organisation Industry Collaboration Strategy is to 'promote sustainability and social responsibility'. We take a broad definition of 'industry' and when we refer to industry partnerships and collaboration this includes work with private, public and third-sector organisations of all sizes. From our inception, we have been committed to providing real world solutions to tomorrows challenges and the application of our research through partnership working and our current research and knowledge exchange strategy continues this tradition.

For example, our Applied Buildings and Energy Research Group (ABERG) is focused on establishing an evidence base to better understand and address the issues of energy consumption in buildings. The team covers a wide range of disciplines including building performance, electrical engineering, construction management, design and social sciences. ABERG is also the home of the unique Salford Energy House, a full sized two-bedroom terraced house built inside an environmental chamber that can replicate almost any weather conditions. With Energy House 2.0 coming soon we are looking to do even more in the area of smart living.

Whilst this strategy is focused on the environmental sustainability of our organisation operations we aim to support and encourage teaching, learning and research through this. We aim to 'practice what we preach' by using our campus as a living laboratory. We are already partners in a multimillion pound European contract for a project to develop 'green' resilience to extreme climate hazards. The IGNITION project backed by £4 million from the EU's Urban Innovation Actions initiative will see the organisation host a 'living laboratory' for urban green infrastructure including roofs, walls and gardens. The project is led by Greater Manchester Combined Authority and supported by 11 partners, including the Environment Agency, and comes in response to increasing flooding, heat hazards and worsening air quality.

In our teaching and learning, one of our principles of a curriculum for industry collaboration is 'Education for ethical behaviour', which presents an approach to student education that seeks to balance human and economic well-being, knowledge and competencies with cultural traditions, respect for other people and for the earth and its natural resources. In practice this requires design of academic programmes with consideration for sustainability including all the Sustainable Development Goals.

We will support our organisation strategy in providing real-world experiences preparing students for life and supporting our research to provide real-world solutions to tomorrow's challenges.

We will achieve this by

- Continuing to share our environmental sustainability experience and challenges through guest lectures
- Offering our environmental sustainability challenges as live briefs for students
- Encourage and support the implementation of a living laboratory on the campus showcasing sustainability solutions

We will monitor this by

- Number of environmental sustainability guest lectures delivered
- Number of environmental sustainability live briefs/student projects supported
- Number of demonstrator environmental sustainability projects on campus

Relevant documents

- Organisation Strategy 2016–2021
- Programme Design, Approval, Amendment, Review and Withdrawal Policy

SUSTAINABLE DEVELOPMENT GOALS

We have the ability to contribute to all the Global Goals through our teaching, learning and research.

Our Governance

We manage and control environmental sustainability on our campus and in our operations through our Environmental and Energy Management System (EEMS) which is certified to ISO 14001:2015 and ISO 50001:2018. This provides the framework for continual improvement in our environmental performance. We will deliver, review and monitor this plan through our EEMS.

Environmental sustainability is embedded within the organisation through high-level commitment from organisation leadership and our policies. A number of organisation groups provide strategic direction, support and oversight for implementation of the organisation policies.

The Environmental Sustainability Team lead the delivery of this plan, directly reporting to the Environmental Projects Board, chaired by the director of Estates and Facilities, who reports to organisation senior management groups via the Estate Programme Group and Sustainability Strategy Group.

The Environmental Sustainability Team manage budgets for energy management, waste management and sustainability communications and engagement. Business cases and requests for further funding are submitted via the Environmental Projects Board through the central organisation financial process.

We have created a number of policies and detailed plans to support our continual improvement aims and progress is reported quarterly to the Environmental Projects Board and publicly annually

Environmental Sustainability Policy

Energy and Water Policy Sustainable Construction Policy Estates Sustainable Procurement Policy Sustainable Food Policy	Energy, Water and Carbon Management Plan Sustainable Procurement Plan Waste Management Plan Landscape Management Plan Sustainable Travel Plan Communications and Engagement Strategy Sustainable Food Plan

Index